Production, Places and Environment

Production, Places and Environment
Changing Perspectives in Economic Geography

Ray Hudson
University of Durham

Routledge
Taylor & Francis Group

LONDON AND NEW YORK

First published 2000 by Pearson Education Limited

Published 2014 by Routledge
2 Park Square, Milton Park, Abingdon, Oxon OX14 4RN
711 Third Avenue, New York, NY 10017, USA

Routledge is an imprint of the Taylor & Francis Group, an informa business

ISBN 13: 978-0-582-36940-5 (pbk)

British Library Cataloguing-in-Publication Data
A catalogue record for this book is available from the British Library

Library of Congress Cataloging-in-Publication Data
A catalog record for this book is available from the Library of Congress

Typeset by 35 in 11/12pt Adobe Garamond

Contents

Contents

Figures

Plates

Tables

Preface

Over the last three or four decades human geographers have explored a variety of different conceptions of and approaches to their part of the discipline. Others have written extensive accounts of these changes and it is not my intention to seek to emulate these accounts here. The aims of this book are both more specific and less ambitious. It sets out to explore the changing character of economic geography in terms of the way in which one economic geographer's work – my own – has altered over this period, seeking to establish lines of continuity as well as points of change and to see these continuities and changes in the context of the ways in which economic geography more generally was changing.

Economic geography has often been at the forefront of debates about the most appropriate way to conceptualize and practise the discipline, investigating different theoretical and methodological approaches, and forging links with different disciplines within the social sciences. In the course of this, geographers have developed new conceptions of space. In parallel, more generally within the social sciences spatial differences have come to be seen as increasingly central to the ways in which economies and societies are constituted. Equally, the relationships between nature and the economy, and the necessary grounding of the latter in the former, have become more prominent both within economic geography and more widely. Increasingly, economic geographers have looked beyond the orthodoxies of mainstream economics to political economy, institutional and evolutionary economics, and to parts of cultural studies, sociology and politics, in the process problematizing the meaning of 'economic' (see, for example, Lee and Wills, 1997).

This shifting character of economic geography forms the context for this book. An interpretation of these changes forms its first chapter. This chapter seeks to elucidate more fully the changing character of economic geography, outlining the way in which the rich mix of epistemological, theoretical and methodological approaches that currently co-exist came to evolve in the ways that they did. Differing approaches have developed in a complex way, as putative new orthodoxies have become subject to critique, and this in turn has stimulated the exploration of further options. The net result is a discipline characterised by a multiplicity of approaches, some competitive, others complementary. While there are dangers in such eclecticism, there are also advantages to be gained from it. The remaining contents of the book consist of 13 chapters drawn from over a period of 25 or so years of my own research and writing, together with a brief introduction to each part that summarizes its contents and the links between the individual chapters and the broader pattern of changes within economic geography. Some of these were originally jointly written, a consequence of their origins in specific, externally funded research projects and more general programmes of on-going collaborative work. Quite a bit of this work has been across disciplinary boundaries. The chapters are organized around four broad, and to a degree interrelated, themes. Because of this, there is a degree of overlap between them at times. While this overlap could have been edited out, to have done so would have run counter to one of the aims of the book – to document how approaches to a subject evolve, with threads of continuity as well as turning points in approach. I have also resisted the temptation to re-write material, especially some of that from earlier years, again at the price of some overlap. It is also worth pointing out that there is another price to be paid in again committing to paper things in a form that, were I writing from scratch now, I would not always want to repeat in quite the same way. It may also lead some to see evidence of inconsistency in approach and analysis at differing points in time, but perhaps others may see this more positively as evidence of learning and seeking to refine explanatory perspectives. However, as one of the aims of this collection is to illustrate to students of the discipline that perspectives change – individually as well as collectively – it would be counter-productive and intellectually dishonest to pretend that such changes in perspective and viewpoint do not occur. Given the links between geography and other social sciences, the book hopefully will also be of value to students beyond the boundaries of geography.

There is, therefore, a serious point in wishing to present things in the way that they are presented here. The chapters of the book can be seen as a sort of autobiographical account, one person's meandering route through the changing map of the discipline of economic geography. It is worth emphasizing that it is therefore unavoidably partial and selective in its engagement with the broader trajectory of change, and at appropriate points some indications of the absences from the text are given. This journey began as an undergraduate in Bristol in the late 1960s with my being introduced to locational analysis and location theories by Peter Haggett, David Harvey and Barry Garner. After a school diet of descriptive approaches of varying sorts, finding geographers engaged seriously with questions of explanation was a source of genuine intellectual curiosity and excitement. A subsequent doctoral thesis set firmly in the framework of behavioural geography developed from a recognition of some of the problems posed by the assumptions of location theory approaches. A move to Durham in 1972 led to an enduring shift of substantive interest to issues of regional development, geographies of industries and the relevant economic and political processes through which uneven development is to be explained (a point elaborated in the introduction to Part 1). In turn, the explanatory limits of behavioural approaches, in particular their partial and emaciated conception of social processes, led to a search for more powerful explanatory accounts based on more meaningful abstractions as to the character of social processes. This began with explorations of Marxian political economy and the more general political economy literature and, somewhat later, moved on to encompass evolutionary and institutional approaches. More recently still, things have – in a sense – almost gone full circle. Issues of knowledge and learning have returned to the centre of the explanatory stage, albeit cast in a very different mould to that of behavioural geography, as part of an attempt to bring together considerations of agency, institutions and structures in explaining geographies of capitalist economies and of uneven development.

Much of this search for more adequate explanation centred around particular sorts of places and the industries that were constitutive of them. Part of the reason for overlap between chapters is that these places and industries themselves have changed, often dramatically, over a 30-year period. Explanations of more recent changes and current structure, however, typically must be constructed with reference to what went before – in terms of either the continuation of past trajectories or breaks from them. In this regard, all geographies are necessarily historical geographies. Another part of the reason for this overlap is that these chapters represent successive attempts to find more satisfactory explanations of the changing geographies of these industries and the changing character of these places, seeking to interpret them from differing but complementary theoretical perspectives. No doubt some will find this overlap irritating, but the point is to try to trace out the way in which approaches to the 'same' set of issues and problems have altered over the years rather than to present these in an edited and sanitized 'definitive' final form. One of the aims of the book is to suggest to students the ways in which perspectives and views develop. In part, this evolution reflects a dialogue within the geographical discipline and academia more generally as to the relative merits of varying, sometimes competing, sometimes complementary, approaches to understanding. In part, this developmental process reflects the way in which intellectuals seek to bring their theories to bear upon the varied characteristics of historical geographies of economies, with a view better to understand them, and, on occasion, maybe even to seek to change them.

While the final responsibility for what appears here is mine, I also need to acknowledge the help of numerous people in producing the materials that form the contents of the book. Several of the chapters were jointly written, originating in collaborative research projects: Chapter 9 with David Sadler, Chapter 12 with varying combinations of David Sadler, Huw Beynon and Andrew Cox, and Chapter 14 with Paul Weaver. These and several other chapters draw upon research financially supported by a number of organizations: the Centre for Environmental Studies (Chapters 2 and 8); the Northern Ireland Economic Council (Chapter 4); the UK Economic and Social Research Council (Chapters 5, 6, 7, 8, 9 and 12); the European Science Foundation (Chapter 10); the UK Department of Education and Science (Chapter 11); Friends of the Earth (Chapter 14); and The International Social Science Council (Chapter 13). Finally, I would also like to acknowledge the generous award of a Sir Derman Christopherson Foundation Fellowship by the University of Durham for 1998–99 that allowed time to complete work on the book.

As well as those noted above as co-authors, to whom I owe particular debts, I would also like to acknowledge the contributions of a number of colleagues and friends over the years in commenting upon various chapters: Ash Amin; Bjorn Asheim; Mike Crang; Mick Dunford; Costis Hadjimichalis; Roger Lee; Alain Lipietz; Anders Malmber; Peter Maskell; Doreen Massey; Linda McDowell; Martin Osterland; Joe Painter; Viggo Plum; Eike Schamp; Ian Simmons; Mike Taylor; Nigel Thrift; Adam Tickell; Henrik Toft Jensen; Dina Vaiou; and last but not least Allan Williams. Matthew Smith of Addison Wesley Longman deserves a particular mention, as it was through conversations with him that the idea of this book emerged and he has been particularly helpful in seeing the project through to completion. Finally, a special word of thanks to Joan Dresser at Durham, who again got the unenviable job of converting a diverse set of files into a clean manuscript, and to Arthur Corner and his colleagues in the Cartography Section of the Durham Geography Department for the maps and diagrams.

Ray Hudson,
Durham,
November 1998

Acknowledgements

We are grateful to the following for permission to reproduce copyright material;

David Fulton Publishers for Chapter 2; The Royal Dutch Geographical Society for Chapter 3; Sage Publications Ltd for Chapter 4; Pion Ltd for Chapters 5, 8, 9 and 14; The Royal Geographical Society for Chapters 6 and 11; Edward Arnold Publishers for Chapter 7; Oxford University Press for Chapter 10; Ashgate Publishing Ltd for Chapter 13.

Part 1

Setting the scene

Chapter 1

Continuity and change in analysing geographies of economies

1.1 Introductory remarks

Why are geographies of production constituted in particular ways in different companies, industries, times and places? How do geographies of consumption relate to those of production? Why do patterns of urban and regional uneven development take the forms that they do? What are the relationships between state policies, geographies of production and territorially uneven development? Why are relationships between the production system and the natural environment configured in particular ways? These are typical of the questions of interest to economic geographers. Accordingly, the objective of this first chapter is to provide an introduction to some of the ways in which economic geographers have sought to answer such questions. A corollary of so doing is to establish a context in which the rest of the book can be situated by selectively introducing some of the ways in which approaches to geographies of economies and uneven development have changed over the last three or so decades.

The particular focus here is upon understanding different approaches to capitalist production and uneven development within capitalism. This is not because other forms of production organization do not exist, but rather that capitalist production is the dominant form over much of the world, either directly or indirectly (for example, in shaping relationships between waged work and unwaged work). Equally the focus upon production rather than consumption should not be seen as implying that the latter is of less significance than the former but simply reflects the fact that the main focus in the remaining chapters of the book is upon geographies of production and uneven development. It is important to emphasize at the outset that this does not claim to be a comprehensive review of changes in geographic thought over this period (for which see, for example, Johnston, 1997; Peet, 1997a). This chapter has much more modest aims – that is, to identify and contextualize some strands of thought and substantive issues that are reflected in the remainder of the book. It does so via a review of the ways in which the substantive foci of geographies of production and theoretical approaches (both within geography and the social sciences more generally in so far as these are relevant) to understanding geographies of production, the production of places and uneven development have shifted over that period.

1.2   Changing approaches to economic geography ────────

1.2.1   From location theories to Marxian political economy

After a well-known period in which regional description became the primary focus of geographical scholarship, the 1950s saw a re-awakening of interest in explanation and theorization within – *inter alia* – economic geography. The re-discovery of economic location theories in the works of von Thunen, Weber and Lösch led to an engagement between some economic geographers and emergent regional scientists in search of general explanatory statements about the spatial structure of the economy (see, for example, Isard, 1956; Chisholm, 1962; Haggett, 1965). Re-focusing concerns from description of the unique to explanation of more general classes of events and spatial patterns was a very important and radical break. On the other hand, the ways in which explanation was sought soon became revealed to be deeply problematic. At one level, this involved conflating explanation with prediction; predictive accuracy became the measure of explanatory power. At another level, there were profound problems associated with an approach that sought to deduce equilibrium spatial patterns on the basis of restrictive assumptions about the natural environment, human knowledge and the character of social processes. The limitations of the restrictive assumptions soon became well known. In the 1960s behavioural geographers mounted their critique of location theories precisely upon the point that the behavioural assumptions upon which these theories were based were highly unrealistic (see, for example, Pred, 1967). Their criticisms arose from a recognition that such assumptions are untenable in an economy that exists in real space and time. Behavioural geographers therefore argued the need to investigate what people actually did know, how they came to acquire this knowledge, and *where* they knew about, rather than assuming that they knew everything and everywhere of relevance to a particular type of behaviour. Consequently, they began to conceptualize the knowledge that people had of environments in terms of 'mental maps' or 'cognitive maps' and, drawing on some strands of psychology, began to seek ways of measuring what people knew (see, for example, Pocock and Hudson, 1978). There were moves to introduce concepts such as 'bounded rationality' (Simon, 1959), recognizing that people could not know everything in a real time and space economy while holding onto the privileging of individual actors and their knowledge. Within the context of geographies of production, such approaches focused on the knowledge that key corporate decision makers possessed about alternative locations in an attempt to explain why economic activities were located in some places rather than in others. As a result, they yielded at best a very partial and imperfect grasp of the relations between knowledge, production and its spatial organization. Such behavioural approaches typically amounted to little more than descriptive accounts of the locational strategies and behaviour of particular firms.[1] Having set out to refine an explanatory approach, they unfortunately slipped into the descriptive trap that neo-classical location theories had set out to escape. As a result, the behavioural approach soon became marginalized. While often interesting in themselves, behavioural studies did little to address the explanatory weaknesses of location theories but, in abandoning them, economic geographers also pushed important questions of agency and action from the research agenda for a decade or so. Others sought to get around the problems of ignorance about individual knowledge and motives in a very different way. This involved adopting probabilistic macro-scale modelling procedures, which in due course became more rigorously theorized via entropy-maximizing approaches which sought to predict the most probable distributions of activities and behaviours in space, subject to any known constraints (Wilson, 1970). These resulted in some circumstances in more accurate predictive models but they pushed questions of explanation and understanding of social processes still further down the research agenda.

The point of raising these questions about the lack of realism in the assumptions of location theories is not simply to criticize them on that score but rather to emphasize that such assumptions were both a precondition for and a symptom of an impoverished and partial view of the social processes of the economy. Assumptions of static equilibrium deny the fact that economic processes are chronically in a state of dynamic disequilibrium, set on open-ended and unknown trajectories of change and development rather than inevitably and mechanistically circling around a point of static equilibrium. As such, they ignore issues to do with the social construction, reproduction and regulation of markets as institutions. Assumptions of perfect knowledge equally deny the fact that economic decisions are chronically made in a condition of ignorance. Seeking to finesse this difficulty via assumptions of bounded rationality fails adequately to address this problem. Assumptions of the environment as an isotropic plane denied the significance of the grounding of the economy in nature and of the chronically uneven character of economic development. They also reduced the significance of spatial differentiation to variations in transport (and sometimes other production) costs within a pre-given space. The net result is that while such approaches placed questions of explanation firmly back upon the agenda of economic geographers, they did so in a way that was based upon unhelpful abstractions. Consequently, they resulted in inadequate theory, which failed to grasp the essential character of the key processes that produced geographies of economies.

The immediate response of economic geographers to the limitations of both location theories and also the behavioural critiques of them was to search for more powerful conceptualizations of the processes that generated geographies of economies. In this, they became more closely linked with broader changes in the social sciences. It became clear that established orthodoxies in the social sciences were incapable of explaining the increasingly problematic character of capitalist development. As the long post-war boom stuttered to a halt, uneven development became more marked and it became clear that poverty and inequality had not been abolished but were again on the increase, the limitations of the orthodox nostrums of the social sciences became increasingly visible. Faced with this, the social sciences began to re-discover the Marxian heritage, as well as exploring other heterodox positions. In their search for more powerful explanations, economic geographers also began increasingly to engage with these strands of thought, and in doing so became more engaged in debates with other social scientists. Increasingly, they turned to Marxian political economy as a source of theoretical inspiration (see, for example, Harvey, 1973). Marxian political economy offered powerful concepts of structure, of the social structural relations that defined particular types of societies (Harvey, 1982; Smith, 1984) and offered a powerful challenge to the spatial fetishism (Carney et al., 1976) of location theories and spatial science. Production in general involves the application of human labour to transform elements of nature into useful products. Different types of economy are, however, characterized by different social structures, differing rationales for production and differing ways of organizing human labour. The driving force of capitalist production is making profit; things are produced as commodities, not as use values for immediate consumption by their producers but for sale in markets. The rationale of capitalist production is exchange and production is validated *ex-post* via sale in markets.

The concept of mode of production seeks, at a highly abstract level, to catch the essence of particular types of economic organization, characterizing these in terms of specific combinations of forces of production (artefacts, machinery and 'hard' technologies, tools – in short, the means of production) and social relations of production. In the capitalist mode of production, the defining social structural relationship is that between the classes of capital and labour. This is a dialectical and necessary (in a critical realist sense: Sayer, 1984) relationship in that the definitions of capital and labour presuppose the existence of the other. Capital owns the means of production, labour owns only its capacity to work (its labour-power). Capital needs to purchase labour-power in order to set production in motion since living labour is the only source of new value (surplus-value) created in production, while labour needs to sell its labour-power for a wage in order to survive and reproduce itself. One consequence of this is a recognition that workers are alienated from the product they produce.

The concept of capitalist mode of production implies thinking about production in terms of values, and this has quite specific connotations within Marxian thought. The value of a commodity is defined as the quantity of socially necessary labour time required to produce it. The qualification 'socially necessary' is critical as it denotes the amount of undifferentiated abstract labour needed under average social and technical conditions of production. Deviations in practice from these averages become critical in shaping historical geographies of production and uneven development. The key factor differentiating capitalism from other modes of production organized for the purposes of exchange is that labour-power becomes a commodity and the social relations of capitalist production become structured around the wage relation. Consequently, capitalist production is both a labour process and a valorization process, as human labour is applied to material inputs to create newly produced commodities. Profits are created because the value created by labour in production exceeds the value of labour-power (thus creating surplus-value which, along with existing values transferred in the production process, is realized in money form on successful sale of the commodity).

It is important to emphasize that the concept of mode of production is a high-level abstraction, designed to reveal the essential defining relationship of a capitalist economy, and not a description of social reality as experienced by people living in capitalist societies. For example, the reproduction of labour falls outside the 'logic of capital' as it can never be fully commodified (with much of the costs of reproducing labour borne by unwaged work within the family) and so a recurrent problem for capitalists is to persuade workers to offer their labour-power on the market. The concept of social formation moves matters one step nearer to the experienced reality as it denotes the ways in which capitalist and non-capitalist, class and non-class social relationships come together in a particular time–space context. This creates conceptual space for different forms of capitalism, allowing that capitalist relations of production may be socially and culturally constituted in differing ways. It also brings increased attention to the role of the state in capitalist societies in ensuring the reproduction of these relationships. Initially, the state was seen as non-problematically meeting the 'needs of capital' so as to ensure societal reproduction. It soon became clear that such State Monopoly Capital formulations (see, for example, Baran and Sweezy, 1968) were too simplistic. Crisis theories focused upon the problems encountered by capitalist states in seeking to ensure the successful reproduction of crisis-prone capitalist economies and societies (see, for example, O'Connor, 1973; Habermas, 1976; Offe, 1975a). Another strand of state theory sought to derive the existence of the state in a political sphere formally separated from the economy from the fundamental characteristics of the capitalist mode of production (Holloway and Piciotto, 1978). This in due course revealed that the project of deriving a theory of *the* capitalist state was a flawed one and that what was required was a historical geographical theory of capitalist *states*, cast at a somewhat lower level of abstraction.

Moreover, moving to a lower level of abstraction involves recognizing that market transactions in a capitalist economy are conducted in terms of prices, not values. Companies declare profits in terms of money, not values. Workers' wages are likewise paid in money. For some this raises the question of how quantitatively to transform values to prices, reflected in the history of the 'transformation problem', and more generally the issue of the validity of value analysis (see, for example, Steedman, 1977; Rankin, 1987; Roberts, 1987; Sheppard and Barnes, 1990, 31–58). Rather than seeking quantitatively to equate values to prices, a more fruitful way of approaching the relationships between them is to recognize the qualitative differences between values and prices and the fact that these are concepts of different theoretical status. Values and prices are indicative of the way in which the capitalist social relationships unite a wide range of qualitatively different types of labour in the totality of the production process. Massey (1995, 307) trenchantly argues that the law of value is useful for thinking through the broad structures of the economy and for forming the 'absolutely essential basis for some central concepts – exploitation for instance' but attempts to use it as a basis for empirical economic calculation are misconceived and doomed to failure. Indeed, 'the byzantine entanglements into which the "law of value" has fallen . . . make it . . . unusable in any empirical economic calculus'. It is therefore important not to confuse values and prices conceptually or seek to equate empirical data measured in prices with theoretical constructs defined in terms of values. The significance of value analysis lies in the way in which it focuses attention upon class relationships and the social structures that they help to define.

Marxian approaches began to find their way into economic geography during the 1970s. The Marxian tradition, however, encompasses a variety of approaches, with differing emphases and aims. It is therefore vital to acknowledge the pivotal role of Harvey's work in bringing Marxian approaches into geography, since he retained a strong commitment to classical Marxian political economy. He likewise retained a commitment to a scientific geography and to constructing more powerful explanations of the structure of the space economy than those that emerged in the 1950s and 1960s (Harvey, 1973). Marxian political economy was, however, much more concerned with temporal variation than with spatial variation in capitalist development. As a result, there was initially a rather limited and one-sided consideration of the links between social process and spatial form. Some sought directly to deduce trends in the organization and geographies of production from the deep inner structural relationships of the capitalist mode of production (see, for example, Läpple and van Hoogstraten, 1980).[2] Others went even further and sought to deduce forms of political action from the structures of capitalist relations of production (see Carney, 1980). More sophisticated approaches recognized that structural analyses could define the limits to and constraints upon production organization and its geographies. Harvey (1982) remains the most elegant and thorough analytic re-statement of Marx's historical materialism into a historical–geographical materialism that sets out the necessity for territorially uneven development in relation to the structural limits to capital. Such approaches could not, however, specify how particular industrial trajectories or patterns of spatially uneven development would evolve within these limits, or indeed explain how (as opposed to why) these structural limits were reproduced. So, for example, while some cities and regions grow and others decline, explaining which grow and which decline is a matter for empirical investigation rather than deduction from immanent structural tendencies. Structural conditions and limits are not, however, pre-given and natural but are socially produced by human actions, intended and unintended. As a consequence, a pivotal issue is the ways in which such structural limits are socially (re)produced within particular institutional arrangements. These processes of structural (re)production thus need to be problematized rather than taken for granted. This is especially so as capital accumulation is an inherently crisis-prone process and this was recognized – inter alia – within Marxian political economy by the increasing attention given to the role of the state in keeping these crisis tendencies within 'acceptable' limits. To begin to understand such issues requires a rather different approach and level of theoretical analysis from that of the law of value and the definition of class structural relations within the capitalist mode of production (see the Afterword to Harvey, 1982). 'Intermediate'-level theoretical constructs therefore need to be interposed between the abstract conceptions of structures and the empirically observable forms of production organization and its geographies, recognizing that the structural relationships of capitalism permit a variety of develop-mental possibilities and trajectories. This intermediate-level theoretical bridge can be built by drawing on concepts from modern social theory (for example, Giddens, 1981, 1984) and from evolutionary and institutional approaches in the social sciences, especially economics and sociology (see, for example, Hodgson, 1988; Metcalfe, 1997, 1998). Seeking to link the structural relations of capitalism with uneven development and geographies of economies also requires a more sophisticated and nuanced view of the relationships between society, space and nature, not least because institutional forms are often territorially defined and demarcated.

1.2.2   Reciprocal relations between the natural, the social and the spatial

One of the problems of location theories is that they adopt a very one-sided conception of the relationships between social process and spatial pattern. Spatial patterns are deduced from (an impoverished conception of) social processes. While drawing on more powerful conceptions of social process, some Marxian approaches initially fell into the same trap. There is no recognition that spatial pattern could not only influence the ways in which those processes operate but could also have a formative role in the ways in which they were constituted in the first place. There is no pre-determined one-to-one correspondence between the structural relationships of capitalism and spatial patterns of the economy but rather reciprocal indeterminate relationships between the social and the spatial. Different analysts place rather different emphases upon the strength of these relationships between the social and the spatial. For Sayer (1984) the relationships are limited to spatial form influencing how causal powers are realized. For Peck (1996) spatial difference plays a rather more active, though vaguely specified, role in shaping social relationships. For Massey (1995) and Urry (1985) the character of social processes may itself be shaped by the way that such processes operate through time and space. Not only are spatial forms a product of social processes but those same spatial forms in turn shape the ways in which those processes are constituted and evolve. These latter stronger views as to the character of social–spatial relationships emphasize that space is socially produced, and so dynamic, rather than being a static container into which social relationships, including those of production, are poured. Indeed, social space is produced by the distanciated stretching of social relationships. Consequently, while the spatial patterning of the economy is an outcome of the social relationships of production, spatially uneven development and the characteristics of specific places help shape the particular form that these relationships take.

   Recognition of the spatiality of production, and of socio-spatial reciprocity, entails a view of production, its organization and its geographies as contingent and contested. Particular forms and geographies of production organization are a product of simultaneous struggles between capital and labour, between companies, and between groups of workers, with states implicated in such struggles both via their regulatory role and, on occasion, as participants. Different social classes and groups seek to shape the anatomy of production and the spaces and spatiality of capitalism to further their own interests. While this is a determinate set of struggles, however, it is by no means one that is (pre)determined, as agency and structure combine in and through place to generate contingent outcomes. There is a considerable range of concrete socio-spatial practices and strategies through which the social relations of production can be realized and reproduced and so there is always a variety of potential resultant geographies (Cox, 1997, 183). The strategies of capital are undoubtedly of great significance in shaping the landscapes of capitalism, although the ways that they do so can vary greatly. Due weight must also be given to the significance of firms as organizational entities in the organization of production, rather than the organization of production being regarded as simply a response to the general requirements of capital. Production is organized by business enterprises operating within extremely complex dynamic networks of internalized and externalized transactional relationships of power and influence (Dicken and Thrift, 1992, 287). While there is considerable force to this argument, it is also important to acknowledge the influence that national states, trades unions and other labour organizations and a variety of groups in civil societies may have on the changing organization and geographies of production. Whilst companies seek to shape space to meet their requirements for profitable production, other social forces seek to shape space in relation to differing and varying criteria as they seek to stretch other social relationships over space and form it in different ways. Organized labour can influence these landscapes in a variety of ways via its socio-spatial practices (see, for example, Herod, 1997; Martin et al., 1994; Wills, 1998b). Furthermore, people seek to create and reproduce meaningful places – understood as complex condensations of overlapping social relations in a particular envelope of time–space (Hudson, 1990) – in which to live and learn as well as work (Beynon and Hudson, 1993). The way in which place is conceptualized has also been problematized. The extent to which places are spatially continuous or discontinuous, have permeable or impermeable boundaries and have singular or multiple meanings and identities is recognized as an issue for empirical investigation rather than as amenable to treatment via an *a priori* assumption (Allen et al., 1998; see also section 1.3 below).

   As well as recognizing reciprocal links between the social and the spatial, it is also important to take account of relationships between the social and the natural (see, for example, Smith, 1984). Paralleling the increased cross-disciplinary importance of spatial difference, relationships between people and their natural environments also became the focus of a growing cross-disciplinary concern. One consequence of this is that, from one point of view, the natural is socially constructed and this can set limits upon forms of production organization and geographies. Beyond that, however, production is unavoidably materially grounded in the natural world (as a source of raw materials and as a repository for wastes, for example). The laws of thermodynamics and conservation of energy and matter governing chemical and physical transformations of matter in production set quite definite limits on the production process. These laws also make it clear that production has unavoidable impacts upon nature and the natural world (for example, see Jackson, 1996). The precise ways in which this mediation between nature and society takes place are structured by the dominant social relations of production. Relationships between the economy and the environment are thus shaped by the specific requirement of capitalist production for profits. This has myriad implications for the organization and geographies of production as well as, in the last analysis, for the sustainability of the production process (Taylor, 1995). While knowledge of natural processes is itself socially constructed, there are natural limits to production that cannot be overcome.

1.2.3 Agency, structures and power relationships

Structures are both enabling and constraining as regards action and agency. Structures thus influence human behaviour in these ways while people reproduce those same structural limits and constraints via their behaviour, albeit often unintentionally. For example, radical communist trades unionists may oppose the class relations of capitalism yet they nevertheless help reproduce them via their everyday behaviour of going to work, even if at the workplace they seek to disrupt production as a perceived way of furthering the immediate interests of those working there. Marx was well aware of the importance of human agency but insisted that people made histories (and geographies) but not necessarily in circumstances of their own choosing. A concern for the relationships between structures and agents was revived in geography as a result of the discovery of Giddens' (1981, 1984) 'theory of structuration'. Structuration emphasizes the reciprocal relations between agency and structures – the 'duality of structures'. Giddens lays considerable stress upon individuals as agents but he also recognizes institutions and other forms of social collectivities as possessing powers of agency. Giddens offers valuable insights in recognizing that agents are both shaped by and help shape structures, although the way in which he conceptualizes these links is in some ways problematic (see, for example, Gregson, 1986). While seeking to re-connect agency and structure, Giddens did so in a way that problematizes both. As a result he has been criticized both for weakening the concept of structure and for obfuscating the concept of agency. Rather than take concepts of structures and agents, and the links between them, seriously, it has been suggested that Giddens tends to blur and dissolve the distinction. Others, working from feminist and post-structuralist perspectives, have emphasized the complexity and significance of agency *vis-à-vis* structure, deepening and transforming the concept of agency by notions of multiple identities. However, this does not progress beyond a concern with 'the complexity of agency' or 'the multiplicity of identity'. As a result, the problem of adequately relating agency and structure remains, since rationalized actions create through repetition the systemic logics of economic forms (Peet, 1997b, 37–38).

In contrast, yet others would, however, claim that the problem of the structure–agency distinction has been abolished by the way that work in certain service occupations has been re-defined to make the service or product indistinguishable from the person providing it. Workers with specific social attributes are disciplined to produce an embodied performance that conforms to idealized notions of the appropriate 'servicer'. In this normalization, the culture of organizations, in the sense of the explicit and implicit rules of conduct, has become increasingly important in inculcating the desirable embodied attributes of workers, as well as in establishing the values and norms of organizational practices. The coincidence of embeddedness and embodiment overcomes the separation of structures from agents (McDowell, 1997, 121). This suggests that this separation is a problem that is to be overcome by dissolving the one into the other, based on a weak sense of structure and a strong sense of agency. It thus circumvents the problem of specifying the ways in which agency and structure are dialectically related in such deeply exploitative forms of work and represented in specific ways in certain circumstances to legitimate these new forms of exploitation in the workplace.

The concept of agency poses more general problems, especially in terms of the relationships between individuals and collectivities of various sorts. There are certainly dangers of reifying organizations in a way that suggests that they have powers independent of the people who constitute organizational activity. Organizational change is animated, resisted or modified by the action of organizational members. Consequently, organizations must be seen as a terrain on which their various members can mobilize. Corporate change therefore is interpreted, sometimes fought over and resisted, both by individuals and by groups of people who may have very different assumptions and agendas about what changes should occur and how they should occur. Organizational members have differential powers but organizational change remains a human process (Halford and Savage, 1997, 110–111). This raises important questions as to the extent to which organizational action is more than the sum of the parts of the actions of organizational members. It also raises key questions about the extent to which the actions of individuals can be understood outside the organizational contexts in which they occur. Organizations clearly do have cultures and histories which both shape and are shaped by the actions and understandings of organizational members, but the precise forms of these relationships are contingent and indeterminate.

This, however, in turn raises the question of how these indeterminate and contingent relationships in practice hold together to create organizations that successfully reproduce themselves. Actor-network theorists argue that their approach speaks to this question, offering a non-dualistic perspective that focuses on how things are 'stitched together' across divisions and distinctions. Actor-network theory seeks to connect the social and the material, in contrast to structuration theory which neglects the material components of both action and structure and is seen by the proponents of actor-network approaches to be overly dependent upon social interaction. Actor-network theory thus opens up but seeks to bridge a fresh divide between the social and the material while seeking to bridge the divide between action and structure. Consequently, actor-network theorists investigate the means by which associations come into existence and how the roles and functions of subjects and objects, human and non-human, are attributed and stabilized. Moreover, they acknowledge that there are distinct asymmetries in power in the position of actors within networks. Actors organize associations or networks while intermediates are organized (Murdoch, 1997, 331). However, this distinction between the organizers and the organized is seen as coming at the end of the network construction process, which is shaped by the actors. The same person can, however, be an actor in one network, an intermediate in another. The 'radical symmetry' that lies at the heart of actor-network theory thus stems from the belief that power and size are not immutable. Actor-network theorists seek to uncover how associations and networks are built and maintained. There is thus much that is attractive in actor-network theorizations.

There are also some major problems, however. Despite its name, actor-network theory in fact has little to say about why agency is exercised in the ways that it is or why structural limits exist in the forms that they do. Furthermore, while the recognition of power inequalities is vital, these are seen as arising only *within* the structure of a given network. There is no concept or theory of power outside the network, which raises problems in seeking to deal with the social sources of power and with the ability of some to control the position of others within a network.[3] Indeed, actor-network theory is characterized by a methodological agnosticism and as such slips from detailed descriptions of particular actor-networks to quite abstract prescriptive methodological statements as to how to analyse networks. It has little to say about what forms of 'stitching together' of networks are more probable than others, and why this should be the case. In this sense, it is probably erroneous to refer to it as a theory as it is somewhat under-theorized and, as a result, problematic as an approach for understanding agency–structure relationships (Murdoch *et al.*, 1998, 14–15).

A concern with power and inequalities in power also arises from another direction. There is clearly a case, as suggested above, for not reifying companies, governments and other organizations, and to argue that these organizations *per se* do not make decisions but that decisions are made by people who are members of these organizations. Such a position is reasonable to a point. Arguments to the effect that all that matters are individuals can, however, easily slide from a concern with individual psychology into a reductionist physiological argument that in the final analysis all behaviour is to be explained via electrical brain activity. It is, therefore, important to remember that individuals exist not as asocial atoms but as social beings whose patterns of thought and action are conditioned by the social relationships in which they are enmeshed. Firms, governments and other organizations have a collective memory *beyond* that of any given individual or group of individuals; and in this sense such organizations can be said to have their own cultures. Moreover, they can exercise power (see Allen, 1997) and some can clearly exercise more power than others, both in specific situations and in general.

1.2.4 Re-discovering the significance of motives, knowledge and learning

A corollary of acknowledging that people are active and thinking subjects, but are not endowed with perfect knowledge, is a revived recognition of the significance of knowledge and learning and the importance of uncovering the rationalities and motives that actually underlie and inform behaviour. The basis of the behavioural geography critique of neo-classical location theories was a recognition that people had imperfect knowledge and varying motivations, but this progressed little beyond descriptive studies of particular cases. More recent research in the cognitive and behavioural sciences emphasizes the ways in which different actor rationalities generate different forms of economic behaviour. For example, substantive or scientific rationality leads to rule-bound behaviour, procedural rationality favours behaviour that seeks to adjust to the constraints imposed by the environment in which people operate, while recursive or reflexive rationality is linked to strategic behaviour that seeks to shape that environment (Amin, 1998). The re-discovery that knowledge and what people actually know are central to production has also been reflected in the burgeoning literature about forms of knowledge and production, learning firms and learning regions (see, for example, Maskell *et al.*, 1998).

Learning and knowledge are also foregrounded via the recognition that organizations develop a 'collective memory' and their own cultures of production, linked to an acknowledgement of the importance of modes of internal organization, competencies and capabilities of firms themselves (see, for example, Foss, 1996). This emphasizes the centrality of certain types of knowledge and competencies to competitive success and of the economy as dependent upon knowledge, learning (in various ways, such as by doing, by interacting, by imitating and so on), adaptation and evolution. For some, knowledge is now the most important resource, learning-by-interacting the most important process (Lundvall, 1995). There are therefore strong links to strands of evolutionary economics and a view that economic development trajectories (corporate and territorial) may be strongly path-dependent (Nelson and Winter, 1982) – thereby explicitly rejecting the validity of neo-classical orthodox views of the economy as oscillating around some point of static equilibrium. There is no doubt that in one sense all production depends upon knowledge of various sorts; the more interesting questions concern the links between particular sorts of production and particular sorts of knowledge and learning process (Hudson, 1999a). Equally there are some important limitations to the ways in which these questions have been framed and answered (Odgaard and Hudson, 1998).

1.2.5 Institutions, instituted behaviour and social regulation of the economy

Recognition of the importance of the cultural constitution of the economy, and its path-dependent developmental trajectories, necessitates acknowledging the importance of institutions, and of conceptualizing the economy as constituted via 'instituted' processes. As institutions are typically place-specific, this also involves taking seriously the territoriality of the economy, the organization of production, and the production of spatial scales. Conversely, it explicitly denies the validity of conceptions of the economy and markets as naturally occurring, governed by natural processes. This institutional perspective draws heavily on the legacy of Polyani and the 'old' institutional economists (Mulberg, 1995). For Polyani (1957, 243–245) all economic processes are instituted processes. As a result, all economic structures are in various ways socially embedded. This is linked to the recognition within economic sociology that the economy is socially embedded in networks of interpersonal relations and so heavily influenced by the presence or absence of mutuality, cooperation and trust (Granovetter, 1985; Dore, 1983). Embeddedness denotes the contingent character of economic action with respect to cognition, culture, social structure and political institutions (Zukin and di Maggio, 1990, 15). Thus capitalism may be constituted in differing ways and economic relationships may be grounded in, and dependent upon, non-economic ones. 'Traded inter-dependencies' may depend upon 'untraded inter-dependencies' as relationships based upon particular conceptions of trust assume a critical importance in some types of economic transactions (Storper, 1995, 1997).

Recognition of the economy as regulated and governed in particular ways follows, logically and historically, from a recognition that economic processes in general and those of production in particular are embedded and instituted in various ways. There is a range of institutional forms, varying from formal institutions such as the state to the informal and tacit institutions of norms, habits and routines that shape the way in which production takes place and the economy is organized. Such varied institutions provide a degree of stability in the face of uncertainty as well as constraints upon, or templates for, future economic developments (Hodgson, 1988, 1993). This resonates with the notion of the creation of distinctive 'worlds of production' associated with particular ways of organizing the production process, which draws upon convention theory (Salais and Storper, 1992). Conventions are defined as practices, routines, agreements and their associated informal and institutional forms that bind acts together through coherent and taken-for-granted mutual expectations. Conventions are sometimes manifested as formal institutions and rules, but often are not. As such, most conventions are a kind of halfway house between fully personalized and idiosyncratic relations and fully depersonalized easy-to-imitate relations (Storper, 1997, 38). Worlds of production are defined as distinctive sets of practices which come together in particular ways as 'bundles' or 'packages', bound together via the glue of conventions and the mutual expectations and shared ways of understanding that they entail.[4]

In some circumstances conventions can be economically advantageous, in others less so, producing a form of 'conventional lock-in' which undermines competitive success. For example, Schoenberger (1994) explains the failures of US companies to adapt particular restructuring strategies because of the ways in which managers' own interests and identities, and *their* sense of the corporations' interests and identities, were embedded in established institutionalized forms of organization. These acted as cognitive filters, effectively locking in companies to established ways of doing things and foreclosing options as to other ways of operating. This exemplifies a broader point: that there are often decisive reciprocal links between human behaviour and institutional form.

The renewed recognition of the significance of institutions has focused attention upon the regulatory role of the state (at national but also supranational and sub-national scales: see, for example, Jessop, 1994). Beyond that, however, it has also highlighted the significance of mechanisms and institutions in civil society as a concern with government has been increasingly replaced by one with governance (see, for example, Painter and Goodwin, 1995; Goodwin and Painter, 1996; see also section 1.3 below). Furthermore, there is now a recognition that different forms of regulatory and governance regimes may result in relations between structures and agents, and between the varying agents in the production process, being stitched together in different ways and taking varying forms as cultures of economies vary. Echoing strands of the agency–structure debate, Gertler (1997, 57) stresses that the relationship between institutions and practices is 'fundamentally dialectical' in nature, with practices having the potential to shape institutions over time. Cultural embeddedness may be defined as the role of shared collective understandings in shaping economic strategies and goals, with culture providing scripts for applying different strategies to different classes of exchange (Zukin and Di Maggio, 1990, 17). Taken-for-granted ways of thinking and behaving – which may variously be described as informal institutions or as conventions – can be a critical component of establishing ideas and ways of doing things as unquestioned, unquestionable and so hegemonic in the Gramscian sense. They may, however, be less a product of some timeless cultural traits than a result of particular regulatory regimes and institutional forms. Gertler (1997, 55) suggests that traits and attitudes commonly understood as being part and parcel of inherited cultures of individual firms are themselves produced and reproduced over time by day-to-day practices that are strongly conditioned by surrounding social institutions and regulatory regimes. Consequently, the very practices taken as signifiers of distinctive cultures are themselves influenced by a set of institutions constituted outside the individual firm.

1.2.6 A cultural turn in political economy?

Much of the discussion in the previous section alludes to a strong theme in the ongoing debate about the emergence of a 'new economic geography' – that there is a (re)recognition that the 'economic' is culturally and socially grounded and embedded (Crang, 1997; Thrift, 1994; Thrift and Olds, 1996). For example, the conceptualization of agents and agency has been influenced by the encounter between economic and cultural approaches. The focus has shifted towards approaches that conceptualize both organizations and employees as actors with sets of cultural attributes which are constituted in, affected by and affect the huge range of interactions that take place at a variety of spatial scales in any form of economic interaction (McDowell, 1997, 119). While it may be overstating the case to claim that this is so in any (and every?) form of economic interaction, it is more difficult to challenge the notion that (what we understand by) 'the economy' is culturally constituted (Crang, 1997). Thus economic rationalities are culturally created, take diverse forms and have distinctive geographies (Peet, 1997b). As Williams (1983) forcefully points out, however, 'culture' is a very tricky concept, imbued with a variety of meanings. Albert (1993) suggests that the assertion that there can be a 'culture of the economy' may seen suspiciously vague to some, while to others it may seem a tautology. But Albert insists that if there is one word that designates a body of individual behaviour patterns shared by a whole population, enshrined within institutions, subject to agreed rules and a common heritage, then it is indeed a culture.

Culture must be seen as a product of social interaction rather than as some pre-given way of seeing the world. If economic practices are culturally embedded, this reflects ongoing and active social processes. It is therefore important to examine the processes by which cultures are actively produced and reproduced by social practices and institutions over time (Gertler, 1997, 51). As such, culturally embedded economic action should be seen as dependent upon collective understandings that shape economic strategies and goals. More radically, it can be claimed that no purely economic, social or cultural relations are distinguishable. In contrast, each is already embedded within the other (Halford and Savage, 1997, 109). This latter claim raises an awkward question, however, for if each is already embedded within the other, it may be difficult to distinguish them from one another – in which case, it is not at all clear what cultural, economic and social would denote.

There is much of value in seeking to problematize the economic in these ways. It is also important to emphasize, strongly, that a lot of what is claimed to be novel actually is not so new. Although the 'cultural turn' in economics and economic geography may seem a radical and hotly disputed shift, cultural analysis has a long history in economics and sociology, from Marx's analysis of commodity fetishism and reification and Durkheim's comments on the non-contractual elements of contracts (McDowell, 1997, 120). An emphasis upon cultural specificity in economic and social relationships is central to Marxian political economy and strands of classical sociology. This is an important corrective to those views that seek to counterpose political economy to cultural perspectives. A recognition that economic and social relationships take different forms in different times, places and cultures, and also are represented in different and competing ways, is central to Marxian political economy. Recognition of different modes of production, of different social formations within the structural limits of a particular mode of production, and of the ways in which these do or do not articulate and relate, is central to a political-economic understanding of the historical geographies of capitalism. Marxian analysis of capitalism precisely challenges competing representations that seek to deny this. Indeed, this is so central to Marxian political economy that it is difficult to understand why it is seen as insensitive to cultural difference in this way.

1.2.7 Pulling the threads together

The basic (but certainly not original) point to establish is a recognition that we need to go beyond simple description and mapping of the spatial patterns of geographies of economies and extend beyond describing and mapping geographies of production and of uneven development. The key issue is how best to conceptualize the social processes of production and the ways in which these both use and produce space and spatial differentiation. The objective is to seek to conceptualize the relationships between individual and collective agency, structure, nature, space and place in a way that helps understand the complex and shifting geographies of capitalist production.

Consequently, it is necessary to construct an adequate conceptualization of the underlying social processes (including those that link people and nature), and of their relation to spatial differences and differences between places. This raises a pivotal question: what is adequate? An adequate conceptualization is one that penetrates beneath the surface appearances of geographies of production, uneven development and the processes of exchange in a capitalist economy and reveals the inner mechanisms of the capitalist production process. A good theory is one that offers a detailed description of the key causal processes and mechanisms.[5] Marxian political economy offers such a window onto the world of production, the economy and uneven development. The nature of the view from this window can be further clarified by reference to critical realism (Sayer, 1984). In epistemological terms, critical realism argues that an adequate theory must reveal necessary causal relationships and mechanisms and recognize the existence of a variety of relatively autonomous causal structures (not just those of the social relations of capital). The realization of these causal powers is contingently dependent upon the ways in which they come together in spatially and temporally specific contexts, in terms of both the combinations of causal structures present and the specific local circumstances of that time–place. Such a position is broadly consistent with the conception of tendential laws within Marxian political economy, with the empirically observed trends reflecting the relative weights of tendencies and counter-tendencies (some of which may be located in non-capitalist social structures endowed with causal powers). Whether a particular set of causal powers is realized in empirically observable form, and to what extent, depends upon the balance between them and counter-tendencies, in the form of both competing structures of causal powers and particular local circumstances. An important corollary of such a view of an adequate conceptualization is that a theory is defined in terms of a description of causal structures and mechanisms. The criterion on which the 'goodness' of a theory is judged is therefore the extent to which it aids interpretation and understanding of a process rather than predictive power *per se*.[6]

Such a perspective, informed by critical realism, suggests that while Marxian political economy provides one window on production and forms a powerful point of departure, it is necessary further to elaborate the theoretical approach. This elaboration must identify and specify lower levels of intermediate theoretical constructs and locate them *between* structural relations and individual behaviour. This is not to deny the importance of individual agency, but rather to insist upon placing individuals in their social context and avoid privileging the methodological individualism of neo-classical economics and behavioural geography. Nor is it to deny the importance of structures, but rather to insist that these are socially (re)produced and that it is an illegitimate move to seek to read off or deduce in a mechanistic and deterministic way everything, including the geography of the economy, from such structures. The aim is to avoid collapsing agency into structure or structure into agency, in the process producing pale shadows of each. At the same time, it is necessary to insist that space is more than simply a pre-formed container in which social processes unfold and that spaces and meaningful places are both socially produced and constitutive of social processes. Equally, it is necessary to recognize that the economy and production are unavoidably grounded in nature, with complex interactions between the social and natural worlds as a result. As geographers have sought more satisfactory answers to these questions arising from a recognition of these interrelationships over the last three or so decades, the ways in which they have sought to construct explanations and deepen understanding have evolved and changed, involving a judicious blending of different intellectual traditions.

Some seek to reject Marxian political economy on the grounds of 'essentialism', that it privileges class relations to the exclusion of all other social relations, everywhere and at all times, and offers a totalizing account of capitalism and its geographies. As a result, economic geography is seen to have been dominated by landscapes of class that foreground class relations between capital and labour and competing capitals. It is claimed that the multiplicity of possible class relations has been reduced to a single key relationship. This is largely because class is seen to have become associated with a solidified and singular categorization of social and economic activity: a capitalist social formation, a capitalist world economy or some other version of capitalist hegemony (Gibson-Graham, 1997, 149). Thus the essence of Gibson-Graham's critique would seem to be that economic geographers analysing capitalist economies have focused attention upon the key class structural relations that define them *as* capitalist (see also Gibson-Graham, 1996). As such, it is a perverse critique. Moreover, it is one that deliberately or inadvertently ignores the extent to which economic geographers (amongst others) have distinguished social formations from modes of production. In doing so, they have acknowledged that while capitalist social relations define societies *as* capitalist, there may be important relationships between capitalist and non-capitalist class relations and between class and non-class relations (including those of ethnicity, gender and territory) in the constitution of geographies of capitalist societies. That the concept of social formation precisely allows for a multiplicity of class and other social relations seems to slip by unnoticed. While wishing to avoid an interpretation of capitalist social relations as totalizing, it is equally important to recognize that there are broad social structural relations (of class, gender and ethnicity, for example) that have determinate (though not deterministic) effects, 'most especially if at the same time their multiplicity and complexity is recognised. Recognition of such broad structures is not the same as the commitment to, or adoption of, a metanarrative view of history. None of the structures . . . need to be assumed to have any inexorability in their unfolding . . . outcomes are always uncertain, history – and geography – have to be made' (Massey, 1995, 303–304). While Marxian political economy can be read or used in an essentialist and totalizing manner by those who wish to do so, for whatever reason, there is no necessary reason why this has to be the case. As Massey demonstrates, more sophisticated analysts are at pains to avoid this charge (see also Jessop, 1990, 4–7). If the charge levelled by Gibson-Graham is that a focus upon capitalist class relations is misplaced because they are no longer dominant and defining, then clearly the object of analysis is no longer capitalist societies. While it is easy to agree with the commendable wish for a more humane economy guided by different social relationships, simply wishing away the realities of capitalist power, material and discursive, is not a very helpful step towards attaining such a goal.

Gibson-Graham argues for an 'anti-essentialist' definition of class as a social process involving the production, appropriation and distribution of surplus labour 'in whatever form' (1997, 91, emphasis added), but concede that 'to empty class of much of its structural baggage [sic] and prune it down to one rather abstract process concerning labour flows might seem rather reductive'. While rejecting the pejorative concept of 'structural baggage', one can only agree that this is indeed 'reductive', as defining class in this way robs the concept of any analytic specificity. Gibson-Graham argue that an anti-essentialist position allows the conditions of existence of *any* class process to assume specific importance in the formation of class societies and subjectivities, without presuming their presence or role. Gibson-Graham thereby deny that there are *any* necessary conditions or relations defining particular types of society, and as such this position is particularly unhelpful. A more helpful perspective, drawing on critical realism, is to acknowledge that there are necessary causal structures which define particular types of society but that societies encompass multiple causal structures, not all of which in this sense are necessary and not all of which are equally powerful. Moreover, in any case the causal powers of such structures can be contingently realized only in specific time–space contexts. Consequently, the causal powers inherent in the social relationships of capitalism may be pre-eminent and must be present in the sense that they define capitalist societies *as* capitalist, but it does not follow that they have a determinate (let alone deterministic) influence on each and every occasion in shaping the geographies of capitalist economies.

Marxian political economy thus offers a valuable window onto the world of production, the economy and their geographies and provides a way of constructing a structured but non-deterministic and admittedly partial account of that world and its geographies. It does not claim to give an over-arching general and deterministic account. It does not claim to grasp and account for all aspects of production in capitalist societies, let alone everything about those societies and their varied geographies. It reveals much, but by no means all, of the explanation for such geographies. To reveal more, it is necessary to gaze upon production through other theoretical windows. Given the changes that have taken place in the character of capitalism over the last 150 or so years, it would be surprising if Marx's own writings were able to deal with its contemporary complexities, but those working in and from a Marxian tradition have sought to accommodate such changes (for example, see Lash and Urry, 1994; Thrift, 1994; Williams, 1989). In recognition of these varied and interrelated changes, in much of the humanities and social sciences, including human geography, we have entered an era of 'epistemological relativism and methodological pluralism' (Gregory *et al.*, 1994, 5). Methodological and conceptual pluralism is therefore 'no bad thing' (Ward and Almå, 1997, 626). A continuing, constructive dialogue between different perspectives is clearly preferable to a continued search for a single, new, all-encompassing paradigm. While there are certainly dangers in an indiscriminate theoretical eclecticism (Fincher, 1983), a search for all-encompassing paradigms may very well be doomed to failure. This is not an argument against metanarratives but rather recognition of the need to pay careful attention to what any given metanarrative includes and excludes. Since no theoretical system can provide a complete and satisfactory explanation, 'we need to find ways of living – critically and creatively – with theoretical dissonance' (Gregory, 1993, 105).

Nonetheless, it is clear that some theoretical approaches undoubtedly have greater explanatory power than others. Not all theories are of equal utility. What, then, are the criteria to be employed in deciding in any given set of circumstances which theoretical perspective should be accorded priority? How is the prioritizing of one theoretical framework over others to be defended? How are sceptics to be convinced of its validity and value? This cannot be done within the terms of the theory itself. Justification of a theoretical position inevitably involves reference to the assumptions upon which the theory is founded and the values and norms in which it is grounded, and in that sense is political. Accepting the truth of any one claim about the world is problematic (Clark and Wrigley, 1995, 207). Even so, Clark and Wrigley argue that accepting that there cannot be universal agreement about the terms and conditions of truth and accepting the plurality of worlds and ways of world making does not deny the validity of theoretical frameworks. While it might perhaps the more reasonable to refer to claims about 'valid knowledge' rather than 'truth', it is difficult to dispute the claim about the validity of theoretical frameworks. We just have to accept the inevitable incompleteness of such frameworks. Theory making is a process of construction rather than deconstruction and in this respect theory making is doubly contingent: contingent upon the predicament in which we find ourselves and contingent upon the theoretical context in which we work. At best, theory making is a process of persuasion and argument both within and outside academia, and choice of theoretical framework is to a degree a political choice.

1.3   Changing substantive foci of interest in economic geography

A corollary of the changing ways in which economic geographers have sought explanations and understanding is that there have been shifts in the substantive foci of their studies. The emergence of political economy approaches certainly led to a strong emphasis upon production, and in particular upon capital–labour relationships, competitive relationships between companies, geographies of uneven development, the relationships of capital and labour to the state, and state industrial, urban and regional policies. These remain issues of central importance to economic geography, although the ways that economic geographers study them have altered and the relative weight of emphasis placed upon them has shifted. It is also important to acknowledge that the substantive foci of interest have also changed in response to changes in the economy, its forms of organization and its shifting geographies. A good example of the latter is the strong growth in interest in issues such as the new international division of labour and of globalization (Dicken, 1992a). The shifting explanatory trajectory of economic geography is thus a product of a complex interplay between perceptions of changes in the economies of actually existing capitalisms and intellectual debates and struggles as to the relative merits of competing and contested theorizations of economies and their geographies. In the remainder of this section some of the newer areas of concern, as well as some of the changes in emphasis in established areas of work, will be outlined. By way of conclusion, some absences and areas of neglect will be noted.

As the Taylorized mass production systems of Fordism slid into crisis, companies first sought a series of intra-national (Lipietz, 1980b; Massey, 1995) and then international (Fröbel *et al.*, 1980) 'spatial fixes' (Harvey, 1982) to preserve the competitiveness of mass production. As it became clear that this was at best a limited and partial solution to their problems of profitable production, they began to experiment with new forms of production organization and re-working capital–labour relations in the labour market and at the point of production. In part, the search for fresh ways of profitably organizing production involved new forms of high volume production such as 'lean production' (see, for example, Womack *et al.*, 1990), in part the (re)discovery of small firms linked into competitive industrial districts (Amin, 1989). For some, this heralded a transition from 'Fordism' to 'post-Fordism', a simple and clear dichotomous break from one set of organizing (and regulatory: see below) principles to another. In contrast, others argued that what was involved was a much more uneven and nuanced set of changes, generating novel and more complex forms of uneven development (and that is very much the position advocated here).

These new forms of production organization involved re-shaping relations (competitive *and* co-operative) between capital and labour in various ways. For example, it involved the introduction of new ways of selecting workers, new forms of work, new ways of organizing the labour process, new forms of industrial and labour relations and often new geographies of employment and production. These radical changes were often facilitated by high levels of unemployment, spatially concentrated, which tilted the balance of power between capital and labour even further in the direction of the former. There was considerable debate as to the meanings of these changes and the beneficiaries of them. Some, for example, saw these new forms of production as heralding the emergence of re-skilled and multi-skilled work for empowered workers and opening up the possibilities of a 'high road' for territorial development strategies (for example, Florida, 1995). Others in contrast saw them as new ways of intensifying the labour process via multi-tasking and the self-regulation of workers by workers bound together via discourses of teamwork and further evidence that the areas that became the recipients of such workplaces were firmly locked onto a 'low road' economic developmental strategy (see, for example, Garrahan and Stewart, 1992). At the same time as Taylorism was allegedly on the wane in manufacturing, however, it was becoming increasingly prevalent in many service occupations (Beynon, 1995) and also becoming defined in new ways in terms of time and uncertainty in the labour market (Beck, 1992).

The recognition of a growing variety of forms of production organization also led to renewed interest in the diverse forms of co-operation as well as competition that shape inter-firm relationships. For example, shifts to just-in-time production methods were seen to involve deeper and more long-lasting relationships between companies, based around trust and mutually shared interests rather than price competition in markets. Others emphasized the asymmetries in power between partners in such networks and pointed to the persistence of price competition as the primary mechanism regulating inter-firm relations. For example, in the automobile sector relationships between assemblers and first-tier suppliers were increasingly structured around longer-term (but never unconditional) contracts, but relations between component supply companies at lower levels of the supply chain continued to be regulated via fierce price competition (Hudson and Schamp, 1995). As well as these changes in the form of contractual relationships linking companies, there was also an increasing recognition of the emergence of longer-term strategic alliances and joint ventures, which often had important direct and indirect consequences for geographies of production (Dicken and Oberg, 1996).

Changes in forms of production organization have also had implications for relationships (competitive *and* co-operative) between workers (waged and unwaged). New forms of organizing production and new forms of individualized pay and remuneration created fresh divisions within workforces, which became intertwined in complex ways with established divisions along the cleavage planes of class, ethnicity and gender (Yates, 1998). Such changes posed major challenges for trades unions and those seeking to represent the collective interests of workers. On the other hand, the growing prevalence of just-in-time production methods and single-source plants as part of 'lean production' strategies created potential opportunities for organized labour as companies became acutely sensitive to interruptions in the flow of output from one plant in the production chain (Wills, 1998a). This potential was often unrealized, however, precisely because of aggregate labour market conditions that militated against industrial action, as workers feared that if they lost the job that they had, they would be unable to get another one.

For many commentators, the capitalist economy has undergone a fundamental structural sea-change from an industrial to a 'post-industrial' or 'service' economy, revolving much more around the commodified production of intangible services than tangible goods. There is no doubt that to a degree such changes have occurred and that there has been growth in services. In this sense, it is correct to speak of a further deepening of the social division of labour (Sayer and Walker, 1992). Equally, it is also important not to overstate the extent of fresh growth, nor to be seduced by official statistics as to the sectoral composition of employment and output. As companies responded to crisis in various ways, they increasingly began to focus upon identified core competencies and to contract out a range of activities that formerly were performed within the company. Thus intra-company occupational diversity was reduced and reflected in the emergence of new service sectors and the expansion of existing ones (across a range of activities as diverse as accountancy, contract cleaning and information technology and computing). Furthermore, there is often confusion as to what actually constitutes a 'service' and its relationship to the material production processes of the industrial economy (Walker, 1985). The performance of many services depends upon the availability of material goods and artefacts, for example. Claims as to a transition to a 'post-industrial' or 'service' economy are thus seen to be based in conceptual confusion and a failure to grasp the significance of the deepening social and technical divisions of labour within contemporary capitalism.

The claims about a transition from Fordism to post-Fordism do not simply impinge upon the production process in a narrow sense but extend to broader questions about the ways in which production is made possible via the actions of states and other institutions involved in regulation and governance. This is because such claims are situated within a regulationist perspective that argues that for stable growth to be possible, modes of regulation must be discovered that are compatible with a particular economic growth model or regime of accumulation. Others argue that in fact it is misconceived to search for stable modes of regulation and that capitalist regulation chronically involves the interaction of a set of regulatory practices, many of which are incompatible and in conflict with one another (Goodwin and Painter, 1996; Painter and Goodwin, 1995). Moreover, there is much more involved in a successful mode of regulation than simply state activity and this is associated with an acknowledgement of the importance of focusing upon governance rather than just government. In part, this recognition is linked to claims as to the 'hollowing out' of the national state as the dominant capitalist state form and more generally the reorganization of the state (Jessop, 1997). Successful modes of regulation generally involve a range of other formal and informal institutions, extending from habits and routines, through the family to various formally constituted institutions in civil society (see, for example, Théret, 1994). This in turn raises questions about the links between the variety of forms of work and employment relations through which economy and society are constituted. For instance, it raises questions about relationships between different types of waged work in the formal and informal sectors of the economy (see, for example, Mingione, 1985). It also raises questions about relationships between waged work and unwaged work (both within and outside the family home) in reproducing the conditions that make wage labour possible, especially in relation to guaranteeing labour supply and the availability of appropriate quantities and qualities of labour-power (Peck, 1996). Furthermore, it focuses attention upon the growing interest in various forms of 'third sector' employment, the growth of socially useful work that would not be validated by the 'normal' workings of the labour market (see, for example, Amin et al., 1998; Hudson, 1995a; Lee, 1996). Capitalist economies are thus seen to be constituted via a variety of forms of work and employment relationships, not all of which fall within the scope of capitalist wage relations.

A further point that can be made in this introductory section is related to the re-conceptualization of space and place. There has been a growing emphasis upon territory (city or region) rather than company as the key basis of competitive success. This has led to a growing focus upon the internal characteristics of regions, their cognitive and institutional assets (Amin and Thrift, 1994) and their capacity to become 'intelligent', 'learning' regions (Morgan, 1995). Regions are regarded as 'nexuses of untraded inter-dependencies' (Storper, 1995) that underpin the relationships of traded dependencies that signify success in the market. This emphasis upon cities and regions as providing the bases of corporate economic success implicitly at least depends upon a particular conception of places as (at least relatively) bounded, closed and sharing a common set of objectives and goals. There are, however, other ways of thinking about place and space that problematize this view of territorially grounded economic success. The work of Massey in the 1970s and 1980s was seminal in bringing geographers face-to-face with a recognition of the need to realize that relations between social process and spatial forms were reciprocal ones. Social space became seen to be a product of social relationships stretched over space and constituted by social relationships rather than as existing as a pre-formed container into which those social relationships were deposited. More recently, she and others have begun to develop more sophisticated conceptualizations of place. Places are seen as condensations of different social relationships coming together in the 'same' time–space location, with the density, variety and types of social relations that intersect there helping to define different types of place. Places are seen as possibly discontinuous, open and with permeable boundaries rather than as necessarily continuous, closed and with impermeable boundaries

(Allen *et al.*, 1998), although in practice it is unlikely that any place was ever completely closed. Places are seen to be complex, endowed with multiple and contested identities and meanings. As a result there are typically struggles to resolve which of these identities and meanings will or should become dominant. The ways in which a place does and should intersect with different geographies of economies are also often contested. Such struggles raise important questions not just about policies but also about politics. If particular places project a dominant, even hegemonic, image of what they are and what they are meant to become, then this has not arisen naturally but is a product of political debate and struggle.

The penultimate point to be made in this section is that there has been a growing recognition of the importance of the links between production and the natural world. Marxian political economy certainly conceptualized the production process as involving the human transformation of elements of nature, but the implications of this were not really pursued for some time (though see Smith, 1984). As fears of the economic growth process grinding to a halt because of the exhaustion of supplies of key raw materials were replaced by a fear that the pollutant effects of capitalist economies would endanger human life on the planet, economic geographers became increasingly interested in such issues. The totality of the production process, from extracting elements of nature, through transforming them in production, to the dissipative wastes generated via their consumption and dumping at the end of their socially useful lives, was increasingly seen as critical. Economic geographers have begun to explore concepts such as that of industrial metabolism and methodologies such as those of Life Cycle Analysis to elucidate these issues (Taylor, 1995). While companies could seek 'spatial fixes' to localized problems of pollution via relocating polluting processes or exporting pollutants (Leonard, 1988), there was a recognition that such tactics simply shifted the location of the problem rather than solving it. Moreover, the impacts of pollutants often returned in unanticipated and unwanted ways to the areas that initially generated them (see, for example, George, 1992). Consequently, there was a growing interest in the requirements for and implications of a transition to clean and ecologically sustainable production and places, especially given the prospect of rapid industrial expansion in countries such as China. Acknowledging the grounding of production in nature has far-reaching implications that imply conceptualizing production in terms of relations between the social, the natural and the spatial rather than just the social and the spatial. Thus there is an interplay between what is emphasized and how conceptualization is undertaken. Moreover, introducing environmental concerns raises issues of equity and social justice as well as environmental sustainability (Harvey, 1996).

Finally, it is important to emphasize that there has been a very significant shift in the interests of economic geographers over the last decade to issues of consumption and identity which is underplayed in what follows (see, for example, Crang, 1998; Wrigley and Lowe, 1995). For some 20 or so years the legacies of location theories that involved discrete attempts to deduce geographies of production and geographies of market areas undoubtedly had a debilitating effect on economic geography. This is not to deny that in other ways there are important connections between the ways in which economic geographers sought to understand links between changes in production and consumption. Much of production involves the production of commodities by means of commodities (Sraffa, 1960), as one company's outputs become another company's inputs via processes of productive consumption. Certainly the logic of the capitalist production process, validated *a posteriori* by the sale of commodities in markets, presumes the consumption of what is produced. Marxian political economy has long been aware of the importance of culture and meanings. There is, however, a considerable difference between conceptualizing the significance of meanings in terms of ideology and false consciousness and in terms of multiple meanings and identities. Nonetheless, the significance of the growth and processes of advertising to try to ensure that products are successfully sold has long been recognized (see, for example, Williams, 1960). For some, the emergence of more product-differentiated flexible production systems for consumer goods reflects a response to growing consumer sovereignty and the increasingly sophisticated demands of consumers. For others, such changes reflect more the aggressive advertising and marketing strategies of companies seeking to segment markets in order to enhance their competitive position. Such strategies have been enabled by the regressive income distribution policies in the late modern world that greatly increased the disposable incomes of a small minority of middle-class consumers (Curry, 1993). These changes in consumption patterns clearly have implications for individual identities, both of those who can afford to consume the new niche commodities and those who cannot. Equally, the growing emphasis upon places competing with one another for investment and jobs within an often global marketplace has involved policies and strategies to (re)create the identities of places. To a degree, this has helped shape geographies of production and become intertwined with the (re)construction of the identities of people who live in these places. The relative neglect of this revived interest in consumption and identity in what follows should not be seen as indicating that it is seen as unimportant but rather simply a reflection of the fact that these are not issues that have featured prominently in my own work.

Notes

1 In the context of geographies of consumption and studies of consumer behaviour, behavioural approaches typically sought to discover the knowledge that consumers possessed about the retailing environment and how they came to acquire such knowledge (for example, see Hudson, 1974).
2 Such approaches bore more than a passing resemblance to the deductive strategies of neoclassically inspired location theorists.
3 While all forms of social relations can be defined as actor-networks, this still leaves unanswered the question of why some actor-networks dominate over others.

4 While the four-fold typology of ideal–typical worlds of production advanced by Salais and Storper is useful for expository purposes, it fails to capture the nuances of production organization in its practical complexity. Nonetheless, the concept of worlds of production remains a valuable way of thinking about the organization of production.

5 It therefore follows that neo-classical accounts, predicated upon assumptions of certainty and static equilibrium, fall at the first hurdle in the theoretical stakes.

6 Although a 'good' theory could certainly yield predictive insights in many circumstances, these would not be the criterion on which the theory should be judged. Moreover, failure to predict would not be a criterion for abandoning the theory.

Part 2

Re-thinking regional change

Introduction

In 1972 I returned to northeast England and to a job in the University of Durham. I was struck by the extent of changes in the region since I had first moved as a student to Bristol in 1966 and intrigued to understand more about the causes of change (this autobiographical context is set out more fully in the Preface to Hudson, 1989a). These are questions that have continued to interest me over the intervening years. The focus of the chapters in this part of the book is therefore upon an evolving understanding of the changing position of northeast England in the wider national and international political economies, the changing socio-economic geography of that region, and the links between changes within and changes outside the region. More generally, it is concerned with grappling with the character of regional uneven development within capitalist societies.

In 1972 I was completing a doctoral thesis which was very firmly grounded in the behavioural geography framework of environmental knowledge and learning (for example, see Hudson, 1974) and was increasingly aware of the explanatory limitations of such approaches. As a result, in seeking to come to terms with what had happened and was happening in northeast England, I, along with others, began to search for more powerful explanatory frameworks that could help comprehend the evident changes in economy and society in the northeast and in its changing position within a wider world.

Initially, this greater explanatory power was sought in analogies with dependency and underdevelopment theories, seeing peripheral regions within the territories of advanced capitalist states as analogous to countries on the periphery of the global capitalist system (Carney *et al.*, 1976). The limits to such an approach soon became clear, however (not least because of the qualitative differences between national states, however peripheral, and regions within national states, however powerful). Reflecting the spirit of the times in much of the social sciences, as well as more specific moves within human geography (Harvey, 1973), this search for analogies in the Third World was soon replaced by a more direct engagement with Marxian political economy and its conception of modes of production and social formations. Concepts of combined and uneven development became a source of explanation and theoretical inspiration (Carney *et al.*, 1997). At the same time, there was a concern to avoid overly simplistic mechanistic and economistic accounts through beginning to engage with issues of ideology and politics as well as the structural relationships of the capitalist mode of production. Chapter 2 is a product of the sort of approach that I was then seeking to develop in order to understand the trajectory of change in the historical geography of northeast England. This approach was grounded in Marxian political economy but sought to avoid 'reading off' this developmental trajectory and the associated pattern of territorial development from structural relationships by means of a recognition of the (collective) strategies of capital, labour and the national state in moulding this trajectory and geographies of uneven development. In retrospect, this certainly sought to forge too close and direct a link between empirically observable tendencies and value categories, adopted a too restricted and unsubtle a conception of both agency and region, and incorporated a rather poorly specified periodization of change (although this could perhaps in part be defended on the grounds that when this chapter was first written in 1980, the full significance of the switch from a national 'One Nation' to a 'Two Nations' political project for and in the northeast were yet to be revealed).

Chapter 3 seeks to offer a more up-to-date and thorough analysis of changes in northeast England and of the position of the northeast in the national and global orders of things. It seeks to do so in two ways. First, empirically, it offers an account of changes in the economy, politics, state policies and geo-graphy of the northeast in the 1980s and 1990s. This is a story of the destruc-tion of the industrial economy of 'carboniferous capitalism', much of it a direct consequence of national state policies, and the attempts to construct a more diversified, 'modern' and robust regional economy via a mixture of inward investment and endogenous growth of small firms in both manufac-turing and services. Since this chapter was first written, a series of plant closures and job losses in the summer of 1998 have confirmed the persistent vulnerability of the regional economy, despite claims as to fundamental eco-nomic transformation in the 1980s and 1990s. The closure of the new Fujitsu and Siemens factories producing integrated circuits came as a sharp reminder that the 'branch plant syndrome' was far from dead. It also indicated that claims as to the transition to a 'high tech' knowledge-based economy were based on a failure to comprehend the character of such inward investment. Job losses at the Barclaycard back-offices on Teesside came as a sharp re-minder that the service sector equivalents of the manufacturing branch plants were also vulnerable to job loss as a result of events and decisions outside the region. Secondly, this chapter seeks to offer a more rounded and nuanced explanation of the changes in and to the regional economy and society over the last half century or so. The central explanatory thesis of this chapter is that the transformation of the regional economy is reciprocally related both to the changing position of the UK in the global economy and to the restructuring of the UK state. Regional changes are seen as both a product of and a condi-tion for wider projects of restructuring the UK state and its relation to economy and society within and beyond the UK. In constructing such an interpreta-tion, it draws upon regulationist accounts of capitalist development and its geographies, critically utilizing concepts such as regime of accumulation and mode of regulation, 'the hollowing out of the national state' and 'globalization' (for example, see Dunford, 1990; Jessop, 1994; Hudson, 1999b). It thus seeks to relate changes within the northeast to profound changes in the character and geography of contemporary capitalism, changes reflected within the UK in the transition from a 'One Nation' to a 'Two Nations' political project (Hudson and Williams, 1995). At the same time, the increasingly deep socio-spatial divisions within the region, alongside the changed character of relations between the region and other parts of the world, raise questions as to the most appropriate way to conceptualize regions, emphasizing openness, porosity of boundaries and spatial discontinuities.

One issue to emerge in Chapter 3 is the growing emphasis upon endogenous growth as the 'solution' to problems of regional economic development. In the northeast, as in places elsewhere that are also under the sway of neoliberal regulatory regimes, the responsibility for urban and regional economic development strategies was pushed back onto cities and regions. In seeking to devise local economic development strategies, places on the margins, such as those in northeast England, looked to those cities and regions that were economically more successful to see what lessons might be learned from this success. Such regions seemed to be characterized by particular sorts of institutional structures, especially those linked to the production and transmission of knowledge and learning, shared beliefs and goals and a sense of common purpose. As a consequence, in peripheral regions such as northeast England there was an increasing concern with creating the conditions in which such so-called 'soft infrastructure' could emerge as the basis for a transition to a knowledge-based, high-skill economy. Chapter 4, the third in this part of the book, summarizes and sympathetically critiques the literature on learning (economies, firms and regions) not simply within the context of the northeast but more generally. The growing interest in knowledge and learning in the practical context of devising territorial development strategies is in fact simply one facet of a broader theoretical engagement between economic geographers and regional analysts on the one hand and evolutionary and institutional economists on the other. In many ways this has been, and continues to be, a fruitful engagement. The focus on knowledge is often presented as a dramatic breakthrough, promising radical theoretical re-appraisal and opening up exciting new policy vistas. Recognizing the importance of knowledge and innovation to economic success is hardly novel, however. The chapter first summarizes the claims made by the proponents of 'learning' and makes some links between the emphasis placed upon knowledge and learning and other theorizations of the ongoing changes in the organization of work, production and territory. It seeks to situate and contextualize claims as to the significance of 'learning' and place them within the context of continuities and changes in processes of uneven development within capitalism. It ends with some concluding remarks about the limits to learning, and questions of learning by whom, for what purpose, in the context of the politics and policies of economic, social and territorial development.

Chapter 2

Capital accumulation and regional problems: a study of northeast England, 1945 to 1980*

2.1 Introduction

This chapter describes and accounts for the post-war restructuring of the economy of northeast England (Figure 2.1, overleaf; see also Carney *et al.*, 1976, 1977). Its central explanatory thesis is that the observed pattern of changes must be understood both as the outcome of, and to some extent a precondition for, the process of capital accumulation, that is, not merely the expansion of value but also the extended reproduction of capitalist social relations. This entails a consideration of politics and policies (for example, those of capital – individual companies – labour and the state), of economic change in the narrow sense and of the crucial question of securing the legitimacy of those social changes that form an integral part of the accumulation process. (The thesis poses epistemological and methodological problems which are no less troublesome as a result of being well known – for instance, see Massey, 1978.)

Such a process is uneven, with an inherent tendency to crisis which, on occasion, actually occurs. As Mandel (1975, 438) suggests, accumulation, the central driving force of the capitalist mode of production, follows a course of 'successive phases of recession, upswing, boom, overheating, crash and depression' which is, as it were, genetically endowed. This arises because accumulation depends upon an increase both in the mass and in the rate of profit and because the capitalist mode of production is characterized by a tendency for the rate of profit to fall, which then sets in motion counter-tendencies either to avoid crises or (temporarily) to overcome them and restore profitability (hence the expanded reproduction of capital) by a process of restructuring, reducing the value of the component parts of capital. The onset of crisis and this restructuring at the level of 'capital in general' are brought about through, and made possible by, competition between individual capitals. These deploy strategies to try to guarantee or boost their profits, to attain surplus profits (that is, profits that are above average for a branch of production). Several strategies are open to individual capitals striving to raise profitability (Mandel, 1975, 77–78). They can switch investments between branches or sectors of

* First published 1983 in FEI Hamilton and GJR Linge (eds) *Regional Industrial Systems*, John Wiley, Chicester

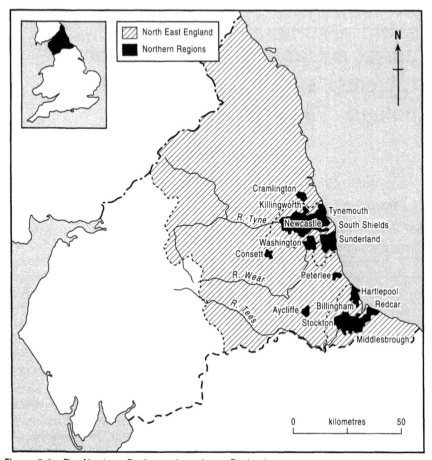

Figure 2.1   The Northern Region and northeast England.

the economy or alter the spatial allocation of investment, changes which, increasingly, are made possible because of the interrelated combination of the centralization of capital and an increased size of individual capitals (itself a product of competition), the larger scale of production associated with the concentration of capital and concomitant changes in the labour process and division of labour as a result of technical progress which centralization and concentration both permit and require. Moreover, the pattern of spatial differentiation resulting from earlier periods of uneven capitalist development is an important precondition for such changes, with the state frequently and increasingly intimately involved through policies designed generally to reduce the cost of elements of the component parts of capital or selectively to lower the price of some of them to some units of capital, while legitimizing these actions as socially progressive.

Thus capitals' restructuring to avoid or overcome crises and maintain, increase or restore profitability requires a redefinition of capital–labour relations in favour of the former and this frequently, if not inevitably, implies the creation of crises (such as increased unemployment or cuts in living standards) when seen from the viewpoint of labour and the working class. At the same time, these changes must be justified, if not as legitimate, then at least as the outcome of natural economic processes, to guarantee the reproduction of capitalist social relations; in this the state is inextricably involved. The post-war history of the northeast is considered against this brief background.

2.2 Recovery and recession: 1951 to 1962

In sharp contrast to the inter-war years and the latter part of the 1940s, the 1951–58 period can be characterized as a 'Golden Age' for the region. The working class, a high proportion of whom had experienced prolonged unemployment in the inter-war years, perceived this to be a time of full employment (although this was only maintained by the out-migration of a considerable surplus population to meet the expanding demand for labour in the South East and the West Midlands) and rising real wages (Figure 2.2, overleaf, and Table 2.1, pages 37–38). It was also a period favourable to the interests of capital. In manufacturing, for example, the 1948–51 profitability crisis was overcome: the mass and rate of profit both rose markedly (Table 2.1). For a time, then, it appeared that the objectives of capital and labour were complementary: how this apparent reconciliation of fundamentally incompatible interests could temporarily be maintained is not without significance.

Of considerable importance was the way that working-class and trades union demands were formulated and pursued, these being greatly influenced by the region's inter-war experience. Even during a time of full employment, and no doubt reinforced by rising average real wages, memories of the 1930s served to shape these demands. As a result, the attainment and maintenance of full employment became an end in itself rather than being used as a platform from which to advance further and more radical demands. A recovery of profitability, therefore, was eminently feasible and, given a rising level of effective demand for labour both within this region and elsewhere in the UK, was compatible with the maintenance of full employment within the northeast. Several aspects of this process can be identified.

First, effective demand for many commodities 'traditionally' produced in the region recovered strongly. The British economy during this period remained essentially based on one fuel: the nationalization of coal mining in 1948 reflected the need to guarantee production of a key element of circulating constant capital, the provision of which had ceased to be sufficiently profitable from the viewpoint of individual capitals. Since national economic recovery and expansion depended upon coal as an energy source, there were compelling pressures to maximize national coal output. Consequently, coal mining employment in the northeast was relatively stable (Table 2.2, pages 40–41). Simultaneously and, to some extent relatedly, effective demand for some of the products of the region's 'traditional' industries – shipbuilding, marine engineering, mechanical and electrical engineering, metal manufacture, armaments – also recovered strongly.

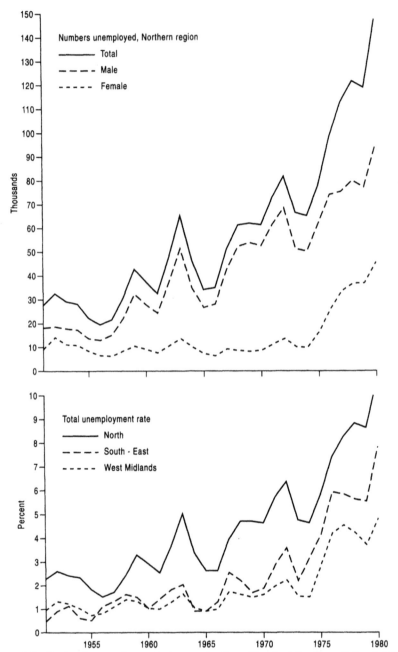

Figure 2.2 Regional unemployment changes in three Standard Regions of Great Britain, annual averages 1950–80.

Table 2.1 Changes in total manufacturing and selected industry groups, Northern Region, 1951–70

Industry group	1951	1958	1965	1968	1970
(a) Employment (thousands)					
Chemicals	40.6	59.2	57.3	49.1	55.0
Metals	58.9	58.3	52.2	50.3	57.3
Shipbuilding, marine engineering	51.7	60.0	39.9	36.2	34.4
Textiles	14.1	15.5	19.2	21.6	23.7
Paper, printing, publishing	12.7	13.0	14.4	18.4	19.7
Total manufacturing	374.7	410.1	393.2	413.7	448.6
(b) Capital stock (£ million, 1970 constant prices)					
Chemicals	242	455	660	957	1 072
Metals	197	311	441	580	612
Shipbuilding, marine engineering	149	162	171	166	165
Textiles	33	38	40	89	112
Paper, printing, publishing	33	39	46	63	72
Total manufacturing	1 085	1 554	2 008	2 647	2 910
(c) Output (£ million, 1970 constant prices)					
Chemicals	53.1	123.0	158.6	170.8	233.5
Metals	93.4	96.3	89.4	95.6	137.1
Shipbuilding, marine engineering	57.6	81.3	54.0	58.9	55.3
Textiles	8.5	14.5	41.3	56.8	40.6
Paper, printing, publishing	10.5	16.3	23.3	27.8	45.0
Total manufacturing	408.4	588.0	680.3	903.7	1 057.8
(d) Profits (£ million, 1970 constant prices)					
Chemicals	13.5	57.9	88.0	98.2	n.a.[a]
Metals	38.8	34.8	34.7	35.6	n.a.
Shipbuilding, marine engineering	14.3	21.5	12.4	15.4	n.a.
Textiles	0.2	4.3	24.8	33.3	n.a.
Paper, printing, publishing	1.9	5.9	10.0	17.7	n.a.
Total manufacturing	115.2	209.2	281.1	431.9	n.a.
(e) Wages (£ million, 1970 constant prices)					
Chemicals	39.6	65.1	70.6	72.6	n.a.
Metals	54.6	61.5	54.7	60.0	n.a.
Shipbuilding, marine engineering	43.3	59.8	41.6	43.5	n.a.
Textiles	8.3	10.2	16.5	23.5	n.a.
Paper, printing, publishing	8.6	10.4	13.3	20.1	n.a.
Total manufacturing	293.2	378.8	389.2	464.4	n.a.
(f) Capital per employee (£, 1970 constant prices)					
Chemicals	5 975	7 689	11 511	19 497	19 491
Metals	3 350	5 338	8 446	11 523	10 636
Shipbuilding, marine engineering	2 880	2 695	4 278	4 575	4 796
Textiles	2 375	2 458	2 109	4 101	4 738
Paper, printing, publishing	2 583	2 985	3 187	3 408	3 640
Total manufacturing	2 896	3 789	5 110	6 398	6 486

Table 2.1   (cont'd)

Industry group	1951	1958	1965	1968	1970
(g) Output per employee (£, 1970 constant prices)					
Chemicals	1 308	2 078	2 768	3 479	4 245
Metals	1 586	1 652	1 713	1 901	2 384
Shipbuilding, marine engineering	1 114	1 355	1 353	1 627	1 608
Textiles	603	935	2 151	2 630	1 713
Paper, printing, publishing	827	1 254	1 618	2 054	2 284
Total manufacturing	1 090	1 434	1 730	2 184	2 358
(h) Profits per employee (£, 1970 constant prices)					
Chemicals	332	978	1 536	2 000	n.a.
Metals	659	597	665	708	n.a.
Shipbuilding, marine engineering	277	358	311	425	n.a.
Textiles	142	277	1 292	1 542	n.a.
Paper, printing, publishing	150	454	694	962	n.a.
Total manufacturing	307	510	715	1 044	n.a.
(i) Wages per employee (£, 1970 constant prices)					
Chemicals	975	1 100	1 232	1 479	n.a.
Metals	927	1 055	1 048	1 193	n.a.
Shipbuilding, marine engineering	837	997	1 043	1 202	n.a.
Textiles	589	658	859	1 088	n.a.
Paper, printing, publishing	677	800	924	1 092	n.a.
Total manufacturing	782	924	990	1 122	n.a.
(j) Profits as percentage of output					
Chemicals	25.4	47.1	55.5	57.5	n.a.
Metals	41.5	35.3	38.8	37.2	n.a.
Shipbuilding, marine engineering	24.8	26.4	31.1	42.5	n.a.
Textiles	2.4	29.7	60.0	58.6	n.a.
Paper, printing, publishing	18.1	36.2	42.9	46.8	n.a.
Total manufacturing	28.2	35.6	42.8	48.6	n.a.

[a] n.a.: not available

Sources: Northern Region Strategy Team, 1975; Hudson, 1976b

This revival partly reflected the impact of the Second World War, the destruction wrought necessitating a period of reconstruction and leading to a relative lack of international competition facing northeast producers. In addition, the upturn also reflected the demand for armaments during the Korean War of the 1950s. Output was boosted by increasing the size of the workforce (especially in shipbuilding) and by some increase in labour productivity (Table 2.1). In general, however, the restoration of profitability in these 'traditional' activities was based on expanded output resulting from fuller use of existing capacity. Productivity also improved, mainly from 'normal' incremental growth, as a result (essentially) of long-overdue replacement investment in

Plate 2.1　Colliery and village: Easington, County Durham.

existing technology rather than as a result of major outlays of fixed capital in new and qualitatively different modern technology and production methods. The recovery in profits largely resulted from an expansion in the mass of surplus value via increased outlays of variable capital at a time when, internationally, there was a tendency towards a sharp rise in the organic composition of capital (that is, the ratio of constant to variable capital advanced in production) and in the rate of surplus value. This pattern of accumulation was to assume considerable significance for the development of these branches in the northeast, although in the conjuncture of their contemporary favourable position in the international division of labour and of the regional balance of class forces, and in particular in view of the way in which working-class demands were formulated, it provided a short-term strategy that was extremely rational when observed from the viewpoint of the units of capital involved.

Second, effective demand for some products 'new' to the region also grew strongly, although from a low base level. These products were associated with activities (for example, in instrument engineering and parts of electrical engineering; clothing; and paper, printing and publishing) that had located there in response to the opportunities the northeast offered for profitable production, allied to the newly emergent regional policies of the 1930s, the decentralization of wartime production, and the strengthened regional policies of the 1945–48 period. From 1948 the Labour government effectively sacrificed regional policy as part of the state's response to the balance of payments and sterling crises of the late 1940s while, during the 1950s, the Conservative government effectively suspended the implementation of formal regional policy and dispensed with

Table 2.2   Employees in employment, Northern Region, 1952–75 (thousands)[a]

Composite Industrial Classification order groups	1952	1956	1960	1966	1971	1973	1975
A. Agriculture, forestry, fishing	39.0	36.5	34.0	24.8	19.1	19.2	17.6
B. Mining and quarrying	179.9	175.6	156.2	104.9	63.7	55.2	49.3
C. Food, beverages, tobacco	33.5	34.8	34.3	34.5	38.4	36.7	36.9
D. Chemicals, allied industries	44.1	50.1	59.8	57.3	58.7	52.9	51.1
E. Metal manufacture	58.7	59.8	62.1	57.9	47.6	44.6	45.0
F. Mechanical engineering	54.0	56.0	58.5	66.6	72.2	71.0	75.1
G. Instrument engineering	1.1	1.1	1.3	1.9	3.4	3.8	3.9
H. Electrical engineering	30.3	36.3	39.6	53.5	56.0	61.5	55.0
I. Shipbuilding, marine engineering	61.4	64.1	60.2	42.3	36.0	34.7	32.3
J. Vehicles	14.5	17.5	15.5	11.6	13.6	14.7	13.6
K. Other metal goods	12.7	13.7	12.1	14.4	13.8	15.0	15.3
L. Textiles	13.6	15.9	18.8	20.6	22.9	24.3	27.9
M. Leather, leather goods, fur	2.9	3.1	2.3	1.7	2.4	2.2	2.9
N. Clothing, footwear	28.1	30.0	31.7	34.7	35.5	36.4	35.5
O. Bricks, pottery, glass, cement	19.1	18.8	18.6	19.3	18.8	17.9	16.9
P. Timber, furniture, etc.	14.9	14.0	12.7	14.6	13.0	15.2	14.0
Q. Paper, printing, publishing	12.7	13.3	14.4	17.2	19.6	21.8	24.5
R. Other manufacturing	7.2	8.7	12.0	15.2	13.5	17.1	14.9

S. Construction	74.9	85.1	87.0	108.6	87.5	102.0	100.1
T. Gas, electricity, water	18.9	19.5	19.4	22.7	20.0	18.1	19.4
U. Transport, communications	101.2	98.0	92.1	86.6	70.5	67.2	68.6
V. Distributive trades	131.6	145.2	158.6	166.3	145.6	155.0	157.3
W. Insurance, banking, finance	12.2	15.3	16.9	20.5	23.6	25.1	25.8
X. Professional services	83.2	91.4	104.6	133.2	153.6	165.9	182.5
Y. Miscellaneous services	83.0	85.0	87.0	103.4	109.1	123.7	126.5
Z. Public administration, defence	71.7	68.2	69.3	74.1	80.0	84.4	95.7
Sub-total manufacturing industries (C to R)	408.8	437.2	453.9	463.3	465.4	469.8	465.2
Sub-total production industries (A to T)	721.5	753.9	750.5	724.3	655.7	664.3	652.6
Sub-total services industries (U to Z)	482.9	503.1	528.5	584.1	582.4	621.3	656.4
Total	1 204.4	1 257.0	1 279.0	1 308.0	1 238.1	1 285.6	1 309.0

[a] These data are not strictly comparable with those in Figures 2.2 and 2.3 although the general tendencies are broadly similar

Source: Fothergill and Gudgin, 1978, 43–48

Plate 2.2   Maurice Gerard, South Moor: a clothing factory in an abandoned colliery building, west Durham.

programmes designed to steer new industries to the northeast (Regional Policy Research Unit, 1979, Part 2). Displacement of surplus labour via migration to the South East and the West Midlands helped to meet the labour power requirements of capitals located there (so supporting national growth) while established employers in the northeast had no interest in further tightening pressures on the labour market; in fact, they continued to vigorously oppose proposals for the introduction of 'new' (especially male-employing) industries into this region, as they had successfully done in the immediate post-war years (hence the spatial and sectoral concentration of such new industries as did invest there). In this, they were supported by the state. As the Northern Regional Controller of the Board of Trade informed the House of Commons Select Committee on Estimates, the 'last thing' his department would want to do would be to encourage firms to go to places with 'no unemployment' (cited by Bowden, 1965, 29). Increased output in the new industries was for the most part achieved not by expanding labour forces or fixed capital stock but by gains in labour productivity. An important factor was that much of the workforce in these activities was female, often married and non-unionized, and being introduced to the regime of factory work for the first time (see, for example, North Tyneside Community Development Project, 1978b).

However, the major focus of dynamism in the regional economy in this period was undoubtedly chemicals, associated not only with the continuing development of existing branches of production but also with the switch to, and development of, petrochemicals (notably at the Wilton complex of Imperial Chemical Industries Ltd) and the introduction of new technologies leading to great increases in productivity and profits per employee (House and Fullerton,

1960; Semmens, 1970; Taylor, G., 1979; Hudson, 1981b). This and subsequent developments there were closely linked to general accumulation imperatives to produce new, synthetic raw materials as a means of reducing the value of elements of circulating constant capital. Thus, the development and expansion of the chemicals sector became a central component of the post-war 'long wave with an undertone of expansion' (Mandel, 1975), which had important repercussions on the overall pattern and pace of accumulation (Hudson, 1981a). Within the northeast, much of the expansion of the manufacturing sector was associated with the development of chemicals – over 50 per cent of that in employment and over 45 per cent of that in capital stock, output and profits.

Third, in addition to net growth in the factory workforce there was also an expansion in service sector employment (Table 2.2, pages 40–41), especially for women. In part, this reflected greater purchasing power from increased employment and wages in other sectors, enabling the further extension of capitalist social relations into the spheres of services and distribution. It also reflected the expansion of job opportunities in educational and medical services, which followed the establishment of the Welfare State in the immediate post-war years.

Thus for a time in the 1950s a generally profitable restructuring of capital co-existed with conditions of full employment within the region. It appeared that harmony prevailed between the interests of capital and labour. But 1958 saw a rupturing of this harmonious relationship and a turning-point in the region's social history, for it marked the beginning of the end of the brief period of full employment in the northeast (and in other peripheral British regions) which led to pressures for increased state intervention to encourage the transformation of the region's economy, in the belief that this would provide the route back to full employment.

Employment loss in this period was, however, concentrated in two branches – coal mining and shipbuilding. This reflected processes of restructuring set in motion in response to crises which had different origins. The crisis in shipbuilding and related activities (including steel production) reflected the changed status of northeast producers as their world market share was sharply eroded by increasing competition from other producers using more modern production techniques (Cousins and Brown, 1970). Consequently, output and total profits fell sharply. In an attempt to maintain the rate of profit, outlays on wages were reduced equally sharply so that by 1963 profit as a proportion of the value of output had risen slightly but employment had fallen (Table 2.1, page 38). In coal mining (and related activities, notably rail and water transport), employment began to fall quickly, although for different reasons. The regional decline in coal mining employment reflected a decision – made possible by the availability on the international market of large amounts of cheap oil – at national level by the British state to switch to a multi-fuel economy in an attempt to cut energy costs, for these had implications to a greater or lesser extent for the international competitive position of all British manufacturing capitals. In the 1950s the drive to maximize coal output had resulted in increased unit production costs; among the implications of a switch to a multi-fuel economy were that, in general, total coal output be reduced and, in particular, high-cost collieries be closed (Regional Policy Research Unit, 1979, Part 4).

A corollary of this sectoral concentration of job loss was a spatial impact, especially in Tyneside and Wearside: the male unemployment rate in South Shields, for instance, rose to over 10 per cent in December 1958, reviving the spectre of the 1930s. In turn, this led to increased political pressure within the labour movement for stronger measures to combat unemployment. In response, the Board of Trade took a rather more active role in steering new industry to the region. It was able to do this because labour forces associated with coal mining and shipbuilding began to disband and the labour market in the South East and the West Midlands remained inflexible, as full employment conditions continued there (Figure 2.2, page 36).

Moreover, these specific sectoral crises and male employment losses in the northeast were overlain by broader, national 'stop-go' cycles of increasing amplitude, largely induced by state macro-economic demand management policies. The combination of these general and specific factors led to a particularly sharp unemployment increase in the region in December 1962 which continued into 1963 (Figure 2.2, page 36). In turn, heightened political pressure to combat this from the Labour Party and trade unions within the region, together with the desire of the Conservative government for re-election, led to the appointment of Lord Hailsham as Minister for the North East and the adoption of the 'Hailsham programme' (Board of Trade, 1963; Carney and Hudson, 1974; Regional Policy Research Unit, 1979, Part 3). This represented a major switch in both Conservative ideology and practice, consisting of a combination of short-term measures to alleviate unemployment together with longer-term, more fundamental measures. The latter amounted to an expanded programme of public sector infrastructure investment which, taking over elements of the prevailing planning ideology in the region (Pepler and MacFarlane, 1949; Durham County Council, 1951), was to be concentrated in certain key locations possessing 'growth potential' (growth points – such as new towns – and a growth zone, lying between the rivers Tyne and Tees and to the east of the A1(M) motorway). It was asserted that a once-and-for-all sharp increase in public sector infrastructure investment would set in motion processes of regional industrial 'modernization' and 'diversification' which, in turn, would provide a route back to 'full employment'.

While at national level the switch to intervention represented a marked change in the stance of the Conservative government, within the northeast the regional modernization programme was seen as a consensus measure. The roots of such policies as a solution to the region's problems can be traced back to the response to crisis evolved in the 1930s by the region's bourgeoisie in an attempt to protect their own interests (Carney and Hudson, 1978). Crucially, they subsequently came to be accepted as a legitimate, even inevitable, solution by the Labour Party and working-class organizations. Thus it was possible in the early 1960s for modernization policies to become the basis of a consensus, or a class alliance, as to how best to solve the region's problems. Given the

Labour Party's political hegemony in the northeast, this has meant its advocating and implementing policies which objectively meet the requirements either of 'capital in general' or of some individual capitals while legitimating these as being directed towards the attainment of declared social-democratic goals (for example, reducing unemployment and improving working-class living standards).

Parallel with this renewed emphasis on modernization policies, important alterations were occurring in the remainder of the regional economy in response to declining employment, especially in coal mining and shipbuilding. In manufacturing, apart from shipbuilding, the pattern of change between 1958 and 1963 exhibited little evidence of crisis from the point of view of capital; on the contrary, gross output, fixed capital stock, labour productivity and profitability all rose sharply. Thus, while there may have been a crisis in terms of falling employment in shipbuilding and coal mining (which became generalized at a political level), the manufacturing sector continued to be restructured in a way that was favourable to the interests of capital. While the recorded increases on these indicators tended to be smaller than those in the 1951–58 period (although in some branches, notably textiles, the reverse was true), of particular interest are the more rapid rises in fixed capital stock and in the capital: employee ratio. While the increase in the latter to some extent reflects employment loss, of greater significance was the accelerated pace of fixed capital investment in the early 1960s (especially in chemicals and metals). In chemicals, the growth of fixed capital investment was related to both capacity expansion, as better capacity utilization was achieved, and the adoption of more sophisticated technologies. In metals, it reflected the recovery in demand that had come about by 1960 because, in response to this, several modernization and expansion schemes were initiated. Even so, this investment was, by international standards, in outmoded technology: for example, the South Durham Iron and Steel Company's integrated South Works at Hartlepool opened in 1962 based on open-hearth steel production.

Moreover, while manufacturing employment (with the exception of shipbuilding) continued to expand slowly during this period, employment in the service sector, in aggregate, grew more rapidly. This was particularly so in distribution and professional services, as in earlier years, and also in miscellaneous services: these gains offset losses in transport associated mainly with the run-down of coal-mining. Despite the slight upward trend in numbers of jobs (Table 2.2, pages 40–41 and Figure 2.3, overleaf), unemployment, particularly for men, continued to grow (Figure 2.2, page 36), not least because many of the new jobs in services, the major source of employment expansion, were for women. Thus from the viewpoint of labour, if not of capital, this was a period of continuing employment problems which culminated in the sharp unemployment increases of December 1962 and 1963.

2.3 Modernization, 1963 to 1970 ─────────────────────

The 1963–70 period in many ways saw the reinforcement of previously established restructuring tendencies in the region's economy and the refinement

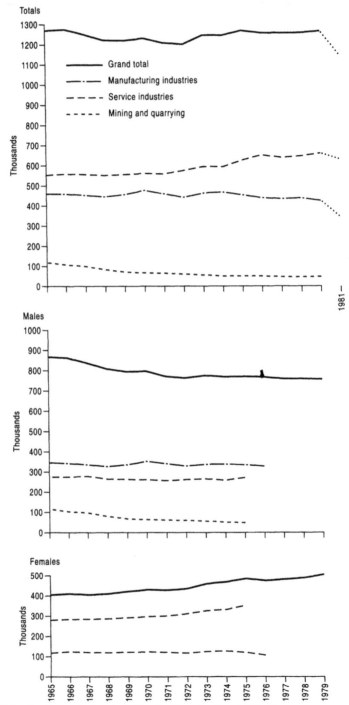

Figure 2.3 Employment change in the Northern Region, 1965–81.

and implementation of the modernization policies which emerged during the early 1960s. In manufacturing there were substantial increases in output, fixed capital investment, labour productivity and profitability while employment and wages also increased, although at a much more subdued pace. However, the rate and character of restructuring varied among different branches of manufacturing. These variations and their relationship to the overall pattern of change in the regional economy require a consideration of the specific conjunction of circumstances that made particular locations within the north-east especially attractive to capitals involved in certain types of production. It is necessary to examine why, for a time, the region provided an attractive destination for a part of some capitals' contemporary 'round of investment' (Massey, 1979). Several elements may be identified.

First, enhanced central government financial incentives were made available to capitals investing in the region. This was particularly so after 1966 (Moore *et al.*, 1977), following the election of a Labour government in 1964. This was initially committed to strengthening regional policy in response to social-democratic aims of narrowing spatial inequalities but, increasingly over time, became committed to pursuing policies intended to restructure the national manufacturing base as an end in itself, or as a means to attaining macro-economic policy objectives, rather than as a way of narrowing regional inequalities. One symptom of this, which was carried over into the 1970s, was the concentration of state aid for R&D in the South East while subsidies for production were channelled to regions such as the northeast (Table 2.3, overleaf). Such policies thus encouraged the emergence of a new spatial division of labour within Great Britain (a point amplified later). State subsidies served to cut production costs by reducing the price of elements of fixed capital, although without necessarily reducing their value: rather, the state assumed financial responsibility for an increased proportion of the costs of production. Consequently, in general this made the region an attractive location for investment in activities characterized by a high and rising investment in fixed capital (notably chemicals and steel); in particular, the availability of large, flat, stable sites and the possibility of deep-water port facilities led to such investment within the region being concentrated at Teesside (for a comparison with the West Netherlands see Läpple and van Hoogstraten, 1980).

Thus fixed capital investment in new productive capacity, output and labour productivity rose sharply in both chemicals and metals production. Although profitability continued to increase in the former, this was not so in the latter (Table 2.1, pages 37–38): indeed, this was a crisis period for iron and steel production. From 1961 to 1966, effective demand for steel produced in the Northern Region slumped and profitability fell sharply. This collapse in profitability was not restricted to the region and served as the immediate trigger for the renationalization of parts of the industry in 1967. The combination of a recovery in demand for steel and a programme of rationalization and fresh investment by the British Steel Corporation (following a period of relatively little investment), although mainly in obsolete production technology by international standards (Cockerill and Silbertson, 1974),

Table 2.3 Expenditure under the Public Expenditure Survey Committee programme: trade, industry and employment, Northern Region and Great Britain, 1969–74

Programme	Northern Region (£ million, current prices)					Northern Region as per cent of Great Britain total				
	1969–70	1970–71	1971–72	1972–73	1973–74	1969–70	1970–71	1971–72	1972–73	1973–74
Regional support and regeneration	65	56	53	49	85	33.3	30.8	30.4	25.3	26.7
Selective assistance	—	—	—	2	12	—	—	—	18.8	18.8
Regional development grants	—	—	—	3	36	—	—	—	42.7	33.3
Regional employment premium	31	31	31	29	31	28.7	28.6	29.2	29.0	29.2
Expenditure under the Local Employment Acts	34	25	22	14	7	40.1	34.8	33.9	20.4	17.4
Industrial innovation	5	3	4	4	4	3.7	1.7	1.4	2.0	1.9
General support for industry	81	88	82	60	53	13.5	14.5	14.7	16.1	18.5
Assistance to shipbuilding industry	4	3	1	4	8	25.2	35.8	11.4	41.4	29.9
Investment grants	77	85	80	56	43	13.2	14.4	14.3	15.0	17.2
Functioning of labour market	7	9	11	13	12	8.6	8.7	8.6	7.8	8.8
Industrial training	3	4	3	5	6	10.8	10.8	10.8	11.0	10.8
Total expenditure	171	168	160	153	208	15.5	14.3	13.1	12.1	12.8

Source: Northern Region Strategy Team, 1976b, 108–21

Plate 2.3   Urban redevelopment in the 1960s: high-rise blocks in Scotswood, Tyneside.

led to a sharp rise in output and a continued increase in fixed capital investment in metals production between 1968 and 1970 (Table 2.1, pages 37–38).

Nonetheless, net employment in chemicals and metals tended to stagnate
or decline. Employment growth associated with capacity expansion by both
existing companies and those new to the region (for example, US multinationals involved in various types of chemicals production such as Monsanto)
was balanced by the closure or restructuring of existing capacity (Robinson
and Storey, 1979). Thus, the 1960s saw an extensive programme of modernization of ICI's inorganic chemicals and fertilizer production at Billingham;
labour-intensive ammonia-producing plants (dating from the 1930s), which
used coal as a source of process gas, were replaced by new plant using first
naphtha and then natural gas as a raw material (Hudson, 1981b). ICI pioneered this development in a search for technological rents (Mandel, 1975,
192–194). The resulting loss of several thousand jobs (Taylor, G., 1979) was
offset to some extent by employment growth accompanying a considerable
expansion and modernization of organic chemical production at Wilton.

These state financial incentives also helped bring about some restructuring
in other 'traditional' sectors, such as shipbuilding and related marine, heavy
mechanical and electrical engineering. Moreover, on occasion, the state was
directly involved by promoting mergers and amalgamations. But a consequence
of this increased centralization of control was a restructuring of production
and employment decline (North East Trades Union Studies Information
Unit, 1976).

Second, the Hailsham programme was implemented and public sector
infrastructure investment (on industrial estates, houses and roads) sharply

increased. These outlays were designed to provide some of the preconditions necessary to attract fresh manufacturing investment in new (especially consumer goods) industries by reducing or taking over elements of their costs of production. This investment was selectively channelled to a few key locations, such as the various new towns, with Washington (8 km south of Gateshead) perhaps being the supreme example (Hudson, 1976a, 1979a, 1980a). Hence the state became increasingly involved in the coordination and programming as well as the financing of investment, especially on Teesside, in response to the expansion of chemicals production there (North East Area Study, 1975). Associated with this spatially selective modernization programme was a substantial involvement of private capital in speculative house-building and central area commercial redevelopment schemes, which found their fullest expression in the redevelopment of central Newcastle as the regional capital (Burns, 1967; Regional Policy Research Unit, 1979, Part 9). Public sector infrastructure investment was not limited to the areas specified in the Hailsham proposals. In particular, the growing demand for water from the continuing industrial development at Teesside led to the controversial decision in 1967 to construct a reservoir at Cow Green in Upper Teesdale, thereby flooding a unique assemblage of flora (Gregory, 1975).

Such public expenditure policies reflected the formation and maintenance of a consensus around the politics of modernization as being the only solution to the problems of unemployment in the northeast. Briefly, following the publication of the National Plan in 1965, which represented an attempt to increase the national economic growth rate through indicative planning and reducing differences in regional growth rates, this specific programme for the northeast became linked to these broader planning policies and the Keynesian notion of balanced regional and national growth (Shanks, 1977), which attempted to modify the inherently uneven nature of capital accumulation.

Third, the 1960s saw the reconstitution of an abundant reserve of labour within the region, which provided an important attraction to some capitals in manufacturing. This had two aspects. One was that female workforce participation rates in the region, especially for married women (Table 2.4), had historically been relatively low so that female labour could be drawn onto the labour market, often for the first time (Hudson, 1979a, 1980a, b). This was,

Table 2.4 Workforce participation rates, Northern Region and Great Britain, 1961–75

	1961	1966	1971	1975
Northern Region				
Male	86.0	83.1	81.1	80.6
Female	31.4	38.3	40.4	43.6
Great Britain				
Male	85.1	83.8	81.2	79.6
Female	37.1	42.2	43.0	45.8

Source: Central Statistical Office, 1975–81

in part, related to the second aspect, the accelerated and chaotic run-down of coal mining and the break-up of the workforce in this and associated activities which released many men onto the labour market. This was particularly evident after 1967 when the run-down of coal mining was hastened as part of the measures to combat pressure on the exchange rate and balance of payments (Crossman, 1976, 451–452; Krieger, 1979; Regional Policy Research Unit, 1979, Part 4). Significantly, rather than opposing closures, the regional branches of the National Union of Mineworkers and the Labour Party negotiated only about the pace of closure: from within the relevant working-class organizations, mining decline was seen as 'inevitable', an indication of the extent to which they had absorbed the dominant ideology as to the nature of regional change. These various changes provided flexibility for employers on the labour market and offered opportunities to recruit a variety of types of labour power. In particular, there were opportunities to employ many people who were both pliant (because they were non-unionized or recruited into unions which promised trouble-free production in return for the right to negotiate closed-shop arrangements and thus increase their own membership) and cheap (particularly during the 1967–76 period when the state directly met part of the price of labour through the Regional Employment Premium, a subsidy paid to employers depending upon the size and composition of their workforces).

These second and third elements, and to some extent the first, provided a combination of circumstances that capitals in some branches of production found particularly attractive. The growing tendency to centralization and concentration, allied to technical progress and changes in labour processes associated with Taylorist and Fordist techniques of scientific management (Braverman, 1974; Aglietta, 1979, 111–122), enabled some capitals advantageously to locate those parts of their production activities requiring relatively large amounts of poorly skilled labour in branch plants in the northeast. This was part of the process of corporate restructuring which, increasingly, was and remains globally rather than nationally based. Mandel (1975, 324) identified the internationalization of production as a distinguishing feature of late capitalism. Changes in the northeast are clearly related to this changed international division of labour. Thus, in 1971 40 per cent of all manufacturing establishments in the Northern Region were branch or subsidiary operations, many of which resulted from investment decisions made in the 1960s by multinationals (North East Trades Union Studies Information Unit, 1977). One result was an increase from 8500 in 1963 to 24,400 in 1971 in the numbers employed in foreign-owned factories in the region (Northern Region Strategy Team, 1976a).

These changes in the northeast were part of much wider processes whereby the 'traditional' regional specialization of industry in the UK was changing to a new spatial division of labour. For a time, regions characteristically fulfil a particular role in the overall production process in which a widening range of activities are becoming organized on an international basis. Typically, assembly and semi-skilled component manufacturing operations are assigned to regions like northeast England (Massey, 1978; Perrons, 1979; Lipietz, 1980b). Thus the characterization of the region as a 'global outpost' (Austrin and Beynon, 1979) is particularly apposite.

Within the northeast, these changes were manifested in two main ways. First, because public sector investment in physical infrastructure was spatially selective, this led to a geographical concentration of 'new' industry which was also associated with a spatial restructuring of intra-regional labour markets and changes in commuting patterns (Hudson, 1980a). During the 1961–73 period, 46 per cent of all new manufacturing firms moving into the Northern Region located in only eight of the 77 Employment Exchange Areas there. These eight were all in the northeast and were associated with new towns (Aycliffe, Cramlington, Peterlee and Washington) or developments by the English Industrial Estates Corporation (Gateshead, Hartlepool, Stockton and Thornaby) as noted by Hudson (1976a, Vol. 1, 139–140). Second, relatively strong rates of growth of employment, output and profits were recorded in industries such as food, beverages and tobacco, engineering, clothing, textiles, paper, and other manufacturing; in others (notably textiles) capital stock also grew sharply. Greater output tended to be associated with gains in labour productivity as well as with larger workforces, but there is little evidence of sharp increases in fixed capital:employee ratios (except in textiles and other manufacturing where there was considerable fixed capital investment partly financed through state subsidies). This suggests an expansion of capacity using much the same technology but with a more flexible labour market enabling greater productivity by changing working conditions at the point of production.

Thus, the alterations brought about in the region's manufacturing sector in general constituted a successful, profitable restructuring from the viewpoint of some of the capitals involved – even in the shipbuilding sector profitability recovered, if only weakly and temporarily. Conversely, for the working class, these changes, together with those experienced in the remaining sectors of the regional economy, were not successful, particularly in increasing employment opportunities: some growth in net manufacturing employment has to be seen against employment loss in other sectors, notably coal mining. The qualitative aspects of jobs lost and created also require consideration. Women accounted for almost half the net increase in the factory workforce between 1965 and 1970 (Figure 2.3, page 46), a symptom of the drive for supposedly non-militant, unskilled labour power: Lipietz (1980a) suggests that feminization can be taken as an index of de-skilling, although this is not to suggest that women's work is necessarily unskilled but rather that unskilled jobs are frequently occupied by women. At the same time some capitals established new factories because of the availability of ex-miners, but the labour thus absorbed did not equal the supply released by the restructuring of coal mining. In the services sector, too, net male employment fell (largely because of the decline in activities dependent upon coal mining) while net female employment increased. Those reductions in female job opportunities which did occur (for example, in distribution, in part associated with commercial restructuring and central area redevelopment schemes) were more than offset by expansion in activities such as education and medical services and public administration because of increased public expenditure by the Labour government in response to social reformist pressures. Moreover, this growth reflected a general expansion of these sectors as well as a restructuring within the state apparatus and changes in the spatial division of labour in the tertiary sector within the territory of the British state, in response to the possibilities offered by technological change and its impacts on labour processes. This was manifested by a decentralization of routine operations to the northeast.

These various changes resulted in a steep rise in male unemployment, declining male workforce participation rates, a fall in males in employment and continuing net out-migration at the same time as the number of females in employment and female workforce participation rates grew. While to some extent policy objectives were realized (the employment structure, for example, being 'diversified'), state intervention failed to achieve its principal aim of lowering (male) unemployment and was associated with a pattern of changes which exacerbated the problems it was supposed to solve. (For an analysis of why intentions and outcomes must differ, see Offe, 1975a; Habermas, 1976.) This disjunction between intention and outcome could have served to call into question the legitimacy of the state and the progressive or, at worst, class-neutral character of its interventions. But, given the general acceptance by working-class organizations and the Labour movement within the region of an ideology which perceived economic change and employment decline as inevitable and modernization as the only alternative, that possibility was not realized and the consensus on the politics of modernization as the route to social reform remained intact, though not unaffected.

2.4 Permanent depression: the 1970s onwards?

With the exception of a brief period of weak recovery in 1972–73, the 1970s were marked by a deepening depression in the northeast, as unemployment mounted to levels last seen in the 1930s. This was a reflection of strategies adopted both by capitals and the British state to meet various crises. Moreover, the impacts of recession – initially the specifically British recession of 1971–72 triggered by the macro-economic policies of the newly elected Conservative government and then the post-1973 international recession (Mandel's, 1978, 'Second Slump') – were intimately related to the way in which capital, encouraged and supported by the British state, had utilized the opportunities offered by the region in previous years.

Some kinds of production, notably steel and petrochemicals (the latter increasingly related to the North Sea oil developments on Teesside), maintained substantial fixed capital investment and this was reflected in the region's rising share of national fixed capital formation in manufacturing (Table 2.5, overleaf). It was linked both to the continued international tendency for fixed capital investment to assume increasing importance and to the accelerated turnover time of fixed capital (Mandel, 1975, 223–247). Such investment was part of a continued drive to increase competitiveness and retain or raise world market shares for existing products through greater labour productivity, largely

Table 2.5   Fixed capital formation by manufacturing industry, Northern Region and United Kingdom, 1971–78 (£ million, current prices)

Year	Gross		Net	
	Northern Region	Per cent of UK total	Northern Region	Per cent of UK total
1971	206	9.4	—	—
1972	178	8.7	—	—
1973	191	8.1	186.4	8.2
1974	305	9.9	309.5	10.1
1975	452	12.8	461.6	13.1
1976	491	12.4	478.8	11.8
1977	588	12.2	594.8	11.9
1978	630	11.0	—	—

Sources: Central Statistical Office, 1975–81; Central Statistical Office, 1981, 157

Table 2.6   Regional development grants, Northern Region, 1972–80, March (£ million)

	Plant and machinery			Building and works			Total
	Special Development areas	Development areas	Total	Special Development areas	Development areas	Total	
Northern Region	237	394	631	56	64	120	751
Per cent of Great Britain total	33.2	42.7	38.5	33.1	34.8	33.5	37.7

Source: Central Statistical Office, 1981, 116

allied with the introduction of automated production techniques (Aglietta, 1979, 122–130) and the development of new products. Generally, however, restructuring has meant a decline in jobs as demand either fell or increased only slowly, leading to development without employment. This fixed capital investment has been substantially underwritten by the state, not only directly by reducing the cost of elements of fixed capital through industrial and regional policies (Tables 2.5 and 2.6) but also through transforming parts of the region's rural periphery, particularly by constructing major reservoirs at Kielder and Cow Green to ensure water for expanded chemicals and steel production on Teesside.

Investment in metals was mainly associated with aluminium and iron and steel production. The former reflected Alcan's decision to build a smelter at Lynemouth (on the coast, about 25 km northeast of Newcastle) using state subsidies to cut the price of key elements of both circulating and fixed

Plate 2.4 The rural environmental impacts of industrial developments on Teesside: the Cow Green reservoir.

constant capital to encourage the production of aluminium within the UK. Investment in iron and steel was associated with the rationalization that followed the renationalization of parts of the industry, in an effort to cut unit production costs of basic steel by belatedly erecting modern plant. (This was an attempt by the state to reduce the price of a central element of constant capital for many other manufacturing activities which had important implications for their international competitive position as well as that of the British Steel Corporation itself). From 1973 the rationalization programme was particularly linked with the British Steel Corporation's 'Ten Year Development Plan' which entailed scrapping capacity and replacing it with modern plants. This necessitated spatial changes in production nationally and regionally, and the development of integrated, coastal works including that at Teesside.

In the early 1970s steel output fell sharply in the UK and in the Northern Region at a time when global demand for steel was growing, a reflection of the weakening international competitive position of British Steel. The subsequent simultaneous fall in global demand for iron and steel products and the expansion of modern capacity in some countries (Linge and Hamilton, 1981) called into question the intention of the Development Plan to increase output. This was reinforced by financial constraints imposed on the British Steel Corporation by the Labour government because of deepening fiscal problems in the mid-1970s, and then by the strict cash limits prescribed in 1979 by a Conservative government committed to monetarist policies, as well as by the EEC's effort to cut capacity. The Development Plan was severely modified: plant closure was speeded up and schemes to expand new capacity abandoned.

In the northeast this resulted in the closure of works at Consett and Hartlepool, while at Teesside the second blast furnace at Redcar is not to be erected and employment is to be reduced. These steps will have consequences for the mining industry because of the decreased demand for coking coal. Employment in coal mining continued to decline during the 1970s – though at a much slower rate than in the 1960s – as a result, in part, of the introduction of improved mechanized deep-mining techniques and the development of open-cast methods in an attempt to reduce further the unit production costs of coal. Furthermore, financial constraints on the British Steel Corporation have led to policies of importing coking coal for the Teesside works (for example, from Australia), threatening employment in collieries on the County Durham coast.

Other manufacturing activities have also reduced capacity with companies writing-off fixed capital to maintain or boost the rate of profit. This process has occurred both in traditional sectors and in the new growth industries of the 1960s. The limited restructuring of the former was increasingly found inadequate as the recession deepened globally and weaker capitals were eliminated. The nationalization of the shipbuilding industry in 1977 followed a period of increased direct state intervention in response to a chronic profitability crisis and the impacts of accelerating inflation which were felt by capitals that had committed themselves to fixed-price contracts in an attempt to finance belated and inadequate restructuring programmes from the late 1960s. But this was simply a prelude to the state's assuming responsibility for a severe rationalization in the face of serious overcapacity and the weak international position of British shipbuilders, particularly following the election of a Conservative government in 1979 which imposed tight cash limits on the industry as part of its wider monetary policy. Employment in shipbuilding in the northeast between nationalization in 1977 and the election in 1979 declined by almost 20 per cent from 36,000 (North Tyneside Trades Council, 1979) and has since continued to fall.

Moreover, the 'new' industries were also affected by the recession; the earlier generation of plants from the 1940s were affected first, but as the recession deepened, the 'growth industries' of the 1960s also suffered, thus vividly demonstrating that the modern, diversified sector of the regional economy was extremely fragile in the changed world market conditions of the 1970s. Product diversity did not mean diversity of production processes and it was the latter that were most readily changed. Capitals which, in the favourable environment offered by the region in earlier times, had established branch plants to produce specific commodities, either reduced levels of capacity utilization in response to falling levels of demand or truncated the 'natural' life cycle of plants by closing capacity and writing-off the fixed capital investment in an attempt to restore the rate of profit. In other cases branch plants had been established to make commodities that were already technically obsolete and at the end of their life cycle but for which a specific, temporary demand then existed; once this disappeared they were closed.

Damette (1980) has introduced the concept of 'hypermobility' to embrace the accelerated switching of investment between locations as the turnover time

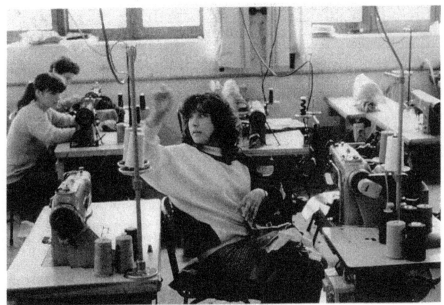

Plate 2.5 New forms of work, new gender divisions of labour: making clothing in northern Portugal.

Plate 2.6 New forms of work, new gender divisions of labour: making leather shoes in northern Greece.

Table 2.7   Notified redundancies, Northern Region, 1970–80

Year	Total	Per cent of UK total
1970	14 000	—
1971	24 506	7.8
1972	12 878	7.8
1973	8 252	10.0
1974	10 672	8.3
1975	20 999	7.9
1976	21 022	12.7
1977	21 931	15.1
1978	18 217	10.7
1979	20 493	—
1980 (Jan.–Jun.)	20 000	—

Sources: Benwell Community Development Project 1978a; Manpower Services Commission, 1981

of fixed capital is reduced (not least because of the state's taking over parts of the costs for individual capitals). Related to this is the possibility that part of the next 'round of investment' by these capitals will not be in the northeast. Given the internationalization of production, there will be an increasing tendency for investment to be switched to other locations offering better opportunities for profitable manufacturing: for example, cheaper labour areas in Mediterranean Europe, especially when the EEC is enlarged, and the Third World (Fröbel et al., 1980; Damette, 1980). This tendency is being reinforced by the policies of states in such areas to enhance the possibilities for profitable production. Such switches in investment strategy, from the viewpoint of the capitals concerned, represent part of a rational response to crisis. The net result of these changes in the northeast was a sudden decline in factory employment after 1975, especially in jobs for women (Figure 2.3, page 46). Total declared redundancies (which understate job losses) in the Northern Region have increased and so also has the proportion of the UK total there. Between 1971 and 1975, some 77,300 redundancies were declared in the Northern Region (Table 2.7), 8.1 per cent of the UK total, but between 1976 and 1980 this rose to 127,863, 10.9 per cent of the UK total (calculated from data in Benwell Community Development Project, 1978a; Manpower Services Commission, 1981). Paralleling the growing tendency towards the internationalization of production is an equally strong tendency towards the regionalization of the associated consequences (Damette, 1980).

At the same time the capacity of the British state to intervene effectively to counter these consequences was severely limited. Macro-economic policies became increasingly monetarist – a response to the threatening fiscal crisis of the British state – and political goals have switched from those of 'full employment' to the containment of inflation and reduction of trade union power. While, for a time, net loss of manufacturing employment was to some extent offset by service sector growth, although the resulting jobs were typically part-time and associated with rising activity rates for married women (Hudson, 1980b), this expansion was curtailed as a consequence of public expenditure cuts after 1975. Similarly, expenditure on regions was cut and the emphasis in regional policy swung even more strongly to restructuring Britain's manufacturing base as an end in itself (Geddes, 1978; Cameron, 1979). This met with scant success as 'deindustrialization' (Blackaby, 1979) accelerated, especially with the advent of the petro-pound and the increasing emphasis on controlling the money supply which was intended to 'shake out' weaker capitals but which further weakened the international competitive position of British manufacturing. Furthermore, restructuring within the British state and the creation of new Development Agencies for Scotland and Wales, partly to meet the perceived threat of a resurgent neo-nationalism precipitating the 'break-up of Britain' (Nairn, 1977), made these more attractive destinations for capitals seeking to invest in the United Kingdom. This had three consequences within the northeast. First, it led to sporadic calls for a comparable Northern Development Agency, particularly from within the Labour Party. Second, in its absence, greater emphasis was placed on encouraging the formation and growth of indigenous small firms as a source of employment growth (Northern Region Strategy Team, 1977); this had a marginal impact in creating new jobs, not least because of the conflict between this goal and macro-economic policies. Third, the awareness of declining manufacturing employment and of service sector expansion meant that there was more emphasis on the latter as a source of future employment growth (Northern Region Strategy Team, 1977), ignoring both the type of service sector jobs provided in the past and the short-term impacts of public expenditure cuts on such activities and the probable longer-term effects of technological progress (Hines and Searle, 1979).

The combination of falling labour demand, both in the northeast and in the South East and West Midlands, and growth in labour supply (Department of Employment, 1978) has resulted in further sharp increases in both male and female registered unemployment (Figure 2.2, page 36), the latter being a specific feature of the later 1970s, although a generalized effect of the world recession rather than one confined to the northeast (Organisation for Economic Co-operation and Development, 1976; Mandel, 1978, 16). Even so, the increases in registered unemployment understate the magnitude of the problem because considerable numbers, especially of young people, have been kept off the unemployment register by various state temporary job-creation or preservation schemes (North Tyneside Community Development Project, 1978a, 94; Manpower Services Commission, 1979). Furthermore, given the sectoral distribution of employment decline, some areas within the northeast have experienced disproportionate shares of the overall regional unemployment increase; in some the registered male unemployment rate is already in excess of 20 per cent (for example, Hartlepool, Sunderland and Consett where it exceeds 30 per cent) and will almost certainly increase further (for example, the male unemployment rate in Consett is forecast to rise to over 50 per cent).

An important issue is the trade union, Labour Party and, more generally, working-class responses to job loss. There have been a few well-publicized campaigns to prevent particular plants closing, such as the British Steel Corporation at Consett and Vickers-Armstrong Ltd at Scotswood (Save Scotswood Campaign Committee, 1979), which have mainly centred on the arguments that such plants were, or could be made, commercially viable or that the social costs of closure were unacceptable. While there may have initially been opposition to closure, this has often dissipated in the face of redundancy payments which, however, are rarely generous (North Tyneside Trades Council, 1979). Perhaps the most significant challenge came early in 1981 when the threat of a strike by the traditionally moderate Durham miners' union led to the abandonment of plans to close five collieries in the region by the National Coal Board, although within a few weeks of this plan being abandoned the closure of one of these collieries had been agreed. In general job losses have been seen, in some sense, as natural and rising unemployment as inevitable. The boundaries to what is regarded as the legitimate scope of state intervention have been reached but, within these, the state is unable to offset effectively the impacts of the law of value. The situation could be described as one of the politics of despair: to borrow Offe's (1975a) evocative phrase, 'the necessary had become impossible, the impossible necessary', in terms of state policy intervention.

2.5   Conclusions: capital, the state and regional crises ─────

The central thesis of this chapter is that the pattern of changes observable in the northeast during the post-war period can be understood only in relation to the process of capital accumulation, being simultaneously a result of, and a pre-condition for, this process. Of decisive importance in this phase of late capitalism has been the growing tendency towards the internationalization of production and the emergence of a new international division of labour in manufacturing. In turn, this has resulted in a reduction in the capacity of the British state to manage and steer the national economy and cope with the problems of the northeast and similar regions. Rather than solving problems at the economic level, the crisis tendency inherent within the capitalist mode of production is internalized into the state apparatus to appear in fresh forms – notably a fiscal crisis (which itself further restricts the state's capacity to manage the economy) and a rationality crisis (a perceived disjunction between the intentions and outcomes of state actions). These could spill over into a sociopolitical crisis, challenging the legitimacy of the state and, indeed, of the capitalist mode of production (Habermas, 1976, 1979).

Carney (1980) has argued that the failure of state intervention in regional problems can give rise to opportunities for the growth of new nationalist and regionalist movements which may, temporarily, precipitate acute disruption of 'normal' political activity. He cites several examples: Belgium (Mandel, 1963); Lorraine and Nord-Pas-de-Calais; and Scotland (Nairn, 1977). However, in northeast England, where increased state intervention, both directly and indirectly, has been a proximate cause of the deterioration in the region's economy, particularly in the declining job opportunities and rising unemployment, this has not been so. Rather, in general there has been a largely passive acceptance by the working class that economic decline, shrinking job opportunities and mounting unemployment are an inevitable part of the natural order, an acceptance made easier by memories of the 1930s and the cushion to consumption levels currently provided by the welfare state. To account for this particular reaction, it is necessary to consider the character and role of the Labour Party and Labourism in the region. Historically, the Labour Party has been the prime representative of working-class aims and aspirations and for some decades has effectively enjoyed a political hegemony within the region. But it has tended to adhere to a view that, in the final analysis, the interests of capital and labour are complementary; hence the emergence of a consensus around the politics of modernization in the early 1960s as the only route to full employment and social progress. The Labour movement held a key role both in establishing the legitimacy of the resultant regional modernization programme and in implementing it. However, when this was at first accompanied and then followed by sharply rising regional unemployment, the limits of the consensus politics of modernization as the route to social reform were clearly revealed. Yet to transcend these limits would require extending both the scope and qualitative forms of state action and, ultimately, challenging the legitimacy of capitalist social relations.

Chapter 3

Re-structuring region and state: the case of northeast England*

... the North is a fantastic success story ... You know that the North East has perhaps the biggest collection of inward investment and Japanese companies bringing the latest technology, superb management. ... The North and the North East is a colossal success story. (Prime Minister Margaret Thatcher, cited in the *Newcastle Journal*, 5 June 1990)

3.1 Introduction

While the mid-1970s were marked by concerns with the limited and partial transformation of the regional economy linked to a restricted conception of regional policy (see Chapter 2), Mrs Thatcher clearly saw a dramatic change in the region's economic well-being in the 1980s. In linking these changes to a new wave of inward investment from Japan, she was alluding to changes in the conception and practice of public policy and in the role of the state. Similar claims about economically successful transformation were made for particular places within the region, such as the former steel town of Consett which became a focus of national media attention in the 1980s (Figure 3.1, opposite). In this case, however, the emphasis was upon the creation of an enterprise culture and indigenous small firms rather than major inward investment projects. Consett was acclaimed by government ministers as an example of public policy successfully nurturing an enterprise culture of indigenous small firms in a former mono-industrial town via collaboration between national government, local government and other new local institutions such as the Derwentside Industrial Development Agency. Within the northeast there was a vigorous promotional campaign prosecuted by the Northern Development Company focused upon the revival of the 'Great North' around a new set of industries and activities, conjuring up images of the region's past successes as one of the 'workshops of the world'.

There have undoubtedly been significant changes over the last 25 years in the economy of northeast England – including a substantial net decline in coal mining and manufacturing employment, capacity and output, fresh inward investment in manufacturing from abroad, creating manufacturing

* This is a revised and slightly extended version of a paper first published 1998 in *Tijdschrift voor Economische en Sociale Geografie*, **89**: 15–30, The Royal Dutch Geographical Society, Utrecht

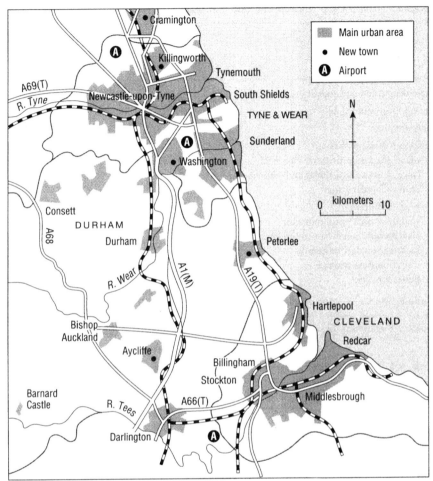

Figure 3.1 Northeast England: selected locations within the region.

employment in new industries, and expansion of service sector employment (Table 3.1, overleaf). On the other hand, concerns have been expressed about the profitability of many sectors of the regional economy and about the character and quality of many of the new jobs. There has also been a dramatic growth in the number of people without jobs (Table 3.2, overleaf). There is, therefore, clear *prima facie* evidence that there have been both winners and losers (in terms of industries, places and social groups) within the region as a consequence of these processes of economic change.

The central thesis of this chapter is that the recent transformation of the economy of northeast England is reciprocally related both to the changing position of the United Kingdom economy in the international division of labour and to the restructuring of the UK state. Regional changes were both a product of and a condition for wider projects of restructuring the UK state and its relation to economy and society within and beyond the UK. These

Table 3.1   Employees in employment, Northern Region, 1975–95 (thousands)

	1975	1995	Change 1975–95 No.	Change 1975–95 %
Agriculture, forestry and fishing	17	11	−6	−35.3
Energy and water supply	78	21	−57	−73.1
Construction	90	59	−31	−34.4
Manufacturing, including	451	249	−202	−44.8
Metal manufacturing and chemicals	109	46	−63	−57.8
Metal goods, engineering and vehicles	201	104	−97	−48.3
Other manufacturing	141	99	−42	−29.8
Services, including	630	747	+117	+18.6
Transport and communication	66	51	−15	−22.7
Distribution, hotels and catering; repairs	212	226	+14	+6.6
Banking, finance, insurance and business services	49	93	+44	+89.9
Other services	302	377	+75	+24.8
Total employees in employment	1 266	1 087	−179	−14.1

Source: *Employment Gazette*, including Historical Supplement, **102**, 10, October 1994

Table 3.2   Unemployment: Northern Region and Great Britain, 1979–97

	Northern Region No. (thousands)	Northern Region % rate	Great Britain % rate
1979	93.9	6.5	3.9
1980	115.6	8.0	5.0
1981	165.6	11.8	8.0
1982	187.2	13.3	9.4
1983	202.6	14.6	10.4
1984	214.5	15.3	10.6
1985	221.1	15.4	10.8
1986	221.5	15.2	10.9
1987	203.9	14.0	9.8
1988	174.0	11.9	7.9
1989	140.0	9.9	6.1
1990	122.7	8.7	5.6
1991	143.4	10.3	7.9
1992	157.1	11.3	9.7
1993	167.8	12.0	10.4
1994	151.4	11.7	9.5
1995	148.2	10.9	8.2
1996	134.9	10.0	7.5
1997 (Jan.)	123.1	9.0	7.1

Source: National On-line Manpower Information System, University of Durham

profound changes in the character and geography of contemporary capitalism have been associated with a tendency for national states to be 'hollowed out', with consequent changes to their regulatory capacities (Jessop, 1994).

The northeast has not, however, been alone in experiencing such changes. Such changes were by no means unique to the northeast within the UK or, more generally, confined to the UK. As processes of globalization have become powerfully inscribed alongside those of internationalization into the political economy of contemporary capitalism, many industrialized regions have experienced severe economic decline over the last two decades. Former economically prosperous and vibrant regions have become the 'rustbelts' of western Europe and North America (for example, see Carney *et al.*, 1980; Rodwin and Sazanami, 1989; European Commission, 1996). In some ways the trajectory of change in northeast England and the restructuring of the UK state can be seen as exemplifying more general tendencies. Equally, there are specific aspects of the UK case that differentiate it from patterns and processes of change in other late modern societies (for example, see Lash and Urry, 1987). These commonalities and differences are important in understanding the particular case of northeast England.

3.2 Regional changes in the national and global contexts ———

In the immediate post-war period there was a significant, though selective, extension of the scope of state involvement in economy and society in the UK as a central element in the Post-War Settlement (PWS) and an emergent 'One Nation' national political project. This embraced the creation of an inclusive social-democratic politics and the narrowing of socio-spatial inequalities. Realizing this vision required constructing a new mode of regulation based on an extension of the boundaries of state activity. This included, *inter alia*, Keynesian economic management to sustain full employment, nationalization of major industries such as coal and rail transport, the establishment of new systems of town and country planning and land use regulation, and the fulfilment of a welfare state, perhaps most powerfully symbolized by the National Health Service.

The expanded scope of state competencies was particularly significant in northeast England, directly impacting on its productive structure. It also involved a significant growth of state involvement into the sphere of social reproduction. This led to expansion in public sector services employment as well as the delivery of educational, health and social services via the state. It thus had a pervasive effect on the lives of the majority of the region's population. Consequently, the northeast became characterized as a state-managed region, with a particular regional form of institutional arrangements and Labourist political culture (see Hudson, 1989a).

This extension of state activity nevertheless involved significant deviations from the paradigmatic Fordist regime of accumulation. Acceptance of Keynesian economic management by both major political parties ensured commitment to a state-managed strategy of intensive accumulation within the national territory, and redistributive policies to raise levels of mass consumption in line with the productivity gains of mass production. However, legacies of Empire, in the form of an unlikely combination of the strength of its organized working class (particularly evident in regions such as the northeast), along with powerful fractions of both financial and manufacturing capital with highly internationalized interests (Beynon *et al.*, 1994), led to a deformed version of Fordism evolving in the UK (Lipietz, 1986). Despite the 1947 Exchange Control Act, which regulated international capital flows into and out of the UK, capital retained sufficient room for manoeuvre to undermine in due course the central premise of a Fordist regime of intensive accumulation within the national territory – national state management of the national economy.

During the 1950s, however, these immanent contradictions were contained as full employment co-existed with profitable private sector production in the northeast. From the latter years of that decade, however, it became increasingly clear that the regional economy was structurally weak and vulnerable in an increasingly competitive international division of labour. It also became evident that there were serious weaknesses in the broader national developmental model; by the mid-1960s these were very visible. There were attempts to address both forms of crisis in a coherent way via linked regional and national modernization policies. These sought non-inflationary national and regionally balanced growth via a stronger central government regional policy and the creation of new, modern regional institutions, coupled with sectoral policies designed to modernize the national economy, especially key manufacturing industries (Department of Economic Affairs, 1965). Government policy responses to these linked regional and national economic crises became increasingly crisis prone. Because the UK economy was relatively open in comparison with other major advanced capitalist economies, it was particularly susceptible to pressures transmitted from international markets. As the performance of the national economy faltered, so too did governments' capacities to maintain full employment and the welfare state, especially with the breakdown of the Bretton Woods arrangements for fixed currency exchange rates in 1972, a key moment in the transition to a more global economy. By 1975 it was clear that government policy initiatives had failed and that the national mode of regulation was no longer tenable as the state faced the threat of simultaneous fiscal and legitimation crises and the economy slid into a deep recession. The transition to a very different neoliberal 'Two Nations' strategy thus began in 1975. It can be seen as a response to the accelerating decline of the UK economy in an emerging era of globalization, and its dominance was confirmed with the ascendance to power of Thatcherism in 1979 (Hudson and Williams, 1995). It had momentous implications for northeast England.

In essence, state involvement – or in Mrs Thatcher's deeply pejorative phrase, 'the nanny state' – came to be seen not as the solution to economic and social problems but as the problem itself. The resultant restructuring of the UK state can be related to the 'hollowing out' thesis (Jessop, 1994) but several aspects of state restructuring as part of the 'Two Nations' project sit uneasily with the canonical model of 'hollowing out'. While there was some transfer of state power upwards to the EU, within the UK the dominant tendency was not one of decentralization to regional and local levels but one of centralization of state power at national level. This was of considerable significance in northeast England, for the restrictions on the competencies and financial resources of elected local authorities both restricted their capacity to act and influenced their vision of feasible local economic development policies. Local authorities, predominantly Labour-controlled, had developed as significant actors in local and regional development strategies over the post-war period. Equally significant was their partial replacement by non-elected quangos (Quasi-Autonomous Non-elected Governmental Organizations), such as Urban Development Corporations and Training and Enterprise Councils, which decisively shaped the definition of regeneration programmes over much of the region and also had relatively substantial financial resources with which to pursue their objectives. The most significant shift in the structure of state regulation, however, was in the boundary between state and market and in the ways in which markets were regulated. The political economy of Thatcherism sought to create more space for market forces as the primary mechanism for steering and restructuring the national economy and gave particular prominence to the role of global markets in this. Processes of globalization were clearly emerging prior to the ascendancy of Thatcherism, but they were given a massive boost after 1979. Equally, while in some respects a response to the deteriorating position of the UK economy in the context of emerging globalization tendencies, the 'Two Nations' strategy also became a project to try to influence the sectoral and spatial shape and form of globalization, both within and outside the UK.

The rhetoric of Thatcherism stressed facilitating and supporting the emergence of an 'enterprise culture' via drastically cutting back state involvement in economic and social life, although state policy and practice revealed a more complex pattern, involving selective expansion in some areas, such as law and order enforcement. Sharply reducing direct state involvement in the production of coal, steel and ships, and in the scope of regional policy, while expanding urban policies, along with more muted cuts in the provision of educational, health, housing and social services through the welfare state, had severe consequences in the northeast, and these are outlined below. So too are the ways in which the economic development policy agenda and institutional arrangements within the region were of necessity altered in these new circumstances, although there are strong threads of continuity as well as changes, and these are discussed in the next section. The final section attempts to draw together the implications of these changes for production and employment in the region, and for the lives of those who continue to live there.

3.3 Rolling back the state, unravelling the 'old' regional economy

Three related aspects of the process of rolling back the boundaries of the state are of particular relevance here. The first is the increasing and deliberate reliance upon the global market as the arbiter of the national and regional economies. The second is the rationalization and then privatization of the formerly nationalized coal, rail, steel and shipbuilding industries. The third is the cutting back of the welfare state, coupled with the increasing incorporation of pseudo-profitability and efficiency criteria in the remainder of the public services sector.

3.3.1 The global market as the arbiter of the national and regional economy

There has been a strong neoliberal hue to national economic policy over the last two decades, with a growing emphasis upon liberating market forces within the UK and upon global market forces as the prime economic steering mechanism within the UK. The high exchange rate and counter-inflation policies of the early 1980s intensified the existing pressures on manufacturing in the UK as a result of a severe international recession. These combined effects were particularly severely felt in northeast England. They were expressed in various ways. Firstly, they directly intensified competitive pressures on private sector manufacturing in 'traditional' branches such as heavy engineering and chemicals and in branch plants of the 'new' consumer goods industries that had previously located in the region. There were widespread capacity closures and substantial job shedding in those plants that remained open, sometimes associated with new fixed capital investment, mainly in the chemicals industry. Secondly, the associated squeeze on public expenditure led to severe financial pressures on both nationalized industries and public services, leading to capacity closures and job losses. Thirdly, deregulation of outward foreign direct investment (FDI) from 1979 led to a massive outflow of capital from the UK in the first half of the 1980s. This intensified deindustrialization, both nationally and regionally. Nationally, between 1979 and 1986 the UK's 40 largest manufacturing firms increased employment abroad by 125,000 but cut employment in the UK by 415,000 (Hudson and Williams, 1995, 45). Partly as a result the Northern Region lost exactly one third (138,000) of its manufacturing jobs in those same years. While there were political limits on the extent to which nationalized manufacturing companies such as British Steel could transfer production from the northeast to locations outside the UK, this was not the case for major private sector companies such as ICI (Beynon et al., 1994). Even so, the major causes of job loss in the region were capacity closure and restructuring to increase labour productivity rather than capital flight and relocation abroad. The combined effects of global recession and deflationary national economic policies were that coal mining employment fell to virtually zero while manufacturing employment in the Northern Region fell by almost 41 per cent and the proportion of regional employment in the manufacturing sector fell by some 15 per cent between 1978 and 1993. The net fall in manufacturing employment occurred despite some new investment and employment creation in parts of manufacturing also linked to government policies which are discussed below in section 3.4.1.

3.3.2 Nationalization, rationalization and privatization

The coal and steel industries – which together accounted for 20 per cent of industrial jobs in the Northern Region – were prime targets of Thatcherite policies of rolling back the state. This reflected political and ideological reasons as well as economic ones. State ownership of key industries was central to the PWS, with the nationalization of the coal industry being of particular symbolic significance. It is important to stress that the industries that were then (and subsequently) nationalized were not the 'commanding heights' of the economy, as was often proclaimed, but rather were industries that for various reasons were seen as vital to national economic or strategic defence interests, though in themselves unprofitable to private capital. There were different and specific reasons for nationalization, linked to the timing of the transfer of these industries to the public sector (for details, see Hudson, 1986b).

Equally the rationalization of these industries within the public sector commenced at varying times, as pseudo-profitability efficiency criteria began to diffuse through the nationalized industries from the late 1950s, initially in the coal industry. As a result the process of capacity closure and job loss in these industries has been underway for almost four decades in the northeast, linked to central government policies and changing international market conditions. The pace of decline accelerated in the 1980s in the context of further rationalization in search of profitability as a prelude to privatization. This was associated with the introduction of new terms and conditions of work, but concerns about poorer quality jobs were soon replaced by fears about an absence of any jobs in these industries. For privatization not only returned these industries to the private sector but in the process virtually eliminated coal mining, iron and steel production, and shipbuilding from the economy of the northeast. By the early 1990s, the presence of these industries which had formed the historic core of the regional production system was reduced to one colliery (although privatized and environmentally destructive opencast coal mining remained in the region), one steel works and some sporadic ship repair activity: in all, fewer than 10,000 jobs, compared to almost 300,000 in the mid-1950s.

3.3.3 Slimming down the welfare state and the transition to a workfare state

The expansion of the welfare state had led to a major growth of employment over the post-war period. Indeed, public sector expansion accounted for most of the net employment growth in the service sector of the northeast in the post-war period prior to 1980 – around 130,000 jobs. The growth of public sector provision of educational, health and social services was linked to a changing gender division of waged labour and a major expansion of female employment outside the home, although much of this was on a part-time and poorly paid basis.

One of the main aims of the Thatcherite policies of the 1980s was to rein in and reduce public sector service provision. Such policies had national impacts but they were particularly severe in a region such as the northeast, with a wide

Table 3.3   Workforce in employment, Northern Region, 1975–96, June (thousands)

	1975	1996	Change 1975–96 No.	%
All persons				
Workforce in employment, including	1 353	1 282	−71	−5.2
Employees	1 266	1 130	−136	−10.7
Self-employed	88	124	+36	+40.9
Work-related government training programme	0	20	20	—
Males				
Workforce in employment, including	843	707	−136	−16.1
Employees	774	600	−174	−22.5
Self-employed	69	92	+23	+33.3
Females				
Workforce in employment, including	510	575	+65	+12.7
Employees	491	530	+39	+7.9
— full-time	303	275	−28	−9.2
— part-time	188	255	+67	+35.6
Self-employed	19	32	+13	+68.4

Sources: Office for National Statistics: *Labour Force Survey, Historical Supplement*, 1996; HMSO Government Statistical Source: Historical Supplement no. 4, *Employment Gazette*, **102**, 10

range of effects upon people's lives and living conditions as well as upon patterns of employment. They had three sorts of impact on the regional labour market. Firstly, the growth of public service sector employment ceased from 1979 to 1983 while the proportion of jobs offered on a part-time basis increased still further (Table 3.3). Secondly, many jobs were transferred from the public to the private service sector as a consequence of legislative changes to enforce compulsory competitive tendering and the contracting-out of work from public sector providers to private sector contractors. Thirdly, for those remaining in work within the public service sector, the introduction of pseudo-profitability efficiency criteria led not only to a reduction in the numbers of jobs but also to a deterioration in the terms and conditions on which they were offered in the residual, slimmed-down public services sector (Hudson *et al.*, 1992). In the 1990s growing pressures to cut welfare expenditure were reflected in changes to move from a welfare to a workfare state, with severe and immediate consequences for the unemployed and long-term sick.

3.4   Creating a 'new' economy, creating new forms of state involvement

The accelerating collapse of the regional economy from the mid-1970s raised serious questions as to the region's future, about the sorts of alternative economic activities that could be developed endogenously or enticed to locate there, and about the policies that might be developed to facilitate such a transformation. The increasing disengagement of national government from regional policy, coupled with its growing attraction to and involvement with urban policy (which in practice was a mechanism which switched public expenditure from regions such as the northeast into inner London: Hudson and Williams, 1995), was of great significance for the northeast. It indicated, *inter alia*, a need for new forms of policy initiatives within the northeast and for new institutional arrangements appropriate to the formulation and implementation of such policies.

Four of the strands of the new economy and associated institutional and policy changes are explored in this section: the renewed search for inward investment, especially from outside the UK; attempts technologically to upgrade the regional economy; the creation of new endogenous small and medium-sized enterprises (SMEs) as part of an 'enterprise culture'; and, to a degree overlapping with this, an increasing emphasis upon private sector services of various sorts, in part linked to new forms of consumption within the region.

3.4.1 Inward foreign direct investment in manufacturing industry and a new role for the region in the international division of labour?

In the post-war years, there were powerful pressures to restrict inward investment to the northeast, especially if new factories would compete in the labour market with established coal mining and 'traditional' manufacturing industries for male labour (Hudson, 1989a). Nevertheless, there was a substantial inflow of branch plant investment from southeast England and the Midlands. This was largely in consumer goods sectors employing female labour (in particular, clothing and electronics), especially between 1945 and 1951 (Keeble, 1971). The low female activity rates in the northeast, a consequence of a particular gender division of waged and unwaged labour, were a compelling attraction to companies in other regions experiencing shortages of female labour. Consequently, for much of the period to the end of the 1950s the northeast was a 'full employment' region and there was little or no attempt to attract inward foreign direct investment (FDI).

By 1963 only 8500 people were employed in foreign-owned factories in the north (see also Hudson, 1995e).[1] The accelerating decline of the old industrial base produced a marked change in attitudes to inward investment as the emphasis switched to regional economic diversification and the attraction of new manufacturing jobs. This goal was pursued via a combination of a strengthened central government regional policy, and institutional innovation and social change within the region, as elements of a programme of regional modernization and (re)presentation (Hudson, 1989a). As a result, there was a considerable increase in foreign inward investment and in the number of multinational branch plants, especially from the USA. By 1978 employment

Plate 3.1   A new industry in a new town: the Sumitomo tyre factory, Washington New Town.

in foreign-owned plants had risen to 53,000, 12.9 per cent of manufacturing jobs (Smith and Stone, 1989). By 1988 the absolute total remained the same but the relative share had risen to 20.0 per cent. By 1993, however, employment had grown sharply to over 80,000, some 33.0 per cent of the regional manufacturing total (Bridge, 1993). Clearly the northeast has had considerable success in competing for FDI.

The net growth of employment in foreign-owned factories reflects the cumulative effects of several tendencies. Acquisitions of UK companies added to the stock of employment in foreign-owned factories, while existing ones were lost because of plant closure or *in situ* restructuring. The dominant tendency, however, was of employment growth as a result of new greenfield investment (although, unusually, there were two significant acquisitions by Japanese companies – Dunlop's Washington tyre factory by Sumitomo and Caterpillar's Birtley factory by Komatsu), heavily concentrated both sectorally and by national origin. Post-1985 almost 23 per cent of new jobs were in electrical and electronic engineering, associated with investment by Far Eastern companies such as Fujitsu, Goldstar, Samsung and Sanyo, while 34 per cent were in the automobile industry, primarily associated with investments by Nissan in Washington – perhaps *the* prime exemplar of the new round of inward investment – and some component suppliers (Hudson, 1995b). The overall stock remained dominated by the USA, however, and flows of fresh investment from the USA and Germany were also significant (Bussell, 1990). Such figures suggest impressive growth of investment, employment, output and exports (since many of the new plants were established to penetrate the EU market).

Plate 3.2   New patterns of production and trade: a Nissan automobile transporter leaving Teesside.

While FDI has diversified the range of industries present within the region, it has been suggested that in the 1960s and 1970s it simultaneously led to a greater degree of homogenization of industrial activity, characterized by routine assembly and component production operations. Thus the region became the location of a particular stage in the overall spatial/technical division of labour within companies.[2] This diversification was also seen as problematic in other ways, creating fears of external control, vulnerability and dependency in a branch plant economy (Firn, 1975), engendering dependent rather than self-sustaining development (Turok, 1993). Formerly coherent regional economies became characterized as 'global outposts', situated at the extremities of corporate hierarchies, far removed from key centres of command, control and strategic decision-making (Austrin and Beynon, 1979).

By the 1980s, however, there were also claims that corporate strategies were changing in significant ways in an era of globalization. Transnational companies were said to have switched to more 'regionalized' or 'glocalized' versions of their global strategies, with R&D, production, distribution and sales organized on, say, a European rather than a worldwide basis, more attuned to territorial variations in consumer demand and regulatory requirements (Hudson, 1997a). These changes can also lead to a degree of intra-regional clustering of factories associated with just-in-time production strategies. Component companies have established plants adjacent to Nissan's factory to allow synchronous production, for example (Sadler and Amin, 1994). There is evidence of component suppliers following the new electronics factories from southeast Asia in the 1990s. This clustering of production may reflect a greater degree of

Table 3.4 Employment structure, Northern Region (1971–97) and Great Britain (GB) (1997)

	North						GB
	1971	1979	1984	1990	1994	1997	1997
Workforce in employment (000)	1 321	1 332	1 210	1 305	1 264	1 292	25 985
% self-employed	7.6	6.4	8.4	9.0	10.0	9.2	14.6
% employees	92.4	93.6	91.6	91.0	90.0	90.8	85.4
of whom							
% manufacturing	36.1	32.9	26.2	24.5	23.3	21.9	18.0
% male	63.6	59.4	56.0	55.0	54.3	51.9	52.4
% female	36.4	40.6	44.0	45.0	45.7	48.1	47.6
% part-time	n.a.[a]	n.a.	n.a.	29.5	29.7	29.9	28.0
% female part-time	12.0	16.8	19.7	22.7	23.7	22.5	22.3

[a] n.a.: not available

Source: National On-line Manpower Information System, University of Durham

embeddedness within the region, with potentially more positive developmental implications than previously disarticulated branch plants. On the other hand, Siemens' acquisition from Rolls Royce in 1997 of Parsons Power Generations Systems on Tyneside, so that it could carry out service work and supply parts to Siemens' factories in Germany, suggests that the 'old' branch plant syndrome lives on (although maintenance is the most profitable activity within the power generation industry, those profits will be repatriated to Germany: Wagstyl, 1997).

There also has been, and is, considerable concern about the quantity, quality and permanence of employment. Many branch plants resulting from earlier rounds of investment either shed labour or closed completely in the 1980s, suggesting that fears of 'runaway' branch plants, closing precipitately and prematurely as companies switched investment to more lucrative locations, were justified. On the other hand, Stone and Peck (1996) argue that the foreign-owned branch plants have been a more secure source of employment than indigenous UK-owned plants over the period 1978–93. Nevertheless, while this may be the case and individual factories associated with inward FDI in the 1980s and 1990s do employ considerable numbers of people, many more jobs have disappeared because of closures of coal mines, steel works, shipyards and factories in other sectors of manufacturing (see Tables 3.3, page 70 and 3.4, above). Indeed, a mass of unemployed people was and is a significant and necessary attraction for such inward investment. By the second half of the 1980s, the northeast had become a low-wage region within the EU, in which labour supply greatly exceeded demand. For example, Nissan received 33,000 applications within six weeks in 1992 for 600 new jobs. Such conditions allow great selectivity in recruitment (Hudson, 1995b). Transnationals often prefer to recruit employed rather than unemployed people, and so exclude the vast majority of those looking for work from consideration as part of their potential workforces. There are therefore clear *quantitative limits* to FDI as a solution to problems of regional unemployment, although it can have a variety of other positive impacts upon the regional economy.

As well as concerns about the numbers of new jobs created, however, there is also considerable debate about their quality. Echoing Firn (1975), fears continue to be expressed about the quality and stability of jobs in the new branch plants of the 1980s and 1990s. As Peck and Stone (1993, 66) note, in relation to the new Japanese factories of the 1980s, while '. . . some inward investment projects can be regarded as "best practice" plants, . . . others do not appear to differ greatly from the traditional branch factory'. The debate as to the quality of new jobs may again be exemplified by the case of Nissan. The company projects an image of work at its Washington factory as empowering production workers, offering an enriching form of work for multi-skilled workers with considerable job security (if no longer a job for life), built around themes of flexibility, quality and teamwork. In contrast, its critics see it as disempowering and disabling, characterized by control, exploitation and surveillance, and bound together through team working in the context of low levels of unionization in a single-union factory (see, for example, Garrahan and Stewart, 1992).

Despite such reservations about the quantity and quality of jobs, interregional competition for mobile transnational investment is nonetheless intense. Consequently, attracting such investment required new regional and local institutions such as the County Durham Development Company, but most notably the Northern Development Company (established in 1986), to promote the virtues of the region as a space for profitable production. The NDC is a 'one-stop shop', selling the attractions of the region to transnational companies in terms of the availability of grants and loans, the quality of the built, social and natural environments and, often crucially, the adaptability and flexibility of a co-operative labour force. It includes carefully selected trades union representatives who promote the virtues of the region's workers, usually in the hope of securing the right to represent them in new single-union deals. Without doubt, the NDC skilfully and vigorously sells the 'Great North' (Hudson, 1991b), in fierce competition with other areas. In turn it uses this success as part of its marketing strategies in seeking to create a 'bandwagon effect' (Smith and Stone, 1989) and thereby entice other companies to the region (Peck and Stone, 1993). The NDC's approach to attracting FDI and industrial regeneration has, however, been criticized as based upon 'hype' and the presentation of a partial and one-sided representation of the state of the region (Robinson, 1990).

Increasingly in the 1990s the NDC has switched the emphasis in its activities from competing for new investment (although it remains very active in this area) to 'aftercare' and maximizing the developmental potential of existing investments. It has placed growing emphasis upon the development of supply chains in the region, especially for the automobile industry, and on the development of industrial clusters rather than simply the attraction of isolated branch plants. It is important in this context to recall that there has been a concern to attract 'cluster forming' industries to the region since at least the early 1960s, and similar expressed wishes to upgrade the quality of inward investment (see Hudson, 1989a), but with little perceptible effect over the period until the late 1980s. There is now some evidence that such clusters may be emerging. There are, however, also dangers in developing a regional economy clustered around a small set of sectors and lead companies, which were pre-figured in past debates about the narrow base of the 'traditional' regional industrial economy and the resultant institutional lock-in (Hudson, 1994b). This suggests a need for institutional structures capable of supporting an 'intelligent' or 'learning' region. The development of such an institutional structure remains a major challenge in the northeast.

3.4.2   Enhancing the region's technological capacity

In the nineteenth century the northeast was a region of technological dynam-ism, with a succession of major product and process innovations developed there. For a while, it was a veritable 'learning region', a 'technology district'. By the second half of the twentieth century, however, it was widely regarded as an economically and technologically backward region, reliant upon a set of old 'smokestack' heavy industries, with little indigenous capacity for product innovation and the creation of new growth sectors and industries. Neverthe-less, in terms of process innovations and the application of new production technologies, industries such as coal, steel and bulk chemicals often belied their 'smokestack' image, making considerable use of sophisticated process technologies, often heavily automated and/or computer-controlled continuous production processes.

In the 1960s, the emergence of a series of regional modernization pro-grammes led to a recognition of the need for technological enhancement of the region's economy; the subsequent availability of central government financial assistance was important in financing investment in new process technologies but had little impact in encouraging new product innovation and R&D in the region. Following the establishment of the Northern Economic Planning Coun-cil (NEPC) in 1964 (one of a series of such Regional Economic Planning Councils as new institutions to help co-ordinate national and regional eco-nomic planning), concerns began to be expressed about the quality of new industrial growth in the region. There was a recognition that more modern science-based and capital-intensive industries were needed in the region with 'high productivity and well paid employment with a high skill content'. It was argued that: 'Further positive action is needed to stimulate technical innova-tion in the region's industries . . . The long term solution to the problem (of economic decline) is to increase the amount of basic research done within the region . . . Investment in this must be encouraged' (Northern Economic Plan-ning Council, 1966, 55). Despite such expressions of good intent, however, in practice little was done to enhance the region's technological capacities in these ways and to facilitate processes of transition to become a more 'intelli-gent' region. While there was a succession of EU and national programmes intended to facilitate technology transfer and enhance technological capacities, these had limited impacts within the northeast. Although ICI was a major exception, with extensive R&D facilities on Teesside, the regional economy more generally remained characterized by routine production rather than innovative R&D throughout the 1960s and 1970s. Moreover, ICI's R&D activities were integrally linked into its overall corporate structure rather than networked with other companies in the region. In so far as there was techno-logical upgrading associated with new fixed capital investment, it was over-whelmingly in new process technologies in established industries (often leading to jobless or job-shedding growth) rather than in radical product innovation and the creation of new growth sectors based around new products with growing markets. Moreover, investments in new processes were often heavily supported financially by government regional policy, with up to 50 per cent of fixed capital investment costs met from regional policy assistance.

By 1990 Mrs Thatcher was emphasizing the extent to which the regional economy had been transformed in the 1980s, especially as a result of new inward investment from Japan. While this led to the production of products new to the region these were not, however, new products. Once again, the emphasis was upon process innovation and new ways of working rather than product innovation and major technological advances in production. In many ways, this reproduced the patterns of the past but with respect to a set of industries that were new to the northeast but that were, in global terms, mature. In so far as these plants are associated with innovation and technological advance, these gains are embedded in global corporate structures rather than in the regional economy. There were also some institutional innovations within the region intended to enhance technological capacities. There were attempts to draw upon the research strengths of the region's universities in science and technology through the creation of science parks and inter-university collabor-ative ventures such as HESIN (Higher Education Support for Industry in the North). While important to some companies, these initiatives have had little impact in generally enhancing the technological profile of companies in the region. Links between research in universities and industry in the region remain rather weak and underdeveloped.

One spin-off from the involvement of higher educational institutions with regional industry was the establishment of a Regional Technology Centre in 1989, a joint venture between the higher educational institutions and the Department of Trade and Industry. It has three main objectives: promotion of technology transfer, particularly between SMEs and higher education insti-tutions in the northeast; information transfer through database facilities with regard to new markets, co-operation between companies and governmental aid schemes; and connecting training demand with suppliers and supporting firms in relation to the recruitment of skilled personnel. Although self-financing, in practice the Centre receives about half its funding from the public sector. Its impact in diffusing new technologies has been limited (Hassinck, 1992a, 1992b).

There were also indigenous attempts to stimulate SMEs in 'high-tech' industries via business innovation centres and agencies such as the North East

Innovation Centre. This Centre was established in 1981 in Gateshead. The services of the Centre are mainly used by small, generally low-tech, firms in manufacturing, especially in engineering and electronics. While it has undoubtedly helped such firms, it has had little impact in enhancing the technological capabilities of the region's manufacturing sector. MARI (Micro-electronics Applications Research Institute) was another local initiative which again has had limited effects within the region. Established in 1979, with central government support, by a consortium of local authorities in Tyne and Wear, Newcastle University and Polytechnic (now the University of Northumbria) and a London-based computer applications company, MARI was a software research centre from which firms, especially SMEs, in the northeast could commission R&D work at favourable rates. In 1989 it was privatized, however, and of necessity adopted a more commercial and less regionally focused strategy. Consequently it had much less impact in enhancing the R&D capacities of firms in the northeast. Similarly, the limited development of R&D and service consultancies in the region in response to national and EU policies is largely oriented to external markets. In summary, such changes suggest at best embryonic processes of embedding innovation and technological developments within the region.

There has also been considerable emphasis in local authority economic development policy discourses (which are discussed more fully below) on encouraging high-tech developments through promoting science parks and technology centres. An example of such an initiative is the Cleveland CAD-CAM Centre, established to try to enhance the technological capabilities of small and medium-sized manufacturing firms (Beynon *et al.*, 1994). Such local policy initiatives have had very limited impacts, however. Local authorities have also been partners in schemes to develop science and technology parks, such as the Sunderland Technology Park, the Newcastle Science Park and the Offshore Technology Park in Wallsend on Tyneside. Such initiatives tend to be stronger on imaginative ideas than on bringing them to a successful conclusion. While there has been considerable rhetorical emphasis upon reconstructing local economies around technologically sophisticated SMEs in manufacturing, there are in fact very few examples of such firms being established and becoming successful. Moreover, some of the best-known examples have subsequently been acquired by major multinationals – for example, Integrated Micro-Products in Consett was acquired by Sun Systems, in order to secure access to its technological expertise. This raises fears of a repetition of the 'branch plant' syndrome.

In summary, while there has been a recognition of the need to upgrade the technological base of the region's economy for over 30 years, and some institutional innovations have been introduced to help attain this objective, the region's economy remains heavily dependent on the routine production of goods and, increasingly, services for external markets. The extent to which it has become a successful 'technology district' or 'learning region' (Hudson, 1996b) is to date extremely limited, although there *may* be the seeds of such a transformation in the greater degree of 'embeddedness' of some of the new inward

Plate 3.3   Enhancing the technological capacity of the local economy: the Cleveland CADCAM Centre.

investments of the 1980s and 1990s. The extent to which such a transformation is possible, and the most appropriate institutional and policy arrangements to facilitate it, remain issues of pressing concern. It is important to note in this context that since the 1960s shortages of suitably highly skilled labour have been a recurrent constraint hindering successful implementation of policies intended to facilitate technological upgrading of the region's economy.

3.4.3   Creating an enterprise culture: small and medium-sized enterprises and new forms of local economic development strategies

From the late 1970s, the emphasis in national government policies for urban and regional regeneration in the UK shifted significantly. In part, this was because branch plant policies had failed to deliver sufficient new jobs to the regions. This created space in which other policies of self-reliance, endogenous growth and the encouragement of small firms could evolve within the broader political project of Thatcherism. A similar switch in policy emphasis was observable over much of Europe, reflecting perceptions of the bases of success in Europe's successful regional economies (Garofoli, 1992; Dunford and Hudson, 1996). At the same time, EU regional policies were increasingly emphasizing small firms and endogenous growth as the route to regional regeneration and this has had a pragmatic influence on the conception of regeneration policies within the northeast through the availability of funding.

Plate 3.4 New industries arising in old industrial areas? The Phoenix Workshops on the site of the former Horden Colliery.

For small-firm policies to be successful in the northeast presupposed, *inter alia*, a decisive transformation of its dominant working-class culture to embrace the values of the Thatcherite enterprise culture. People socialized into a world of working for a wage for others (often the state as employer) would have to become self-employed or, better still, themselves employers of wage labour. This decisive cultural transformation was to be facilitated by new mechanisms and institutions, often bridging the private and public sector divide (such as local enterprise agencies, agencies such as British Steel (Industry) and British Coal (Enterprises) and Training and Enterprise Councils), to help create the conditions in which new small firms could be born and then grow. As a result, County and District Councils also had to acquire new skills in place promotion, in formulating local economic development measures in pursuit of these goals and in competing for a variety of sources of public funding to help implement them. In practice, however, much of the content and form of local authority policies was simply regional policy written small, a familiar mix of policies to provide factories and premises, help with training and provide marginal financial assistance directly to companies. What was new, however, was the role that local authorities took on in offering advice on the development of business plans and co-ordinating aid and assistance from a variety of sources, liaising with regional promotional institutions, national government departments and the Government Office for the North East and, on occasion, directly with the European Union. Newcastle City Council set up Business Development Centres and a Business Advice Centre, and North Tyneside Council established a Business Resource Centre, as 'one-stop shops' offering

Table 3.5   Northern Region firms registered for VAT, 1979–94: production sector

Registrations	Deregistrations	Total stock	Net change	Date of return	% change	% change for GB
0	0	3 239	0	1979		
450	358	3 331	92	1980	+16.8%	+9.1%
432	303	3 459	129	1981		
474	335	3 598	139	1982		
552	369	3 782	183	1983		
502	388	3 895	114	1984	+19.2%	+9.5%
620	389	4 127	231	1985		
646	472	4 302	174	1986		
690	483	4 509	207	1987		
719	533	4 696	186	1988		
717	566	4 847	151	1989	+11.2%	+9.1%
700	514	5 033	186	1990		
547	563	5 016	−16	1991		
New series						
424	561	4 515	−137	1992		
380	536	4 359	−156	1993	−6.0%	−1.2%
390	505	4 244	−115	1994		

Source: National On-line Manpower Information System, University of Durham

information and advice to local firms, for example. There is a danger of a plethora of such agencies confusing local firms, especially if they seem to be acting competitively rather than co-operatively. The sophisticated forms of real service provision, intense inter-institutional contact and collaboration, and public–private sector co-operation that characterize the successful small-firm economies of regions such as Emilia-Romagna and western Jutland have not (yet) developed in the northeast.

In general, the available evidence suggests that policies to encourage SMEs in the northeast have had at best limited success. Although net firm formation rates in production industries exceed national rates in the 1980s (Table 3.5), more generally the Northern Region recorded both the lowest new firm formation rate and the lowest net firm growth rate of all UK regions between 1980 and 1990 (Keeble and Walker, 1994; Johnson and Conway, 1995), As a result, the region's share of the national stock of businesses fell (Table 3.6, overleaf). Equally, the actual employment creation effects of SMEs have been, at best, limited (for example, Beynon et al., 1991, 1994; Hudson, 1993; Hudson and Sadler, 1991). There is, then, little evidence that such local economic regeneration policies have had their intended effects, that the new institutional arrangements put in place have had anything more than a cosmetic and superficial effect in bringing about a cultural transformation that embraces the Thatcherite conception of the values of entrepreneurship and enterprise, or that local people have established their own companies. Most of the new small firms established by local residents are in low-tech services rather

Table 3.6   Northern Region firms registered for VAT, 1979–94: all sectors

Registrations	Deregistrations	Total stock	Net change	Date of return	% change	% change for GB
0	0	52 350	0	1979		
5 751	5 676	52 425	75	1980	+6.4%	+8.0%
5 820	4 436	53 809	1 384	1981		
6 087	5 421	54 474	666	1982		
6 575	5 350	55 697	1 225	1983		
6 580	5 681	56 594	899	1984	+5.1%	+8.6%
6 457	6 080	56 971	377	1985		
6 822	6 331	57 463	491	1986		
7 318	6 225	58 555	1 093	1987		
8 323	6 669	60 210	1 654	1988		
9 105	6 755	62 559	2 350	1989	+9.8%	+13.9%
8 336	6 706	64 188	1 630	1990		
6 929	6 839	64 279	90	1991		
New series						
6 189	7 466	60 439	−1 277	1992		
6 269	7 094	59 612	−825	1993	−1.9%	−4.0%
6 433	6 749	59 297	−316	1994		

Source: National On-Line Manpower Information System, University of Durham

than high-tech manufacturing (see Storey, 1990; Hudson *et al.*, 1992). Such new manufacturing companies as have been established have often resulted from the actions of 'entrepreneurial immigrants', attracted by financial incentives and labour market conditions, rather than those of 'local heroes' (Caulkins, 1992). Companies established by local residents more often reflect a defensive fear of unemployment rather than an offensive embracing of the enterprise culture. For those who do set up their own small companies, 'enterprise means . . . a twilight world of hard work, low pay, casual labour and insecurity as . . . people plod along trying to secure a decent living through enterprise' (MacDonald, 1991).

The legacy of a culture of waged labour has thus proved highly resistant to change, not least because unemployed former workers often recognize that the economic climate in the areas in which they live is not a promising one for the would-be entrepreneur. The high failure rate amongst new companies suggests that this is a very rational judgement (Hudson and Sadler, 1991). The failure of an enterprise culture to take root in the region is not a result of the inadequacies of individual psychology or of institutional innovation but rather a consequence of local economic and labour market conditions.

3.4.4   The turn to private sector services

Given the generalized switch towards a private service sector economy, it is no surprise that in the northeast attempts have been made to identify potential service sector activities that might be developed there. The severe decline of coal mining and manufacturing ensured that the service sector became relatively more important within the northeast. The cutbacks to the welfare state added to the significance of private sector services as activities such as care for the elderly were transferred to the private sector and compulsory competitive tendering led to increased sub-contracting from the remaining public sector services to those in the private sector (Hudson *et al.*, 1992). Beyond that, however, there was growth in areas of private sector services. For example, banking, finance, insurance and business services doubled their total employment (albeit from a low base) between 1971 and 1993. Private sector services growth has been considerable (Kirby, 1995), especially in the period 1984–87 linked to tourism and retailing and new forms of consumption and leisure within the region.

In addition, there has been growth in service sector employment in back-offices and telephone call centres, attracted by a combination of physical redevelopment activities as a part of urban regeneration programmes providing appropriate sites and buildings (see below) and a considerable supply of cheap, flexible, reliable (with low rates of absenteeism) and English-speaking labour. Back-offices were established by companies such as Barclaycard in Middlesbrough in the 1980s (see Beynon *et al.*, 1994), in many ways the private sector in the 1980s echoing the decentralization of public sector back-offices to the region in the 1960s. In the 1990s, however, the focus of attention has been upon the establishment of call centres and other customer support activities, serving national and global markets from within the northeast. This tendency is in part related to the intensification of pressures to cut costs associated with deregulation of financial and business services and the Taylorization of work in these activities. Prominent examples include British Airways ticket telesales (set up on the Newcastle Business Park in 1991 as the first such operation in the region and now employing 700), AA Insurances and Matrixx Marketing (also Newcastle), Abbey National (on Teesside), Ladbroke (in Peterlee), Orange (1400 jobs at Darlington), London Electric, Mercury One-2-One and The Insurance Service (on the Doxford International Business Park on Wearside). There is fierce competition for such jobs from other localities outside the South East, notably north Yorkshire (especially Leeds), Merseyside and central Scotland in and around Glasgow, as well as from Ireland. Although providing several thousand jobs in the northeast, in many ways these call centres and back-offices are analogous to the branch plants of mass production manufacturing, providing routine functions in establishments with little autonomy or local control.

Tourism has also become a more prominent feature of the regional economy (North of England Assembly, 1995), a relatively low productivity but labour-intensive activity, now accounting for 83,000 jobs, employment which is likely to continue to grow (Thomas, 1995). Tourist attractions have often been linked to the formation and growth of local small-scale service activities, such as pubs, clubs and hotels. Tourism is focused primarily around two aspects of the regional environment. The first is the attractions of its natural environment, especially in its rural and upland fringes. The second is the

Plate 3.5   Industrial heritage as tourism: Beamish Museum, County Durham.

historical legacy of its built environment, which has more of an urban location. Recognizing the constraints of the environmental characteristics of such
urban areas, often blighted by the scars of industrial production followed by
decay, attempts have been made to transform the legacy of earlier industrial
production into the basis for contemporary employment and economic activity (Hudson and Townsend, 1992). The heritage of deindustrialization has
been selectively preserved and transformed. The past is restored in a partial
and sanitized form as a tourist attraction for people who wish nostalgically to
engage with a typically romanticized version of an earlier industrial era. Indeed, whilst in some cases it is the preservation or reconstruction of actual
industrial landscapes, of old ways of work and living, that is the focal point of
attraction (as in museums such as Beamish), in others the selling point is a
fictional world that itself portrayed life in the industrial past through rose-
tinted lenses (as, for example, South Tyneside becoming Catherine Cookson
country). In short, this is the old industrial areas' particular version of the
heritage industry (Hewison, 1987).
     Representing old industrial areas as attractive destinations for tourists has
also necessitated important changes in local and regional institutional arrangement. A regional Tourist Board has been created and local government economic development departments have developed new skills and expertise so as
to invent and then market the attractions of their areas. The turn to tourism
represents a recognition of the economic marginalization of a region that was
once a focal point in an accumulation process that now increasingly bypasses
it, with deep implications for its future developmental trajectory. Promotion
of such activities might, in some circumstances, appropriately form one element

in local or regional economic development strategies, but to rely upon them as their main, or sole, basis is a very risky course of action. For it places the northeast in competition with the vast array of locations, scattered around the globe, which seek to sell themselves to tourists in various ways. There is no guarantee that the northeast will prove more attractive to consumers than places in southern Europe or than the mystic delights of distant continents as the tourist industry becomes increasingly globalized. Recognition of issues such as these is leading to '. . . some waning of the perception of tourism's value to the local economy' (Thomas, 1995, 71).

In summary the new economy of the private service sector is characterized by rather poorly paid jobs, mostly offered on a part-time or casual basis and taken by women, and accepted on terms and conditions that confer only minimal benefits, typically in non-unionized workplaces. For example, female part-time employment in the region expanded greatly from 148,000 in 1971 to 258,000 in 1987. While the growth of part-time and casual work allows flexibility from the point of view of employers, such work is often precarious and insecure when viewed from the perspective of employees, with wages linked to individual performance or team and group targets (Taylor, 1997); Beck (1992) refers to this new form of individualized and uncertain work as a new kind of Taylorism, emphasizing the continuities with earlier forms of manufacturing branch plant investment in 'global outposts' in the region. Many people in the region are prepared to accept employment on such a basis, however, which is precisely one reason why the region is attractive to companies establishing such activities.

Much of this growth of private service sector activity and jobs has been underpinned by new forms of central government urban policy, such as Enterprise Zones – the Metro Centre, located in the Gateshead Enterprise Zone, is perhaps the most prominent example of the growth of the retail sector and of new forms of retailing activity. It is, however, the operations of the Urban Development Corporations (UDCs) that lie at the heart of the property-led economic regeneration strategies which centre around private service sector growth, business services such as call centres, new forms of retailing, leisure and consumption and expensive housing. The UDCs have been key players is establishing local 'growth coalitions' such as the Newcastle Initiative, Teesside Tomorrow and the Wearside Opportunity which place great emphasis upon the physical transformation and changing appearance of places. The UDCs have been heavily involved in major urban tourist and leisure projects such as marinas, recreational centres and hotels. The UDCs have a brief to transform selected old industrial areas within the conurbations through providing an environment in which private capital will invest in spectacular and speculative projects such as St Peter's Basin and the Royal Quays on Tyneside, or Teesside Park and the Marina and harbour development at Hartlepool on Teesside. This process is most sharply symbolized by such waterside locations within the old riverside conurbations of Tyneside, Wearside and Teesside. The UDCs set the agenda for development, for what is to be defined as development, on Teesside, Tyneside and Wearside in the 1990s. The visual images of former

Plate 3.6 Creating employment? The Hylton Riverside Enterprise Zone, Wearside.

areas of derelict docklands, warehouses, shipyards and associated working-class housing areas converted by a combination of refurbished and brand new buildings are very powerful ones. UDCs have undoubtedly been successful in physically transforming areas in this way, but questions nonetheless remain as to the full range of effects of such property-led regeneration strategies and about the mode of operation of UDCs as regeneration agencies.

Essentially derivative of a USA-based urban redevelopment model (Judd and Parkinson, 1990), UDCs are non-elected bodies appointed and generously funded by central government, one part of a broader tendency towards the centralization of power within the UK state that characterized the politics of Thatcherism. UDCs can be seen as authoritarian corporatist institutions, driven by particular private sector interests, typically with token representation of the interests of organized labour and local residents. While the Teesside Development Corporation can certainly be characterized in this way (Beynon *et al.*, 1994), the Tyne and Wear Development Corporation has

Plate 3.7   Regenerating inner urban areas in the 1990s: St Peter's Basin, part of the area of
the Tyne and Wear Development Corporation.

made greater efforts to involve the local communities in the areas in which it
operates (Robinson *et al.*, 1993). These variations in the mode of operation of
UDCs are important insofar as they lead to meaningful involvement of local
communities in the regeneration process – although there are strict limits as to
how far local community interests can take precedence over those of the
private sector.

Local authorities, typically Labour-controlled in areas in which UDCs
operate, have tended to accept their subordinate position, albeit unwillingly,
and work with UDCs. For they represent a substantial source of extra central
government expenditure in an era in which this has generally been shrinking
and central government control of local authorities has increased. The task for
the UDCs is to cut through the regulatory controls formerly exercised by
elected local authorities, deflecting local criticism that this is anti-democratic
with claims that such projects will result in the creation of thousands of jobs

in the midst of areas blighted by high unemployment and widespread poverty. In practice, implementing their strategies results in increasingly sharp social and spatial differentiation, juxtaposing middle-class affluence with working-class poverty and unemployment, only partially alleviated by relatively small numbers of mainly poor quality, low-wage and part-time new jobs. It is important to acknowledge that UDCs are established for a finite period of time and this raises important questions as to exit strategies and the trajectory of change in their designated areas once the flow of central government funding ceases.

3.5   Conclusions: regional regeneration and polarization? ──────

What sort of regional economy has then emerged from this often conflicting process of economic restructuring and labour market change? There is no doubt that the composition, structure and pattern of ownership of the regional economy have altered significantly. The old coal- and steel-based economy of 'carboniferous capitalism' has all but disappeared completely. The region has attracted large volumes of inward FDI in new industries and activities, especially in the automobile and electronics sectors, which have had catalytic effects on the wider regional economy. They have pioneered new models of production, new ways of working and forms of industrial relations and new structures of relationships between companies within supply chains. Equally, there have been dramatic changes in the built environment of parts of the region, especially in the riverside conurbations that fell within the remit of the UDCs. These have formed key locations in the transition towards a more service-based economy in the region. There undoubtedly is a 'new' regional economy and to observers such as Mrs Thatcher its creation is a major and unqualified success.

Nonetheless, despite the evidence of success and positive transformation, there is also evidence of a number of worrying tendencies. With one or two exceptions (such as chemicals) profitability remains weak (Durham University Business School, 1995). The threat of further significant rounds of job shedding and plant closure remains, especially in clothing and footwear (significantly now the largest manufacturing sector in the northeast in terms of employment and heavily dependent on a few large retailers as customers). The evidence as to the extent to which the new manufacturing branch plants of the 1980s and 1990s actually are more regionally embedded 'performance plants' offering better quality, more-skilled jobs (as compared to the 'global outposts' of the 1960s and 1970s), with regionally based supply chains and strong patterns of intra-firm linkages, remains inconclusive. On the other hand, the new business service investment in call centres and back-offices strongly echoes the characteristics of that earlier round of 'global outposts' within an emergent international division of labour in these service activities.

The aggregate pattern of labour market change in the region over the last 25 years suggests complex processes of labour market change involving an aggregate drop in effective demand for labour alongside a more deeply and complexly segmented labour market marked by relatively high unemployment, much of it long-term. As recently as 1994, almost 40 per cent of the registered unemployed in the region were long-term (more than one year continuously) unemployed, with the recorded incidence of such long-term unemployment depressed by the impact of temporary job schemes, while for each recorded vacancy there were 24 people registered as unemployed (North of England Assembly, 1995). Paradoxically, at the same time, there were fears of skilled labour shortages (for example, of engineers) in sectors such as electronics which had expanded through inward investment, raising serious questions about institutional arrangements for training and skilled labour supply within the region (Cassell, 1995).

More surprising and paradoxical than the regional unemployment rate consistently exceeding the national rate is the fluctuating decline in unemployment in the region from the mid-1980s and again in the 1990s. This tendency of simultaneously falling employment and registered unemployment reflects the unprecedented growth in people registered as permanently sick and so no longer counted in the unemployment statistics. The number of people in the region who have fallen out of the labour market due to long-term illness and permanent sickness now exceeds the number counted as unemployed. Adjusting the definition of unemployment to allow for this (although thereby assuming, contentiously, that all of those registered as ill or sick could return to work) as well as the impacts of people looking for work who are ineligible to register for unemployment benefit, temporary job schemes and early retirement, Beattie *et al.* (1997) demonstrate that the number of people registered as unemployed and claiming unemployment benefit in the North in January 1997 – 124,000 – is less than 40 per cent of the 323,000 'really unemployed'. The Labour Force Survey for the spring of 1997 shows 9.8 per cent of the region's economically active as unemployed but no less than 33.3 per cent of the working age population as without jobs.

The increase in unemployment is a clear indication of the way in which, for many people, the restructuring of the economy has involved a switch from one form of state management to another. Previously, they depended upon the state for employment and for the provision of a range of public sector services provided through the welfare state as citizens' rights that were important for the quality of their everyday lives. As unemployment has increased (in turn increasing pressure on public expenditure), however, the state has increasingly had to invent new ways of managing the unemployed, especially the long-term unemployed and permanently sick. For them, the only foreseeable future is one of dependence upon state transfer payments as clients of a welfare state (Pugliese, 1985) that is being transformed into a workfare state.

In one sense, localized concentrations of unemployment are not new. There have long been marginalized people in marginalized places dependent upon such transfer payments (Hadjimichalis and Sadler, 1995). What is new is the unprecedented concentrations of permanently sick people who are no longer part of the labour force. This has posed serious challenges to the state welfare system which are still adequately to be resolved. What is also new, and especially stark

Plate 3.8 The impacts of industrial decline: empty housing in South Bank, Teesside.

in former one-industry settlements, is the generalized and *simultaneous* occurrence of mass unemployment in *many* such places. This in turn has required new forms of institutional arrangements, new forms of state organization specifically focused upon social containment and control through varied training, re-training and temporary employment schemes. This has been most clearly manifest in the activities of the Department of Employment/Manpower Services Commission and schemes such as the Youth Training Scheme, the Community Programme, Employment Training and Training for Work, and now Welfare to Work. These new national schemes and state organizations have come to play a pivotal role in social regulation over large swathes of the northeast, centred on the allocation of places on training and temporary employment schemes and the management of a substantial surplus population. Many of those who participate in such schemes simply then return to the ranks of the unemployed. In 1993/94, for example, only 47 per cent of YTS leavers found jobs while 31 per cent again became unemployed – the highest figure for any English region and indicative of the depressed state of the regional economy (North of England Assembly, 1995).

As a consequence of a lack of employment opportunities in the formal economy and the low levels of welfare payments, there has often been a parallel growth in other forms of work as part of individual and household survival strategies. In Cleveland County perhaps 1 in 10 of those registered as unemployed are involved in the 'black economy' (Beynon *et al.*, 1994). There has also been an expansion of criminal activities in these areas, as became clear in the Meadow Well riots in 1991 on Tyneside, and in the growth of teenage prostitution in central Middlesbrough on Teesside. The latter developments are particularly stark examples of a broader process of growing use of illegal child labour (Dobson, 1996). In summary, these varied tendencies are an expression of people seeking to supplement inadequate incomes from transfer payments in a situation in which they are effectively permanently excluded from the formal labour market and in which the level of provision via the welfare state has been greatly reduced. Such people are typically consigned to the worst of the remaining stock of public sector social housing, and subjected to high levels of surveillance by police and social security officials, especially so in those locations in which they are spatially adjacent to the transformed consumption spaces of the new middle classes. Socio-spatial segregation has become even more marked as a result. Whether such a deeply divided society is sustainable remains an open question. Whether it is compatible with successful economic regeneration likewise remains an open question, but the evidence seems clear that Europe's economically successful regions are generally characterized more by social cohesion and inclusion rather than by division and exclusion (Dunford and Hudson, 1996). The challenge for the northeast is to continue the positive aspects of the process of change begun in the 1980s while ensuring that economic success for some does not undermine the social bases of economic success for the region overall but rather reinforces the continuing economic success of the region by facilitating and supporting the emergence of skilled and well-paid work in a socially inclusive society.

Notes

1 Most of the available official statistical data refer to the Northern Region, which is a wider region than the industrial northeast. In terms of data on industrial change, however, data for the North(ern) Region accurately represent the position in the northeast.
2 It is, nonetheless, important not to overgeneralize about the branch plant economy. There were, for example, significant differences in the 1960s and 1970s between new capital-intensive chemicals plants (primarily attracted by grants towards the costs of new plant and machinery) and new consumer electronics factories (primarily attracted by abundant supplies of appropriate labour).

Chapter 4

The learning economy, the learning firm and the learning region: a sympathetic critique of the limits to learning*

4.1 Introduction

The context of, and focus of concern in, this chapter is the recent growing interest in – one might almost say obsession with – 'learning' and 'knowledge' as a route, perhaps the *only* route, to corporate and regional economic success. This is one facet of the growing, and generally productive, engagement between economic geographers and regional analysts on the one hand and evolutionary and institutional economists on the other (see, for example, Maskell *et al.*, 1998; Storper, 1997). This focus on knowledge, and the processes through which it is transmitted, is often presented as a dramatic new breakthrough, of epochal significance, promising radical theoretical reappraisal and opening up new possibilities for the conception, implementation and practice of policy (see, for example, Braczyk *et al.*, 1998). It would of course be futile to deny the significance of knowledge, innovation and learning to economic performance. Production as a process that simultaneously involves materials transformation, human labour and value creation, necessarily depends upon the knowledge and skills of individual workers and on the collective knowledge of a range of social and technical conditions and processes that make production possible. This also directs attention to the institutional bases of knowledge production and dissemination, recognizing that these are social processes and 'instituted processes' of a Polyanian type. Competitiveness and economic success are thus seen to be grounded in a variety of types of knowledge and knowing.

The argument in the chapter, however, is that recognizing the importance of innovation and knowledge creation to economic success is hardly novel and that the contemporary focus on learning is in many ways simply a new twist on an old theme that 'knowledge is power'. This is an important insight. It is also a partial one – not least because of the relationship between the possession of power and the capacity to shape the production and/or appropriation of knowledge. Not all economies are capitalist but the historical geography of capitalism cannot be sensibly understood without giving due recognition to the revolutionary impacts of innovation on the what, how and where of

* First published 1999 in *European Urban and Regional Studies*, **6**(1): 59–72, Sage Publications, London. Reprinted by kind permission of Sage Publications Ltd

production. It would be a foolish enterprise to seek to deny the importance of learning and knowledge creation, and the institutional settings and forms in and through which these processes occur, to innovation and to the dynamic of un-even development in capitalist economies. Indeed, one cannot over-emphasize that the creation of knowledge has been integral to the competitive dynamic of capitalist economies since they were first constituted as capitalist. As Marx emphasized well over a century ago and Schumpeter later restated, much of the revolutionary dynamic of capitalism has *always* rested in its capacity to create new commodities and new ways of producing them via a sequence of radical transformations of the forces of production and of the organization of the labour process (see, for example, Aglietta, 1979). So what, one might reasonably ask, is new? What, one might enquire, is all the fuss about?

The first point that can be made in response to these questions is that the current 're-discovery' of the importance of knowledge and the processes through which it is produced brings with it a baggage of strong claims about new and enriching and empowering forms of work (see, for example, Florida, 1995). The alienated and deskilled mass worker is apparently now no more than a subject of history. The emphasis on knowledge is also associated with claims about the possibilities for increasingly egalitarian, more equal and progressive forms of economic and social development. It likewise brings related claims as to new possibilities for urban and regional regeneration strategies, and suggestions of new developmental trajectories for problematic cities and regions (see, for example, Morgan, 1995; Simmie, 1997). There is no doubt that in particular cases, times and places, such claims have a certain validity but they also require careful and critical scrutiny. Such a scrutiny must acknowledge the necessary structural limits to a capitalist economy and the disciplines (if not quite iron laws) that these set on what is both necessary and possible (as opposed to what may be desirable but impossible – to adopt a memorable phrase from Offe, 1975a).

The structure of the remainder of the chapter is as follows. First, the claims made by the proponents of 'learning' approaches are briefly summarized, and some links are drawn between the pre-eminent emphasis that they place upon knowledge and learning and other literatures that analyse ongoing changes in the organization of production and work in contemporary capitalism and which have differing emphases. The aim is to situate and contextualize claims about the significance of 'learning'. Secondly, these claims are placed within the context of continuities and changes within capitalism, and the ways in which these have been understood, as a further step in this process of contextualization and situation. Finally, some conclusions are briefly drawn around the limits to learning, and questions of learning by whom, for what purpose, in the context of the politics and policies of social, economic and territorial development.

4.2 The learning economy: learning firms and learning regions

There has been a growing recognition of the importance of knowledge in the contemporary organization of production in what many commentators see as an era of globalization (see, for example, Strange, 1988; Giddens, 1990). This takes a variety of forms but the central point is that the production, distribution and exchange of knowledge is claimed to have attained an unprecedented significance in the operations of the economy. Much of the discussion about the significance of knowledge takes place under the rubric, and around the theme, of 'learning'. There are, however, several strands to the learning literature, which emphasize different aspects of, and ways of, learning: learning-by-doing (Arrow, 1962); learning-by-using (Rosenberg, 1982); learning-by-interacting (Lundvall, 1992); and learning-by-searching (Boulding, 1985; Johnson, 1992). Perhaps the most influential of these within recent debates has been Lundvall's emphasis on learning-by-interacting, informed by a concern to understand how (predominantly small) companies in open economies can remain competitive in an environment of rapid technological change and uncertainty. It essentially focuses upon companies learning about and adapting to 'best practice' via interaction with other firms and institutions as the route to competitiveness.

Lundvall (1995) has recently remarked that the term 'learning economy' signifies a society in which the capability to learn is critical to economic success. For Lundvall, contemporary capitalism has reached the stage at which knowledge is the most strategic resource and learning the most important process. There is a recognition that this process is to a considerable degree path-dependent, although significant breakthroughs often involve shifting onto new, rather than further along existing, paths. The learning process could thus involve the capability to move from already successful to potentially even more successful new 'state-of-the-art' development trajectories, or to learn how to sustain currently successful trajectories of development, or how to shift onto more- from less-successful paths. Learning both presupposes and produces knowledge (although the learning literature tends to gloss over the different forms and processes of knowledge production: Odgaard and Hudson, 1998) but knowledge is not an undifferentiated entity and it exists in a variety of forms. There is, in particular, a critical qualitative difference between information, which is codifiable (and so commodifiable and tradable) knowledge that can be transmitted mechanically or electronically to others (for example, as bits along the fibre optic cables of a computer network), and in principle can become ubiquitously available, and tacit knowledge in the form of know-how, skills and competencies that cannot be so codified and ubiquified. Foray (1993, 87) defines tacit knowledge as '. . . knowledge which is inseparable from the collective work practices from which it comes'. He goes on to emphasize that '. . . some tacit knowledge is always required in order to use new codified knowledge'. Foray thus emphasizes the asymmetric relationship between these qualitatively different types of knowledge. Acknowledging that knowledge is 'tacit' problematizes its communication and transmission to others who lack access to the unwritten codes of meaning in which such knowledge is embedded and upon which its meaning depends.[1] Such tacit knowledge may indeed be unique to particular individuals rather than collective in character – in which case the problems of communication are, *a fortiori*, problematic – but it *is* often collective rather than simply individual, locally produced and often place-specific. Know-how thus cannot be divorced from its individual, social and territorial contexts and in that sense is only partially commodifiable. It therefore can only be purchased, if at all, via the labour market as embodied knowledge and not in the form of patents, turn-key plant or other forms of 'hard' technology.

Recognition of the uncodifiable aspects of learning and knowledge creation[2] is important since it signifies that these processes are qualitatively different from the simple transfer of codifiable knowledge as information. As a consequence, learning involves more than simply transactions of information exchange within markets or hierarchies. Lundvall (1992) has emphasized the national context of innovation systems and learning and the significance of shared language and culture as well as formal legislative frameworks in shaping trajectories of innovation and learning. This emphasis on the national as a key site of regulatory processes resonates with broader critiques of claims as to the decreasing significance of the national in the face of processes of globalization. The notion of 'globalization' has become increasingly contested and questioned, with a growing number of analysts stressing the continuing salience of the national in terms of the organization and regulation of the economy (see, for example, Boyer and Drache, 1995; Gertler, 1997; Weiss, 1997).

While recognizing that the national territory can and continues to be a crucial milieu in some circumstances, it is also becoming increasingly clear that there is no *a priori* reason to privilege this particular spatial scale, irrespective of time and place. Nevertheless, the renewed emphasis on the salience of the national has implications not just for the proponents of globalization but also for those who wish to privilege the regional over the national in terms of the production of knowledge and learning (see, for example, Storper, 1995, 1997). Regional and locality-based learning and knowledge production systems can, however, be of equal or greater significance (Maskell and Malmberg, 1995; Maskell, 1998), not least in the context of arguments that innovation systems are constituted sectorally – and at least potentially globally – rather than nationally (Metcalfe, 1996). The sectoral constitution of innovation systems globally across rather than within national boundaries emphasizes the significance of the place-specifically local within the global and of the links between corporate learning and territorially embedded knowledge (a point revisited below). Such tacit knowledge and learning capacity is seen as *the* key competitive corporate and territorial asset. While the debates as to the relative importance of sectoral versus territorial bases of learning, and of the relative importance of different territorial scales in processes of learning and innovation, do not challenge the importance of knowledge *per se* to the contemporary economy, there are other reasons for treating such claims which foreground the role of knowledge and learning with a degree of circumspection, however.

A corollary of the renewed emphasis on the generalized significance of knowledge in production is a recognition of the need to move away from a conception of the old 'linear' R&D model dominating the production of knowledge, associated in particular with a Taylorist conception of the technical division of labour. This approach was informed by a perceived need for the separation of mental and manual labour as the key to achieving scale economies and labour productivity growth within mass production systems. Such separate R&D departments and the associated routinization of R&D activities formed one element in a historically specific form of the organization of the labour process within large firms. Such an organizational model has now been recognized as a historically specific one, which in some circumstances remains appropriate and powerful but in others can incorporate crucial weaknesses, especially in an era of rapid shifts in product markets, for there are no necessary feedback loops from the users of and customers for innovations to those within the firm charged with responsibility for producing them. Consequently, new products may not be attuned to consumer tastes and fail in the marketplace while, conversely, opportunities for new products may be missed. Furthermore, the growing significance of the symbolic meanings attached to consumption, *a fortiori* in circumstances in which the commodity is an event or spectacle rather than a material object, places an even greater premium on knowledge of consumer tastes and on the ability to shape them through advertising.

Knowledge creation and innovation is accordingly seen as something that must become all-pervasive throughout the firm, at all levels and in all departments and sections. The ideal is to emulate the (originally Japanese) process of '*kaizen*', continuous improvement through interactive learning and problem solving (Sadler, 1997), a happy state which it is claimed is brought about as a consequence of the existence of an actively committed and engaged workforce within particular types of corporate organization, dedicated to enhancing corporate performance. The emphasis is upon creating dense horizontal flows of knowledge and information within and vertical flows of knowledge and information between the various functional divisions of the company, while opening the ears of those involved within the company to voices from outside its boundaries. The aim is to build a 'seamless innovation process', bringing together everyone in the firm involved in product development, from those who had the initial idea to those who finally took it to the marketplace. Creation of multi-disciplinary and cross-departmental 'concept teams' with responsibility for product development is seen as a way of sharply reducing the socially necessary labour time taken to bring new products onto the market. Increasingly these are organized as 'globally distributed teams' which 'meet' via video-conferencing and other forms of electronic technology. This reliance on such distanciated social relationships of intellectual production, rather than face-to-face meetings, reflects increasing pressures on managerial time and resources but can also create problems as these teams seek to work to very tight deadlines (Miller *et al.*, 1996). While these globally distributed teams are not quite 'virtual organizations' or 'virtual corporations' (Pine, 1993), they do represent a significant change in the organization of the processes of knowledge creation and innovation within companies.

Complexity theory strongly suggests the need to adapt a view of social systems as evolving in a non-linear fashion (Amin and Hausner, 1997). One implication of this in the context of innovations is that revolutionary innovations (organizational, product and process) may be produced in unexpected ways. The emphasis now is therefore upon recognizing that innovation is an interactive process that involves the synthesis of different types of knowledge rather than privileging the formal scientific knowledge of the R&D laboratory over other forms of knowledge. As a consequence, there is considerable emphasis on acknowledging the legitimacy and 'voice' of different types of knowledge (not least as radical innovations may well challenge the dominant 'logic' within an industry), of reuniting the mental and the manual which were torn asunder by Taylorism, on reinventing polyvalent multi-skilled workers, and so enhancing corporate competitiveness by producing higher quality products more flexibly.

Such tendencies are observable both in small, flexibly specialized firms and units and in new forms of high volume production that seek to combine economies of scale and scope, ultimately in mass-customized production with a batch size of one (Hudson, 1994a, 1997a, 1997b). Intensifying competition and shorter product life cycles (which may be part of an aggressive, offensive competitive strategy rather than simply a response to changes in consumer tastes) are necessitating a closer integration of R&D with the other functional sections within companies, with far-reaching implications for the internal organization and operation of those companies. This growing emphasis on the significance of learning and knowledge creation, and new forms of production organization, links in with propositions about the emergence of new forms of more rewarding, satisfying and engaging work than was available to the vast numbers of workers who manned the mass production lines of factories and offices, alienated in deskilled and dehumanizing jobs. It is, however, worth recalling that even at the high point of Fordism only a minority of labour processes were organized on Taylorist principles and that a considerable amount of manual work was performed by knowledgeable craft workers (Pollert, 1988). Innovation and learning have nonetheless become seen as creative processes that must be suffused throughout the entire workforce, capturing the knowledge of all workers to increase productivity and product quality and at the same time enhancing the quality of work. In essence, this amounts to the reinvention of polyvalent skilled craft workers, a return to a pre-Taylorist era prior to the invention of scientific management. Florida (1995) writes approvingly of the emergence of a new form of production organization in a knowledge-based economy, in factories that are claimed to be becoming more like laboratories, with knowledge workers, advanced high technology equipment and cleanroom conditions free of dirt and grime. This does indeed powerfully suggest that the old distinctions between manual and mental workers are being cast aside, that every worker is now becoming an innovative knowledge worker.

Know-how historically was, and in large measure still is, typically a kind of knowledge developed and then kept within the confines of a firm. Furthermore, the boundaries of the firm still remain significant for knowledge that is central to the core competencies and strategic goals of a company. Nevertheless, the increasing complexity of the knowledge base upon which the totality of the production process depends is increasing the social division of labour in knowledge production and resulting in growing numbers of collaborative long-term relationships between firms (see, for example, Kitson and Michie, 1998). A variety of processes, ranging from the growth of out-sourcing and contracting out to the increasing prevalence of joint ventures and strategic alliances between even the largest global corporations, is indicative of a rather different model of shared corporate learning. As a result, 'know-who' is becoming of growing importance in the production of know-how (Lundvall and Johnson, 1994). This growing emphasis on knowledge and learning therefore also links in with claims as to new forms of relations between companies, based on co-operation, trust, and the sharing of knowledge for mutual benefit. These forms of interactive inter-firm relations for knowledge creation are particularly associated with the supply chains of major Japanese manufacturers (which is not to imply that they are in some sense culturally defined and confined to Japan and Japanese companies). Considerable emphasis is placed upon new forms of network relations, both 'horizontal' relationships between small and medium-sized enterprises (SMEs) and 'quasi-vertical' relationships between big firms and their suppliers and/or customers, which stress the sharing of R&D, of knowledge and the products of learning to the benefit of all partner companies in the network. For example, institutional innovations such as placing resident engineers in customers' factories allow efficient channels for inter-corporate learning within networks of inter-firm relationships.

Some of these networks are based upon spatial propinquity, others are not. A useful distinction can be drawn between spatial propinquity and organizational proximities (Bellet et al., 1993). The former may (but does not necessarily) facilitate the latter by increasing the probabilities of encounter between agents within a system but is not necessary for interaction between individuals or groups. Organizational proximity does not necessarily require spatial adjacency or proximity but does presuppose the existence of shared knowledge and shared representations of the environment and world within which the firm exists, although the various units and sections of the firm may be in spatially discrete and distant locations. Such a form of proximity also enables the synthesis of varied forms of information and knowledge via co-operative and collective learning processes between firms within the institutions of an industry. Organizational proximity is therefore a necessary condition for creating innovations and resources through processes of collective learning and is simultaneously a product of these processes. However, the networks through which learning is enabled and expressed are not necessarily territorially defined and demarcated, and in some respects the growing sophistication of IT and communications technologies has weakened this link further. The emergence of 'global distributed teams' in innovation and product development is indicative of this weakened link (Miller et al., 1996). Conversely, the technological facilitation of information flows has simultaneously enhanced the significance of place-specific tacit knowledge within key nodes of command and control and representation in a global economy (Amin and Thrift, 1994).

Some networks are without doubt deeply spatially embedded and recognition of this provides a bridge into more general notions of the significance of territorially based knowledge to economic competitiveness and success. The concept of the learning firm as an institution for the production of knowledge is thereby transposed into the notion of the learning region (Morgan, 1995; for related concepts see Camagni, 1991). This perspective emphasizes that regional economic success is heavily based upon territorially defined assets derived from 'unique', often tacit, knowledge and cognitive assets, and stresses the importance of spatial proximity in collective learning processes. Considerable emphasis is placed upon the pivotal role of regional institutional structures which allow regions (and firms within them) to adjust to, indeed anticipate and shape, changing market demands. Innovation and knowledge creation are seen as interactive processes which are shaped by a varied repertoire of institutional routines and social conventions. This involves not simply inter-corporate collaborative links but also links between companies, the (local) state and institutions in civil society, emphasizing the permeability of the boundaries between economy, state and civil society in the creation of regional competitive advantage.

This notion of a cohesive society, with permeable boundaries between economy, civil society and state, is powerfully captured in the concept of the 'negotiated economy', originally developed in relation to analyses of the specificities of the Danish case (Amin and Thomas, 1996) but of a more general provenance. Within the negotiated economy, the state fulfils a distinctive role as arbitrator and facilitator of relations between autonomous organizations, as well as continuing with its more traditional roles of providing specialized services and defining the legislative framework of rules and regulations. This is a model of state activity which emphasizes enablement and which falls between the concepts of the 'liberal' and 'interventionist' states (Offe, 1975b). The concept of the negotiated economy can thus be linked with that of 'the learning state' and a mode of regulation positioned between market and hierarchy through which an enabling state seeks to create the conditions for a dialogic approach to conflict resolution and policy formation in general and innovation, knowledge creation and learning in particular. This approach rests on discursive, moral and political imperatives rather than formal contracts and legal sanctions in achieving consensus and taking decisions. It thus places the emphasis on shared values, meanings and understandings, specifically territorially embedded, and tacit knowledge and the institutional structures through which it is produced. These emphases are caught in notions such as those of 'institutional thickness' (Amin and Thrift, 1994), 'social capital . . . [those] features of social organisation, such as networks, norms and trust, that facilitate coordination and co-operation for mutual benefit' (Putnam, 1993), or, perhaps most powerfully, as 'regions as a nexus of untraded interdependencies' (Storper, 1995). As Storper (1995, 210) puts it, '. . . the region is a key, *necessary* element in the "supply architecture" for learning and innovation' (emphasis added) while the emphasis on 'untraded dependencies' or 'relational assets' focuses attention upon the necessary territoriality of critical elements of non-market relations and tacit knowledge. This signals a decisive shift in focus from firm to territory as the key economic actor in the knowledge-based competitive struggle, to a collective and territorialized definition of competitive advantage which emphasizes the cultural and social underpinning of economic success. In so far as this represents a growing recognition of the limits of narrow neo-classical and technicist views of the economy based on analogies with the behaviour of physical systems (Barnes, 1996), then this in itself is a very important step forward – the rediscovery and re-emphasis of the economy as a social process. Equally, it is important to be aware of the limits to such an approach and emphasis.

Rather than privilege territorial over corporate knowledge production and learning (or vice versa), the critical point is to explore the *relationships between* these two institutional bases of learning. Camagni (1991, 127) emphasizes that firms seek to combine codified information and tacit knowledge into 'firm-specific knowledge'. More specifically, in so far as globalized forms of corporate organization are emerging, they are predicated upon the integration of fragmented products of local learning to further corporate interests. This may well involve dis-embedding them from the contexts in which they were initially produced, and perhaps therefore finding ways of converting tacit knowledge into codifiable information. Alternatively, in situations in which knowledge is so organizationally and technically specific to a firm and so deeply embedded that it cannot be alienated from its origins, it may simply involve big firms acquiring smaller ones as a way of gaining access to such knowledge (a familiar story within the historical geography of mergers and acquisitions: see Athreye, 1998). Global corporations are, it is suggested, de-veloping organizational forms focused upon ensuring the repatriation of the varied results of different localized learning experiences and their integration within a collective body of knowledge to serve strategic corporate interests (Amin and Cohendet, 1997). The implication is that the processes of seeking to secure access to locally produced knowledge are also processes of inter-corporate competition. The issue is therefore one of the relationships between knowledge production and acquisition and competition and co-operation between various territorial and corporate interests.

4.3   Old wine in new bottles: or another trip around the mulberry bush?

The learning firm is, however, hardly a novel concept in the sense that know-ledge has always been crucial to capitalist development. There are, however, limitations in the way in which learning approaches deal with the production of knowledge. Their emphasis is upon learning as a way of catching up with 'best practice' in a selection environment, and adapting to significant innova-tions in organization, process or product. The issue of how radically new knowledge is produced, and redefines 'best practice' as radical innovations are created, is left largely unexplored (Odgaard and Hudson, 1998). Moreover, the emphasis on the transmission of knowledge *per se* in a 'learning economy' may well lead to an underestimation of the significance of other forms of learning that are more ubiquitous and central to capitalist competitiveness.

Capitalist corporate success in production has always depended on ensuring one or both of two things, either finding ways of making existing commodities more profitably and/or finding or inventing new commodities to produce sufficiently profitably. Companies have evolved a variety of strategies to reduce the costs of production of existing commodities, involving a variety of 'spatial fixes' (Harvey, 1982) to enable the costs of producing with existing technologies to be reduced and for them to remain competitive. Such strategies typically involve search for and learning about locations offering lower unit costs of labour or other material inputs to the production process. Storper and Walker (1989) contrast this approach based on 'weak competition' with a Schumpeterian one based on 'strong competition', a strategy based on the creation of new commodities and products and/or new ways of producing existing commodities rather than on seeking ways of making existing commodities competitively by searching out sources of lower cost inputs within the parameters of existing process technologies. Strategies of strong competition are thus also based upon innovation, learning and the creation of knowledge but of a very different sort from those which underpin strategies of weak competition.

These differing 'weak' and 'strong' competitive strategies and their grounding in different types of knowledge and learning are reflected in the extensive literature on spatial variations in conditions of supply and markets for various inputs to production processes and in the equally extensive literature on process and product innovations, respectively. Both types of innovation are based upon knowledge generated at the corporate level, though typically knowledge of new processes and products soon diffuses through a particular branch of production (albeit with its trajectory of diffusion legally regulated by patent – assuming of course that such regulations are enforceable). This tendency towards the erosion of a temporarily conferred competitive advantage by the diffusion of knowledge as information and of technological innovation underlies the continuous 'hunt for technological (and other) rents' (as Mandel, 1975, graphically and memorably expressed it). This diffusion of knowledge in turn provides the impetus for capital's continuous search to revolutionize the how, what and where of production. One has only to look at the economic history of the 'successful' regions of the nineteenth century in which industrial capitalism was first born and then consolidated to recognize the *key* role of product and process innovation from the very beginnings of the process.[3]

There are also strong grounds for critically evaluating the claims that this renewed emphasis on knowledge is associated with the empowering of workers in satisfyingly enriched – as a result of reducing, if not eliminating, the alienation of workers from their work – and multi-skilled jobs (see Blyton and Turnbull, 1992). It is worth recalling in this context that the rationale of Taylorism was to break the power of the multi-skilled craft worker to challenge the imperatives of capital and disrupt the smooth flow of the production process. Taylorist scientific management therefore sought to disembody knowledge and know-how and to break up the production process into a myriad of separate and deskilled tasks whose pace was controlled by the speed of the line rather than the inclination of the individual worker. The emergence of separate R&D departments and a linear model of learning and innovation within the firm that separated manual from mental labour and privileged the latter over the former, and privileged codifiable formal scientific knowledge over the practical and often tacit knowledge of the skilled manual worker, was equally an integral part of this process. Taylorism was invented precisely as a way of wresting control of the labour process from skilled craft workers, and in so far as the economy remains a capitalist one, there remain pressing reasons why capital should want to retain such control in many types of contemporary production; indeed, there are now frequent references to the growing Taylorization of office work and a range of 'white collar' occupations (see, for example, Beynon, 1995) at the same time as others enthuse about the reinvention of the multi-skilled manufacturing worker.

It is difficult to imagine, therefore, that capital would willingly wish to return control of the production process to workers that it potentially could not control. The search for alternatives to mass production reflects capital's need to break the capacity of the mass worker collectively and spontaneously to challenge its quest for profits – although unintentionally, this may in turn create potential new opportunities and points of leverage for organized labour since the contradictory character of the class relations between capital and labour can be refashioned but they cannot be abolished in an economy that remains capitalist. While the probabilities of strikes and other forms of disruption to the production process are certainly much lower now than in the decades of 'full employment' in the 1950s and 1960s in the territories of the major advanced capitalist states (for reasons which are discussed below), it is clear that any form of disruption to production organized on lean, just-in-time principles can quickly spread along the entire supply chain, bringing production to a precipitate halt (Hudson, 1997a). Given the disciplining context of more or less permanent high unemployment, this recent and ongoing reworking of work can more plausibly be seen as representing a new way of ensuring managerial control and intensifying the labour process, of reproducing in enhanced form the asymmetries of power between capital and labour. Managerial strategies and regimes of labour regulation remain focused on seeking to ensure the continuity of production and the compliance of workers.

Rather than empowerment in new forms of satisfying work built around notions of re-skilling and team working, the new forms of work are based upon multi-tasking and new ways of intensifying the labour process. Workers are enmeshed within disempowering regimes of subordination, characterized by control, exploitation and surveillance, accepting arrangements through which they discipline themselves and their fellow workers, while bound together through the rhetoric of team working (see Garrahan and Stewart, 1992). As a result, these may actually be worse jobs than those on offer on the old mass production lines, increasing stress (see Okamura and Kawahito, 1990) and changing the mode of regulation of the labour process. No longer is it 'us' versus 'them'; 'they' are now part of 'us'. Considerable ambiguities and uncertainties follow from these changes of identity, not least in relation to forms of organization and representation of workers' interests.

A further point needs to be made concerning the number of such jobs and the criteria on which people are selected to fill them. For even when the claims that these are better quality jobs are shown to be true, there is still a savage sting in the tail for labour. The capacity of firms to create these new regimes of work depends upon their ability to exercise great selectivity in whom they choose to employ and the terms and conditions on which they employ them, often in no-union or one-union factories, especially in countries or regions in which neo-liberal regulatory regimes have become dominant. Employees are selected more on the basis of their attitudes, psychological profile, age and physical condition, and personal and family circumstances than on their technical skills. For example, in many service sector occupations, personal appearance and social skills have become key recruitment and retention criteria (McDowell, 1997). Many manufacturing companies typically seek to recruit physically fit young males, with family and other financial commitments, who will be loyal to the company and accept new ways of working on the factory floor. Only a tiny fraction of those who apply and of those who feasibly could fill these jobs are employed, typically after an extensive selection process. Firms can exercise this degree of selectivity only against a background of high unemployment and for this reason are also very careful in their choice of locations in which to introduce these new forms of work and employment (Hudson, 1997b).[4]

There is also some scepticism as to the validity of the notion of new forms of network relations between companies as involving equal partners. It is certainly the case that there has been a considerable increase in outsourcing and subcontracting by many major companies. In this sense one can reasonably refer to a shift from vertical integration within companies to quasi-vertical disintegration, involving a redefinition of the boundary of the firm and a redefinition of the criteria on which to make the 'make or buy' decision. But to argue that these new relations are between equal partners is to ignore the sharp asymmetries in power between companies, and the extent to which such networks involve not co-operation based on trust but often not-too-subtle coercion if companies wish to keep their customers or suppliers. There is no doubt that the systems of relations between automobile or computer companies and their suppliers or between major retailing chains and their suppliers definitely are not relations between equals (Hudson, 1994a, 1997a). Indeed, there is no *a priori* reason why one should expect relationships between firms in a capitalist economy to be between equals; in fact, one should expect quite the reverse. In assessing the claims as to networks of equal partners, it is also important to remember that one of the features of the last couple of decades has been wave after wave of mergers and acquisitions as the centralization of capital has reached renewed heights after a couple of decades in which merger and acquisition activity was very subdued – and it is these massive transnational corporations, the 'movers and shakers' who dominate the global economy, that are frequently at the centre of decisive network relationships.

The concept of the learning region and the proposition that regional economic success reflects specifically regional assets and institutions for the production and dissemination of knowledge is also hardly a new one. Even the most cursory glance at the historical geography of capitalist development indicates that this has long been the case. In the United Kingdom in the latter part of the nineteenth century, for example, 'coal combines' lay at the heart of carboniferous capitalism in the industrial boom regions of the nineteenth century (Harvey, 1917). These combines comprised interlocking intra-regional networks of highly innovative firms, extending across sectors, integrated by physical input–output linkages and various forms of formal economic and financial linkages (such as interlocking directorates and mutual share ownership). More importantly, they were underpinned by non-economic relationships between key individuals and families and by networks of supportive institutions that evolved around, in and through the formal economic relationships between companies. These spanned the boundaries of local civil society and the state, as a dense network of interlocking institutions, attuned to the needs of the dominant regional firms and sectors, emerged there. For many such regions the problems subsequently became those of 'institutional lock-in', an inability to make the change from one development trajectory to another precisely because the institutional bases of the region reflected the past dominance of now-declining firms and sectors (Grabher, 1993; Hudson, 1994b). This is a salutary reminder that institutional thickness *per se* is no guarantee of successful regional economic adaptation and innovation as it can constrain rather than facilitate processes of collective learning and change.

Furthermore, the fetishization of knowledge and learning, and their institutional bases, may lead to a neglect of *other* institutional factors that underlie regional competitiveness. There could be no clearer illustration of the point that in order to understand the historical (including the contemporary) geography of capitalist production, it is necessary to grasp the ways in which such successful regional economies in the nineteenth century were grounded in social relationships that extended far beyond the workplace into home, community and the institutions of local civil society, and in due course the state (Beynon and Austrin, 1994; Carney and Hudson, 1978). Moreover, central to the embedding of industrial capitalism in these regions was the construction of discourses and ideologies that represented this as the 'natural' course of regional and socio-economic development. These representations constituted an attempt to present one view of a particular trajectory of capitalist development as 'natural' and 'unavoidable', to instil this as a hegemonic and uncontested view, subliminally learned and accepted by the populations of these regions.

This attempt to establish such a view as hegemonic was important because it was clear in the nineteenth century that economically successful regions, which were certainly learning regions containing learning firms, were also deeply socially divided ones. They were characterized by enormous disparities in incomes and wealth, juxtaposing extremes of conspicuous consumption with widespread abject and absolute poverty. The contemporary claims that the learning region offers a new and socially inclusive model of development are ones that need to be scrutinized carefully since they tend conveniently to ignore the point that the social relations of capitalism are at least as deeply marked by social inequality now as they were then. There are critical issues related to *who* controls the processes of knowledge production and learning. 'Learning firms' within a region may be successful economically and the institutional structures of a 'learning region' may both be produced by and facilitate the reproduction of 'learning firms'. This, however, does not necessarily equate to an egalitarian model of regional socio-economic development. Such firms remain unavoidably built around antagonistic class relations and may well also presume inequalities in other social relationships such as those of age, ethnicity and gender as an integral part of their strategies for competitiveness and success in the marketplace (*cf.* Massey, 1995, Chapter 8).

4.4 Conclusions and reflections on the limits to learning: learning by whom, for what purpose?

There is no doubt that firms have a great variety of possible approaches to production. Equally, there is a variety of forms of capitalist development model, nationally and regionally, and this indicates that there is a fair amount of room for manoeuvre in seeking to define regional development strategies (see, for example, Albert, 1993; Lash and Urry, 1987). While acknowledging the 'room for manoeuvre' (to borrow the phraseology of Seers *et al.*, 1979), the key point to emphasize is the continued existence of the social structural constraints which set limits to what is possible within a capitalist economy. These cannot simply be conveniently forgotten or assumed away. One implication of this is that capitalist development of necessity remains driven by competition and the search for profit. Another implication is that such development must therefore remain uneven – both between classes, between and within other social groups, and within and between regions, *a fortiori* if it is recognized that regions themselves are constituted as socially heterogeneous and as spatially discontinuous, whatever the claims about social homogeneity and spatial contiguity (Hudson, 1990). Certainly in some circumstances, development models may be based on less rather than more divisive guide rails – but capitalism requires the existence of reciprocally defining classes of capital and wage labour, although of course there may well be attempts to represent the situation in ways which deny this.

'Learning' and the production of knowledge are undoubtedly necessary elements in the processes of competitive commodity production and in some respects can themselves become commodified. The diffusion of information about new organizational, product and process innovations is a central element in the competitive dynamic of capitalism. The greatest competitive advantage is, however, conferred by precisely that knowledge that remains tacit and uncodifiable, not amenable to generalized transmission to others. It is thus the most valuable form of knowledge in conferring competitive advantage, precisely because it cannot have a price put upon it. Successful firms and regions thus guard it jealously. If, however, firms 'learn' via producing and protecting such knowledge, if 'regions' seek to learn in the same way, in the final analysis this is to enhance their competitiveness in a range of markets. As a consequence uneven development within and between regions and their constituent social groups is unavoidable. Knowledge and learning may be necessary for economic success but they are by no means sufficient to ensure it; nor, even more so, are they sufficient to ensure equality, cohesion and social justice.

For some firms, becoming 'successful learners' is the route to competitive success, and these firms have to locate their operations somewhere. Conversely, other firms have to lose as part of this struggle, and they have to devalorize capital, close plants and dismiss workers somewhere. There is no necessary territorial correspondence between these two faces of creative destruction – indeed, there are strong grounds for expecting the production of new commodities often to take place in new production spaces. For some regions, becoming 'learners' likewise offers a route to competitiveness, albeit one characterized by internal social division, if not strife. This may involve developing institutional structures to enable existing firms and sectors to evolve successfully, or new ones to become established in a region, or some combination of the two. It is important to acknowledge that for those regions that do successfully embark on the 'high road' to regional economic success, this very success raises new problems in terms of a requirement continuously to learn and anticipate, if not create, market trends. Moreover, if some regions 'learn' and 'win', many more will fail to do so and 'lose'.

The command and control functions of an increasingly spaced-out global economy will doubtless continue to locate within the few global cities, economic 'winners' but marked by deep social divisions (Sassen, 1991). These global cities are characterized by intensely dense institutional structures for producing and disseminating information globally, but at their heart lie interpersonal contact networks decisively bound together by the ties of critical tacit knowledge (Amin and Thrift, 1994). They are also deeply divided places, with socio-spatial differentiation deeply etched into their urban landscapes as a necessary feature of the ways in which their economies are constituted (see Allen *et al.*, 1998). Moreover, beyond the boundaries of the global cities, a post-mass production, post-Fordist world of specialized regional economies, all on their own successful learning trajectories, and all winning, is not a feasible option within the social relations of capitalism. Equally, a post-mass production, post-Fordist world of product-specialized high volume production regional complexes, all producing just-in-time and in one place in their own unique niche in the global market place, is not a feasible option. There may be some cases in which some regions 'win' by following one or other of these strategies – but there will be many more that 'lose', either failing in the attempt or doomed to failure by the success of others.

There is no doubt that an explicit recognition of the role of the production and dissemination of knowledge in a capitalist economy can help further understanding of uneven development. Cognitive assets can certainly be crucial in defining competitive advantage. Equally, the case for a regional political economy that remembers the lessons of a Marxian political economy and recognizes capitalism as structurally and necessarily, inherently and unavoidably, character-ized by uneven processes of growth and decline remains as valid as it ever did. Acknowledging that capitalism is shaped within particular structural boundar-ies which pivot around class relationships is not to imply that social life can be, is to be, or should be, reduced to such relationships; nor is it to deny that gender, location, ethnicity and much more besides are constituted as separate cleavage planes of social division and at the same time as foci of individual and collective identities. It is, however, to suggest that class relations cannot be ignored – not least in the production and diffusion of knowledge itself.

There is equally no doubt that innovation and learning can be important concepts in understanding why some firms and regions are economically suc-cessful and others are not. Equally, it is evident that the 'learning' paradigm may both legitimate the success of some firms and regions and the failure of others, and seem to hold out the enticing prospect of a more prosperous future to still many others. Nevertheless, it would be as well to recognize the limits that 'learning' entails, both as an explanatory concept and as a guide to territorial development policies. Not least, the political economy of learning implies a need to unlearn – or at least ignore – other concepts of political economy, with different developmental implications. 'Learning' is by no means a guarantee of economic success. Still less is it a universal panacea to the problems of socio-spatial inequality, and in some respects it is used as a cloak behind which some of the harsher realities of capitalism can be hidden. Addressing the problems of uneven development and inequality undoubtedly poses very hard policy and political choices for those who seek to devise progressive development trajectories in such a world, torn between attach-ments to place, class, gender, ethnic groups and no doubt a lot more besides.

Notes

1 There may be a danger of 'tacit knowledge' thereby being invoked as an unknowable residual explanatory variable, in a way analogous to neo-classical growth theorists' treatment of technical change. This is not to deny the existence or significance in some circumstances of tacit knowledge. It is to suggest that there are methodolo-gical problems in revealing its existence and effects.

2 It is important to note, however, that the processes of *producing* new knowledge, knowledge that comes into existence for the first time, are not dealt with directly in learning approaches, as a corollary of their grounding in associationist, stimulus–response conceptions of learning and their concern with outcome rather than pro-cess (see Odgaard and Hudson, 1998). This represents a major problem with the learning approach *in its own terms*.

3 *En passant*, it is worth noting that this creates problems in characterizing the nine-teenth century as a regime of extensive accumulation: *cf.* Brenner and Glick, 1991.

4 This also raises broader questions as to the maintenance of social order and the perceived legitimacy of capitalist relations of production and the state policies that sustain them, especially when such locationally concentrated pools of surplus labour have become a permanent and structural feature of labour markets.

Geographies of changing forms of production and work

Introduction

While the focus in Part 2 was upon the changing character of regions, and the changing ways of understanding them, in Part 3 the emphasis switches to industries. The particular concern in Part 3 is with issues of forms and geographies of production, employment and work and changes in these, acknowledging that such changes are unevenly distributed over time and space. It examines these issues at the level of individual workplaces, companies and sectors. As such, it engages with well-established concerns in the social sciences and also with one strand of a more recent debate within regulationist approaches, that focuses upon changing forms of labour process control and strategies for recruiting and retaining workers. It challenges the notion that such changes are to be understood in terms of a simple dichotomy, of a transition from Fordism to post-Fordism. A recurrent theme running through the chapters in this part of the book, one that has already been evident in Part 2, thus engages with a key issue within the regulationist literature – the presence or absence of connections between changes at the macro-level in regimes of accumulation and at the micro-level in the organization of work within individual workplaces. The argument is that the evidence suggests that there is no necessary connection between changes at macro- and micro-levels. Indeed, it is by no means clear just what it means to speak of 'post-Fordist' production, since many of the forms of organizing production and work that are allegedly 'post-Fordist' in fact have pre-Fordist histories. As such, it has implications for the broader debate about changes from Fordist to post-Fordist regimes of accumulation and modes of regulation,

There are also continuities with the chapters in Part 2 in another sense, and these are perhaps most clearly evident in Chapter 5. For there are clear relationships between the character of particular places, labour market conditions in them, and the ways in which companies use these spaces in their production strategies. There are definite links between labour market conditions, 'hire and fire' strategies and labour control regimes. Chapter 5 focuses upon changing forms of production, employment and work in 'old' industrial regions, drawing its empirical evidence from a variety of regions in Europe, as well from northeast England. This chapter was originally written in the mid-1980s, and in the preceding decade or so it had become clear that significant changes were occurring in the organization of economic activities, work and employment and in consumption patterns and living conditions. It was also clear that there was considerable selectivity in the incidence of these changes, socially and territorially. This complex pattern of changes raised important theoretical questions as to the most appropriate way to interpret and understand them. Many of the answers offered drew heavily, in one guise or another, upon notions of increased 'flexibility' and an alleged change from a Fordist to a 'flexible' regime of accumulation, from a supposedly rigid and inflexible Fordist mass production system to flexible production systems. In this chapter, these theoretical claims of a qualitative shift in the character of contemporary capitalism are set against the evidence of changes in 'old' industrial regions of the advanced capitalist world, such as northeast England, Lorraine and the Ruhrgebiet. Whereas there undoubtedly were (and are) some economic activities and localities that could be validly characterized as exhibiting symptoms of flexible production, the evidence from the 'old' industrial regions did (and does) not point to such a conclusion. Indeed, in the years after this chapter was written, these regions have typically remained blighted by high unemployment and widespread poverty, on the margins of the economy rather than central to it. Moreover, it is not at all clear what the concept 'flexible regime of accumulation' denotes, since capitalist social relationships have always exhibited great flexibility in coping with their inherently contradictory character. In contrast, it is suggested that Fordism (both in the narrow sense of a particular way of organizing production and in the broader sense of a regime of accumulation) had never established more than a tenuous hold in many such regions. Secondly, it is suggested that the changes in the organization of production, employment and work that were then taking place in and through these regions did not constitute a transition to a new post-Fordist flexibility. In contrast, such changes are more appropriately interpreted as evidence of a selective reworking of existing methods of production alongside experiments to find new profitable methods of high volume production. Furthermore, the implication of these changes was that, rather than capitalism being characterized by transition from one stable regime of accumulation and mode of regulation to another, it is more accurately characterized as a mosaic of uneven development and contradictory regulatory practices.

Studying economic change in particular sorts of places involves analysing changes in the specific industries that have been constitutive of these places. Often such places have had a mono-industrial character, or at best have been constructed around a limited set of industries and economic activities (for example, coal, chemicals or steel: Beynon *et al.*, 1991, 1994; Hudson and Sadler, 1989, 1990). As state policies were devised to diversity the economies of such regions, industries that were new to them (though not necessarily themselves new) came to locate within them. One such industry was the automobile industry. Over the years, automobile production has provided the arena for radical changes in the organization of production and work. Not least, the automobile industry was the birthplace of the once-revolutionary changes in production that came to be encapsulated in the term 'Fordism', structured around particular patterns of capital – labour relations and relationships between companies. In Chapter 6 the search for alternative high volume methods of production, incorporating economies of scope as well as of scale, and their variable geographies is explored. This involves consideration of new forms of structuring capital – labour relations and relationships between companies along the production *filière*, and the various spatial structures that these can take. Often old industrial regions, blighted by high unemployment, provided locations in which automobile companies could seek to experiment with new ways of producing and new ways of organizing and controlling the labour process. An important conclusion is that there is no necessary relationship between particular forms of organizing production and geographies of production – depending upon local circumstances and contingencies, the same model of production organization can take a spatially agglomerated or deglomerated form. This has very important implications for the formulation and implementation of local and regional economic development strategies. These conclusions, drawn on the basis of evidence available by the early 1990s, have been reinforced by

Plate P3.1   Monitoring the labour process: numbered workplaces on the line in a shoe factory in northern Greece.

subsequent developments. In particular, the comments (first made in 1993) that 'the Japanese model could itself be on the verge of radical and potentially destabilizing change' has taken on added significance in the light of subsequent events.

The implications of the introduction of new forms of high volume production for individual workers, organized labour and trades unions are then explored more fully in Chapter 7. There is considerable debate as to whether the new forms of work and employment relations that are integral to high volume production methods are an improvement over those of Taylorized mass production or represent new and intensified forms of work organization. In many accounts of high volume production, the prime emphasis is upon reshaping relationships between companies. In so far as the implications of new methods of high volume production for workers are considered, they are assumed to be beneficial, generating reskilled, multi-skilled, enriching and satisfying work in contrast to the alienating jobs of Taylorized mass production. As a corrective to this, this chapter focuses upon an alternative view of the implications of these changes for workers. It argues that in fact high volume production often produces worse jobs, characterized by intensification of work and multi-tasking, with new forms of regulating the labour process which serve to divide workers from one another and increase stress and tension in the workplace (and beyond). Moreover, mass production has not been abolished, but relocated to peripheral locations within the global economy. Workers in the newly industrializing countries of the peripheries of global capitalism have often been unable to challenge such changes because of the character of their dominant political regimes, with political élites encouraging inward investment by multinational companies and ensuring that labour market conditions are consistent with this objective. Workers in the core countries of advanced capitalism have been unable to mount effective resistance to these changes, not least because they have been introduced against a background of high and seemingly permanent unemployment, in part induced by changes in state regulatory approaches with the ascendancy of neo-liberalism. Such changes thus pose great challenges to trades unions and other institutions of organized labour to find ways of organizing effectively on an international scale to contest the undesirable effects of these changes on workers and their communities.

Chapter 5

Labour market changes and new forms of work in old industrial regions*: maybe flexibility for some but not flexible accumulation

5.1 Introduction

As the postwar 'long wave with an undertone of expansion' became transformed into one with 'an undertone of contraction' (Mandel, 1975), it became clear that momentous changes were occurring in the organization of economic activity, in patterns of work and employment, in production, and in consumption patterns within advanced capitalist economies. They raise intriguing questions as to the most appropriate way of interpreting them, and many of the answers that have been offered draw heavily, in one guise or another, on notions of increased 'flexibility'. For some, the changes denoted a 'crisis of Fordism', its imminent demise, and the transition to a new 'post-Fordist' flexible regime of accumulation, a transition which is located at the heart of the myriad observable changes (see, for example, Harvey and Scott, 1987). Others have focused on the more specific, though closely related, issue of the transition from 'ageing' Fordist mass production to 'ascendant' flexible production systems (Storper and Scott, 1989).

Partly because these processes of transformation and change are still ongoing, the extent and generality of the switch to 'flexibility', and indeed precisely what 'flexibility' means ('flexibility' for whom, for example), remain a matter for debate. What I seek to do in this chapter is to explore this issue by setting some of these general claims and propositions against the detailed evidence provided by the diverse forms of recent (1970s and early 1980s) changes in labour markets, in labour processes, and in the organization of production in 'old' industrial regions of the capitalist world. Undoubtedly there are some economic activities and localities that can be validly characterized as exhibiting symptoms of 'flexible production systems' or 'flexible accumulation'. The evidence available on the 'old' industrial regions does not, however, point to such a conclusion. Rather it suggests two things. First, it suggests that Fordism (either in the narrow sense of a particular method of organizing production, or in the broader sense of a regime of accumulation) has never established more than a tenuous hold in many of those regions. Second, it suggests that the changes

* First published 1989 in *Society and Space*, 7: 5–30, Pion Ltd, London

in organization of production, employment and work that are currently tak-ing place in them do not constitute a transition to a new 'post-Fordist flexibility' but rather a selective reworking which reproduces, in modified form, pre-Fordist and Fordist methods of production. In so far as such regions remain loci of major ensembles of industrial production within the capitalist world (notwith-standing their relative, and in some cases their absolute, deindustrialization over the last couple of decades), this second conclusion has broader implications for our understanding of the transformations currently taking place within capitalism.

To put these claims into perspective (and to provide a background to the empirical evidence of the interrelated changes in labour markets, labour pro-cesses, and in the organization of production), it is necessary to define briefly what is meant by 'old industrial regions' (OIRs). OIRs are those areas that formed the cradles of industrial capitalism, and are situated where capitalist production grew rapidly in the nineteenth century around industries such as coal mining, chemicals, iron and steel, and related metal processing industries (engineering, railways, shipbuilding, etc.). From a very early stage they were organized in large oligopolistic conglomerates and were tied into international markets.[1] Despite attempts to 'diversify' their industrial structure and labour markets, especially in the 1960s, via various national state policies, many of the OIRs remained heavily dependent upon their 'old' industries as a source of waged employment. Even so, for reasons that are set out below, changes in economic activities and employment within the OIRs have not been confined solely to these industries.

In the next section the variety of recent changes in labour markets and labour processes in the OIRs is summarized. Clearly the sorts of changes described there are by no means limited to these regions, but they are perhaps most acutely experienced in these areas where 'work' was widely, if errone-ously, defined as full-time regular male wage labour – 'a job for life'. It is also important to acknowledge that such changes are not experienced in the same way, or to the same extent, in each OIR. There are often significant differ-ences as well as similarities between industries, regions and smaller localities within OIRs, differences which reflect local variations in politics, culture or the practices of civil society. It is in this sense that the specificities of localities and regions can be rescued from the dangers inherent in the contemplation of the unique, and be integrated into an understanding of the dynamics of the restructuring of capitalism (in the form of a global system of commodity production and exchange). Indeed, the construction of such an understanding requires this integration. This and other issues central to the theoretical inter-pretation of changes in the OIRs are explored in section 5.3, and in the final section I tentatively explore some of the political implications of these changes.

5.2 Forms of labour market and of labour process change in the old industrial regions[2]

The most obvious effect of the changing geography of production upon the labour market in the OIRs is the rapid increase in and current high levels of registered unemployment. There is a tendency to interpret the growth of unemployment simply as one aspect of a growing dichotomy between those people in full-time regular waged employment and the long-term unemployed, who are solely reliant on state transfer payments. Nevertheless, this considerably oversimplifies the sorts of changes that have been occurring in these labour markets. For, related to the shrinking number of jobs available and the rising unemployment in the OIRs, there has been a series of changes affecting the character and duration of these jobs, employers' strategies for hiring and firing labour and the differential access to different sorts of waged labour for young and old, men and women (see, for example, Osterland, 1986). These issues are explored in turn.

5.2.1 The growth and changing character of unemployment

As unemployment has expanded, there have been important changes in the character of unemployment and in who is becoming unemployed. Female unemployment has grown to a far greater extent, in relative and absolute terms, as the 'new' manufacturing industries of the 1960s (see subsection 5.2.2.2) have shed labour and as the downward multiplier effects of loss of income from industrial jobs have worked through to service activities in those areas. Increasingly, long-term unemployment has ceased to be the exclusive preserve of men aged 50 years or more and has become generalized over both sexes and all age groups, young and old alike, but with a growing concentration of unemployed among the young. This general expansion of permanent long-term unemployment has provoked new forms of state involvement to contain its consequences (see subsection 5.2.8). Also as a result of this, important theoretical and political questions are posed (see section 5.4).

However, even in the 1980s it does not automatically follow that those losing their jobs in the 'old' industries have joined the register of unemployed. It may well be that many of them are re-employed within the region (or indeed elsewhere: see subsection 5.2.5). In other words, the structure of the labour markets within which job losses from 'old' industries occur is important in shaping the distribution of unemployment arising directly and indirectly from these job losses. There is now growing recognition that local labour-market conditions are often much more important than factors such as social class, traditionally thought of as decisive, in influencing the incidence of unemployment (Ashton and Maguire, 1987), although, as Osterland (1986) points out, for those who do obtain alternative jobs it may often mean re-employment in another company, in a less skilled job, on lower wages, and at the expense of someone else who is added to or remains on the unemployment register.

5.2.2 The contraction of core employment and the expansion of peripheral employment in 'old' industries

Historically, the predominant form of work in 'old' industries in the 'old' industrial regions eventually became structured in terms of regular waged work – a (male) 'job for life', with a progression available through a hierarchy of job types within the coal mines, steelworks or shipyards. In recent years there has been a redefinition of the extent of this core workforce, and, as it has shrunk, two related changes have occurred. First, peripheral employment associated with work in these industries has grown. Second, there have been important changes in the terms and conditions on which employment has been offered, associated with a reorganization of the labour process.[3] Both aspects reflect a concern to increase labour productivity, and in this way to increase the international competitiveness of production from plants in these regions. Because of the natural environmental basis of production and/or for technical reasons associated with the organization of production, it is often impractical to decentralize production in 'old' industries, such as coal mining, basic chemical production or steel production, to smaller plants. Indeed in the 1950s and 1960s there were considerable productivity gains associated with the introduction of new and more automated production technologies consisting of bigger production units and complexes into these industries (see, for example, Hudson, 1983a; Läpple and van Hoogstraten, 1980). To increase labour productivity, therefore, one must secure an acceptance, among the workforce, of changed conditions of work and of greater flexibility within big plants and industrial complexes – often under threat of closing them completely and switching production abroad, threats made against a background of global overcapacity and slow or no growth in demand for their output. One result of this has been to promote and enhance interplant competition in terms of labour productivity – something that workers have often acquiesced to because they see it as a way of preserving 'their' plant and community (see Hudson and Sadler, 1983a, 1986a).

5.2.2.1 New forms of work and of conditions of employment in the shrinking core workforces of the 'old' industries

One aspect of the imposition of new working conditions is linked to the introduction of new technology. Although in some economic activities there may be a general tendency to switch from a Fordist to a more flexible method of production, it is also important to acknowledge that in industries such as deep coal mining there are strong pressures currently to attempt to 'Fordize' production as a way of competing with opencast output and with coal imports. In the UK this is particularly evident in British Coal's attempts to introduce the mine operating system (MINOS) (Winterton, 1985), both increasingly to automate production and to tighten management control of labour underground. The considerable relative autonomy that miners had over the pace of production is being eroded as the new mining technologies are being selectively introduced into collieries and indeed onto particular coal faces. This helps to produce divisions between miners within such a colliery and also between them and miners in collieries denied such investment (because of the promise of substantial wage increases to be achieved via productivity bonuses, and because of the promise of long-term employment).

In other cases, the introduction of new conditions of work is associated less with the introduction of new technology than with the intensification of work with existing technology. There are several dimensions to the process of reorganizing work around a more flexible labour force. One is the introduction of new and extended shift systems (see, for example, Hudson, 1986a, 1987). A second relates to the lowering or ending of demarcation barriers. For example, on Teesside, Imperial Chemical Industries plc (ICI) has negotiated an agreement with local unions whereby its shrunken core workforce undertakes maintenance work – with the additional benefit to ICI that it can then shed maintenance workers from this workforce (Beynon *et al.*, 1986a). Elsewhere in northeast England there have been substantial reductions in demarcation barriers within ship repair yards on Tyneside (Hudson, 1986b). In the USA the National Steel Corporation (NSC) and the United Steelworkers of America (USW) agreed upon a deal to reduce craft designations to allow work assignments of greater flexibility, with these assignments allocated by joint management–labour committees; workers were to be dismissed if they refused reassignment (Clark, 1987, 16–24). In the United Kingdom, United Merchant Bar, a new company jointly owned by Caparo Industries and the British Steel Corporation (BSC), agreed a deal with the Iron and Steel Trades Confederation (ISTC), which provided for flexibility between production and maintenance workers and contained a no-strike clause (Hudson and Sadler, 1986b). These are simply four from among many examples of (albeit usually unwilling) trades union acquiescence to the erosion of well-established demarcation lines and working practices. A third dimension to this reorganization of work and working conditions is a move to 'one-union' arrangements in 'old' industries. Particularly in the United Kingdom, there has often been a plethora of trades unions representing the interests of workers in such industries, and to some extent increased flexibility has been associated with a switch to 'one union' deals. In 1986, for example, United Merchant Bar agreed such a deal with the ISTC, the first that involved any company in which BSC had a stake. The effect this will have on workers elsewhere in BSC is likely to be profound, not least because United Merchant Bar's mill is one that BSC had formerly closed and it is physically in the middle of BSC's major integrated works at Scunthorpe.

A fourth way in which working conditions have been redefined involves wage cuts and a switch to decentralized wage bargaining. In the nineteenth century, in what were then 'new' industrial boom regions such as northeast England, capitalists and workers often negotiated 'sliding scale' arrangements, whereby wages rose and fell depending on the level of market demand for a company's output (see Carney and Hudson, 1978). In the 1980s there has been a reluctant acceptance by workers, in what are now often slump localities within 'old' deindustrializing regions, of competitive forms of wage determination, as part of the new flexibility that is the price to be paid for maintaining some capacity and employment in the 'old' industries. For example, Clark (1987, 15–16) details how workers in the USW agreed to a direct cut in wages, as part of a deal with the NSC, in an attempt to enhance the company's competitiveness and therefore their chances (as they saw them) of remaining in employment.[4] In other cases, wage bargaining has been decentralized, with an increasing proportion of the wage linked to productivity at individual plants, or even to productivity at individual coal faces in mines or of production lines in factories. Consequently, workers within an industry, company or individual plant become divided on a territorial basis, as the competition to increase productivity both to raise their wage incomes and to try to guarantee the survival of 'their' plant is intensified. For example, Sheffield Forgemasters introduced new collective bargaining arrangements and altered pay and working conditions after a restructuring of the company into 10 divisions, seven of which were based in Sheffield itself. The changes included decentralization of industrial relations to a divisional basis, and a unilateral abolition of the centralized site-based shop stewards organization in a clear attempt to divide and rule. Somewhat unusually in the political climate of the 1980s, at the Atlas site in Sheffield (where five of the divisions were located), these changes were vigorously contested by the workforce, which came out on strike (ignoring a call, from the ISTC's central headquarters in London, to return to work because strike action was jeopardizing financial aid to the company from the Department of Trade and Industry). The other major Sheffield site, River Don, was then picketed until the workforce also came out on strike. The strike eventually ended after 16 weeks, in February 1986. Although Forgemasters agreed to reinstate workers sacked during the strike and agreed to a joint shop stewards committee to represent the two major Sheffield sites, pay and productivity bargaining was decentralized to plant level in a situation where no full-time trades union officials were permitted on site (for fuller details see Hudson and Sadler, 1986b). In contrast, in other cases, far from contesting attempts to reformulate wage bargaining on this decentralized basis, workers and their trades union representatives have embraced such changes, sometimes eagerly and sometimes reluctantly, in the hope that they might safeguard 'their plant'. In such cases the politics of class and of attachment to place become intricately, and often dangerously, entangled in relation to the prosecution of working-class interests.

5.2.3 New forms of work and of conditions of employment in the core workforces of 'new' manufacturing firms in 'old' industrial regions

5.2.3.1 The branch plant economy

Until the end of the 1950s there was remarkably little change in the patterns of production or in the labour markets of the 'old' industrial regions. Broadly speaking, before the late 1930s the wild cyclical swings in the 'old' industries led to periodic mass unemployment, but there was little political pressure for state involvement to diversify employment opportunities in these areas. After 1945 such pressures began to develop, and regional policy was placed more prominently on the agenda, as part of the national state's contribution to the postwar social democratic settlement, especially in Western Europe. The general increase in state involvement, in economic and social life, was seen as a central element in attempts to impose a new Fordist regime of accumulation in the main capitalist states (see section 5.3). Nevertheless, there were pressing national imperatives, tied to the demands of postwar reconstruction following the demands of the war economy, to freeze the pattern of production and its associated pattern of demand for (male) labour in these areas and to maximize output of coal, steel, ships, and so on.

Prior to the end of the 1950s the extent to which 'new' industries were selectively permitted in these 'old' industrial regions was decisively influenced by the existing major companies that controlled the 'old' industries and/or by national states. Alternative male employment was strictly off the agenda, although some growth in 'new' female employment, in the factories of, for instance, clothing or electronic engineering companies, was accepted and indeed welcomed. Even so, there were strict limits to the extent to which the gender composition of the waged labour force could be altered, precisely because of a working-class culture that had emerged around a strict gender division of labour between non-waged female employment in the home and the requirements of male shift-working in industries such as coal, chemicals and steel.

From the 1950s these limits began to weaken as male employment in the coal mines and shipyards began to fall, a decline that was linked to changes in international energy markets and the international division of labour in ship-building (especially, at this time, the resurgence of the Japanese yards, reconstructed around 'state of the art' technologies). The slackening of these constraints was associated with a variety of labour-market changes in the OIRs, including growth in the employment of 'green' female labour as the core workforces of many of the new factories. This was often a growth in part-time work; thus, women could work for a wage in a factory and still have time to do unwaged work in the home. In contrast, the availability of large numbers of male workers, socialized into the disciplines of working in three shifts, led to other companies seeking to recruit them to their core workforces. For example, Opel, General Motors' subsidiary in the Federal Republic of Germany, set up a car assembly plant in Bochum, in the Ruhr, in the early 1960s and recruited most of its initial workforce of 15,000 from coal miners made redundant in surrounding areas as collieries were closed. Within three years most of these miners had left and had been replaced by migrant workers from southern Europe, because it proved impossible to socialize ex-miners into the disciplines of a very different and archetypically Fordist labour process of car assembly lines.[5] Elsewhere, however, ex-miners seemingly adjusted much more easily to work on the factory production line (see, for example, Bulmer, 1978).

These factories in 'new' 1960s industries in the 'old' industrial regions were often branch plants of foreign multinationals, especially US-based ones locating in Western Europe. They brought with them new industrial relations practices, and new styles of management and of work organization. As the above examples illustrate, in some cases this provoked problems as these management innovations came into conflict with established cultural expectations

Plate 5.1 A new industry in an old industrial region: Opel's automobile assembly plant, Bochum.

of how work should be organized, whereas in other cases, new industrial relations practices and old cultures both found ways of adjusting to one another. In areas such as northeast England, for example, trades unions competed in the 1960s to offer 'one-union' deals to boost their membership, in exchange for promises of trouble-free production. Compared with contemporary conditions of work and levels of pay in the coal mines, such new factory jobs seemed to offer employment in attractive conditions, and the introduction of differently organized production and labour processes proceeded relatively smoothly as new core workforces were, for a time, assembled for these branch plants (see, for example, Hudson, 1986a). It is important to emphasize, however, that the scale of labour-market changes associated with this introduction of 'new' industries and 'new' Fordist labour processes was quite limited and, in many cases, temporary as these branch plants shed labour or closed completely in the 1970s.

If the 1960s round of branch plants in the OIRs of Western Europe was associated particularly with US-based multinationals, that of the 1980s became particularly identified with Japanese direct investment abroad, both in Western Europe and in North America. Quantitatively, the extent of this investment is fairly marginal as a proportion of the total foreign direct investment (FDI) by Japanese companies. It is often more significant as a proportion of FDI in some[6] of the 'old' industrial regions, but, of greater importance, it has had considerable qualitative significance, especially in terms of its 'demonstration effect' to other companies. Through their strategies to recruit core workforces and to organize production within their new branch plants and emergent production complexes,[7] these Japanese companies have begun radically to change the rules.

It is not so much that they have introduced new practices (although they have) but rather that they have intensified existing ones and combined them in novel ways. Not surprisingly, given the high levels of registered unemployment in the OIRs, inward-investing Japanese companies have been able to exercise great care in selecting their new core workforces. In northeast England, for example, Nissan received over 10,000 applications for the first 240 jobs that it advertised. In these circumstances, Nissan was able to compose its new core workforce extremely carefully; using extensive sophisticated psychological testing procedures it was able to ensure that its new employees possessed appropriate attitudes towards work and felt a commitment to the company. The lesson has not been lost upon other new non-Japanese companies, who have often emulated the example set by the Japanese. When PMA (Positive Mental Attitude – now a subsidiary of Coats-Patons) established a carpet-fibre weaving plant in Hartlepool (which is some 15 miles from Washington) in the early 1980s it reportedly selected its initial core workforce of 24 from the 3500 people that it interviewed (Boulding, 1988). Such great selectivity is by no means confined to the United Kingdom. In West Germany, for example, in the late 1970s Daimler-Benz established a major car assembly plant, around a dedicated line, at Bremen (an area badly affected by reductions in shipbuilding employment). By 1987 it had expanded its workforce there to 13,000, carefully selected in terms of its age, ethnic composition and skill levels: the average age of the workforce was around 30 years old (much less on parts of the line); in contrast to many West German assembly plants there were very few southern European or North African migrants; and only skilled manual workers were recruited, irrespective of the tasks they would carry out in the plant.[8] But, like Nissan's assembly plant in northeast England, this does not represent a transition from Fordism to flexible production but rather represents one element in corporate strategies to preserve Fordist production. New plants are located in OIRs, where prevailing high levels of unemployment allow the workforces to be selectively composed of young, physically fit people who are committed to 'their' company, and therefore labour productivity and competitive production in classically Fordist dedicated line branch plants are enhanced.[9]

Japanese companies, especially those in the electronics sector, have often been crucial in introducing not only innovative recruitment practices but also new working terms and conditions (Marsh, 1983) around what remains a basically Fordist production strategy in their foreign branch plants. No-union or one-union deals, often tied to no-strike agreements, have proliferated. In part this is because trades union leaders in the OIRs have extolled the passivity and other virtues of their members, as part of the global competitive struggle to secure new investment and new jobs (see, for example, Inward, 1987). In UK regions such as Scotland and Wales, one-union deals are the norm in the new Japanese electronics factories, and in the northeast Nissan agreed such a deal with the Amalgamated Engineering Union (AEU). There are signs that some trades union leaders, who previously had been actively involved in the enticement of Japanese capital, are now beginning to realize the implications of their earlier stance, even though they now seem powerless to reverse a trend that they had previously encouraged. In the northeast, for example, in response to the AEU's agreement with Nissan, Mr Joe Mills, regional secretary of the Transport and General Workers Union (TGWU), had this to say at a conference at Durham University, on 25 November 1986:

> Because of high unemployment and in desperation to co-operate with inward investment, some unions are ignoring their traditional role to organize workers and are standing back and allowing companies to choose which union they want, similar to choosing washing powder from a supermarket.

Cynics, of course, suggest that he might have adopted a different stance had the TGWU secured the deal at Nissan. This ambivalence on the part of trades union leaders, partly born of a fear that the choice may now be between a one-union and a no-union plant, is central to Japanese (and other companies) being able to get the sort of one-union deals that they want, often the key to and precondition for greatly increased flexibility over working conditions, grading and wages. However, this is not part of a transition to new flexible production systems but is a reflection of a shift in the balance of power between capital and labour, a shift in power which allows management to push through changes that allow the re-emergence, in modified form maybe, of Fordist mass production.

5.2.3.2 New small firms and the growth of self-employment

From the 1970s it became unambiguously clear that the combination of falling fixed capital investment and changes in technical conditions of production meant that the new jobs in branch plants would not replace those shed from old industrial regions. In these circumstances, often with considerable encouragement and rather more limited financial support from national states, the emphasis switched to regional self-reliance through the formation and growth of small firms. In the OIRs, especially where industries such as coal and steel were nationalized, special *ad hoc* state agencies were created specifically to encourage this self-reliance. To some extent this switch to a small-firm strategy reflected the resurgence of regions such as northeast Italy and also a misunderstanding of the possibilities for and limits to growth based upon small manufacturing firms in the OIRs of northwest Europe and North America (see also Sengenberger and Loveman, 1987).

Although the aggregate numbers of jobs created have been minimal, small-firm growth has nevertheless had important effects in these regions. It has often been associated with strident calls for the emergence of a new 'enterprise culture' in these areas; in practice the unemployed are blamed for being unemployed because, it is implied, they lack the appropriate entrepreneurial attitudes. It goes without saying that this is a very partial account of the causes of unemployment. In so far as new small firms have produced jobs, those employed as wage labour are often in a particularly vulnerable position in terms of their capacity to resist pressures to work harder and longer. They may be part of the core workforce of these companies, but if so their position as core workers is a precarious one (especially given the very high failure rates among small firms). For those who have become owners of one of these new small companies for the first time, there is a switch from employee to employer, with all that this signifies for their class position; and more generally, there are implications relating to the increasing fragmentation of a working-class culture based upon a shared experience of life as waged labour.

5.2.4 The growing peripheral workforce

The division between a core and a peripheral workforce is not a new one (see, for example, Friedman, 1977). In many OIRs the distinction between 'secure core' and 'insecure peripheral' employment has long been a grey zone rather than a sharp black-and-white divide. Such areas have often been characterized by a male workforce possessing various complementary skills, variously acquired via working in steelworks or shipyards. Although for some these provided 'jobs for life' with one employer, for others they were the key to a more or less uninterrupted series of contracts, allocated via informal contact networks in clubs, pubs or union offices. But what is distinctive about the last decade or so is the redefinition of the relationship between core and periphery, and the increasing diversity of conditions on which waged employment is offered in this expanding peripheral workforce.

One aspect of this changed relationship is the growth of subcontracting. In industries such as shipbuilding, subcontracting has long been an established part of the production process, but it has increased markedly in recent years. For example, over the last decade the workforce of the Bremer Vulkan yard has been halved as subcontracting has increased significantly. Coal, chemical and steel companies now systematically subcontract services which were formerly performed within the company; the workforces of these subcontracting companies become part of the peripheral workforces of the companies they serve. Some extent of the scale of subcontracting can be gauged from evidence given in a survey from the West Midlands of the United Kingdom, in which it was revealed that 39 per cent of the 370 workplaces surveyed had replaced directly employed labour by contractors (*Financial Times*, 5 May 1987). The economic rationale for this was succinctly spelled out by the manager of a Teesside engineering company (Beynon *et al.*, 1985, 15):

> Every single thing we do for them, they can do for themselves. As activity has reduced so they cut capacity – so people are now sub-contracting at what used to be 80% of their activity level, but they'd reduced capacity to 70%. The thing they're not going to do again is go through all the hassle of recruiting and laying men off. There are now more conscious decisions to sub-contract. Traditionally our business has been dragging people out of holes. Now this is not casual sub-contracting, but planned sub-contracting.

It is these last two or three sentences that catch the decisive qualitative shift in the role of subcontracting.

This growing institutionalization of subcontracting as a planned part of production strategies has also had important repercussions. For example, it has resulted in increased competition between subcontracting companies (many of whom have recruited some of the workers made redundant because of the switch to subcontracting), especially in situations where increased flexibility among the remaining core workers is absorbing some of the work that might otherwise have been put out to contract. As a consequence, wages and working conditions for those employed in the subcontracting companies have often deteriorated, especially in cases where the companies are able to recruit 'off the cards' labour (see also section 5.2.5). Such situations in Hartlepool were described in these terms by two men caught up in them (quoted in Morris, 1986, 149):

> The big companies and the little ones are cutting each other's throat for the work and, of course, it all comes back on the men – half of them are on the fiddle anyway. Keeps the cost down you see, and if you're a bloke on the dole, what choice have you got. If someone offers, you jump at the chance.

The second man put it like this:

> In the end the numbers dropped off and companies didn't want to pay the prices the big firms were asking, so they went to the smaller companies who were prepared to undercut. They could do this because they'd use mostly part-time workers or self-employed (that is, off the cards) and just call them in when they wanted them . . . if you want the work you've got to be prepared to take what's on offer, and if you don't accept it you don't get the work.

Such examples could be found for other places too; they convey a vivid impression of the way in which subcontracting is being reoriented as a way of preserving the 'old' industries and their methods of production. In no sense can this be regarded as evidence of the emergence of new 'flexible production systems', but rather it is evidence of redefined relations between capitals and between capital and labour, a redefinition which enables present production strategies to be reworked in an attempt to restore the competitiveness of existing 'old' industries that make 'old' commodities.

The two examples above also reveal the effects of this on the people who work for subcontracting companies. They illustrate the way in which changes in the role of subcontracting are related to other changes in the organization of production; changes in the conditions on which subcontractors employ people are linked to other labour market changes. In particular they point to a link between the growth of subcontracting and what is perhaps the most dramatic manifestation of the growing peripheral workforce: the resurgence of casualization. This is far from restricted to peripheral subcontracting companies, however. There is emerging evidence that it may be endemic: a survey by the General, Municipal and Boilermakers' Union in the West Midlands of the United Kingdom found that casual or temporary workers had replaced permanent ones in over half of 370 workplaces, employing over 200,000 casual workers in total (*Financial Times*, 5 May 1987). One strategy of big companies in the 'old' industries is to rehire former core workers on a casual basis in response to the ebb and flow of market demand, and in the United Kingdom the nationalized steel industry has been prominent in securing this sort of deal from the major steel unions. In 1983 BSC closed its 44″-bore pipe mill in Hartlepool, subsequently rehiring some of its former workforce on a six-month contract to meet a specific order. Further contracts were also agreed, but BSC refused to recognize continuity of employment across them, claiming that each new contract – although representing continuous employment – signified a break in service. The terms on which waged employment was on offer were also clearly laid down in an agreement taking the form of a 66-point schedule of working practices, which stressed the need for maximum flexibility and which emphasized that holiday entitlement and redundancy payment rights were to be waived, with every employee required to work any patterns of shifts or days that management specified (for fuller details, see Hudson and Sadler, 1985, 30–35). This sort of particularly stark instance of the recurrence of casualized work revives memories of earlier phases of capitalist development in such places. But it is important to stress that this does not point to the emergence of new flexible production systems. On the contrary, it represents the reimposition of old 'hire and fire' strategies as the position of labour in the market has been seriously weakened; therefore workers must accept this sort of flexibility to receive, in exchange, temporary access to waged work.

5.2.5 The growth of the black economy: working 'off the cards'

It is probably fair to say, especially in relation to particular jobs characteristically filled by women, that 'off the cards' employment has been well established over a long period in the old industrial regions. For example, work in clubs, pubs and corner shops has frequently been paid 'cash in hand'. In this sense the notion of a 'black economy' in these regions is not a new one. Indeed, its preservation and expansion in such regions in the United Kingdom have been encouraged by state policies, for if a woman is married to an unemployed man, any amount in excess of a specific sum (currently (1987) £6.00 per week) that she earns formally, is deducted from her husband's supplementary benefit claim. Some work is clearly organized specifically to take account of this and hence to tap a pool of otherwise reluctant potential recruits (Morris, 1986, 151–152).

There is also evidence, however, that this 'black economy' has been increasing in scale in recent years, again not so much in terms of wholly new forms of unofficial employment being introduced to OIRs, but via an expansion of existing ones (though there are limits to the scope for this). For example, it is clear that small-scale subcontracting firms have made increasing use of 'off the cards' labour so as to cut labour costs and to increase their flexibility in the hiring and firing of labour. It is also clear that there has been some growth of 'informal' kinds of activities which provide services to those in work; for example, electricians, painters, plumbers and joiners (often already in 'formal' employment) have found increasing scope for 'off the cards' work as other people engage in house improvement or maintenance. There are various forms through which such services are 'informally' provided (see, for example, Morris, 1986, 45–46), but they share one common limit to their growth: that is, they depend by and large on wages earned in the formal sector, and the prevalence of this form of money income can be severely circumscribed in localities characterized by chronic mass long-term unemployment.

5.2.6   The growth and changing roles of part-time work

In many service sector activities in the OIRs, as elsewhere, the character and timing of the tasks to be performed have meant that it has been usual for employment to be offered on a part-time basis. This tendency has continued and indeed expanded. In the 1960s the growth of part-time employment in manufacturing reflected not only pressures to contain labour costs but also the pressures of labour shortages and hence the necessity to devise production strategies and shift systems (specifically to allow married women to combine waged work in factories or offices with unwaged work in the home: see, for example, Hudson, 1980a, 1980b).

The more recent mid-1980s growth of part-time work is related to, and also cuts across, the redefinition of core and peripheral employees. There is no simple relationship between the growth of part-time employment and this ongoing redefinition. In some cases part-time employment has increased relatively and/or absolutely within a shrinking core workforce, whereas in others the increase of part-time work has been related to a growing peripheral workforce. The first tendency has been particularly marked in relation to part-time work in service activities. Again this can be related in part to specific national variations in state employment legislation; for example, in the United Kingdom reducing paid employment to less than 17 hours a week enables workers' rights to be eroded and employers' costs to be reduced. Townsend (1986, 317) remarked that in the United Kingdom there is 'a distinct group of regions that have suffered an equally depressing loss of full-time jobs but one accompanied by an above-average recorded gain in part-time employment. This group are all associated with heavy job loss in steel, coalmining and/or shipbuilding . . .'. He goes on to note that these areas originally had below-average levels of part-time employment and that the raising of these levels has been associated with the expansion of female employment. In other instances, however, part-time work has grown as part of the expansion of peripheral employment, as manufacturing companies have used part-time employment among a growing peripheral workforce, recruited from former core workers and for short periods (often 'off the cards') to meet temporary surges in demand.

5.2.7   International labour migration

The diverse forms of labour-market change outlined above share one characteristic in common in that they refer to changes in labour-market structures and in individual strategies for seeking waged work within labour markets in the OIRs. But just as these internal labour-market changes relate to shifts in profitable production sites within a changing international division of labour, there have also been, to a degree, international repercussions in relation to labour-market changes.

In the 1960s in many of Western Europe's OIRs, severe labour shortages in the coal mines and steelworks (particularly 'dirty' jobs such as those near coke ovens or hot-metal areas in iron and steel plants), as well as in car and other manufacturing plants, led to a considerable influx of southern Europeans and, to a lesser extent, of North Africans to fill these jobs (as well as jobs in services activities such as catering and cleaning). From the mid-1970s, as employment fell in these industries political pressures rose to stem the flow of new immigrants and to repatriate ones already resident. In France, for example, the national government introduced financial incentives in an effort to 'bribe' migrants to return home. Although the extent to which such migrant workers had become structurally incorporated into particular jobs in the OIRs put limits on attempts both to restrict the flow of new migrants and to accelerate the return of existing ones, pressures to reverse the previous trend continued amidst arguments about migrants taking 'our' jobs.

In other OIRs there has been a different form of growing involvement with international labour migration. This involvement is exemplified in localities such as Teesside, where many skilled manual workers who were made redundant by BSC and ICI initially became contract workers on North Sea oil rigs. Related to this, many have subsequently become international migrant workers.[10] Industries that are diminishing in Teesside are expanding in other parts of the world, and men who formerly worked in them on Teesside are helping to build up productive capacity in chemical and steel plants in Africa, Asia or the Middle East. It would seem that a contact network, established on Teesside in the early 1970s, when it was undergoing major expansion of chemical and steel capacity, has been crucial in providing the mechanism through which men from this area obtain jobs scattered all over the globe. A letter from a group of ex-BSC workers in South Africa to their former works newspaper on Teesside exemplifies this (*Steel News*, 1982, 3):

> You are probably wondering how adventurous have redundant steelworkers become? They meet up in the strangest places. As and at this moment, working for this company, we have Arthur, ex-shift manager Rod and Bar mill – Project engineer; Alan – ex-Mills and Spares Stores – Site Agent; Malcolm – ex-Beam mill electrician – Site Electrician; and E.A. ex-Consett works – Site Technician. We are all working for the SASOL company who are building a project about as big as I.C.I. Billingham out here at Secunda.

Such jobs abroad are usually offered on short fixed-term contracts because they relate to plant construction. However, the high rates of pay, even after agents' fees have been deducted – for the men sell their labour power at wage rates that were formerly unimaginable in Teesside – mean that such work is keenly sought after, especially by younger men. As is often the case with migrant workers, remittances are used to maintain the family and to invest in housing. Family ties, however, can easily be strained, as a 15-year-old boy explained (Beynon *et al.*, 1985, 18):

> My dad's a contract worker. He worked in the Middle East – Saudi Arabia and places like that. A year ago he came back from Oman. It's expensive to take the rest of the family over there . . . so he has to go out for three months at a time and when he comes back he's only back for a week and a half or something – so when he's back you don't see much of him. You get used to it – he's always away.

Given the collapse of the apprenticeship system in Teesside's chemical and steel industries, it is virtually certain that boys such as him will be denied the chance to acquire skills that they could sell on this international labour market; for them the future is one of remaining an unemployed person in their home region.

5.2.8 The expanding and changing role of the state in managing the surplus populations of the 'old' industrial regions

It is salutory to recall that in many localities of the 'old' industrial regions the 1950s and 1960s expansion of production was threatened by labour shortages. State policies were formulated to protect and, if need be, to expand workforces in coal mines, chemical plants, shipyards and steelworks (see, for example, Hudson, 1989a). In the OIRs of France and West Germany, labour-market pressures were so great that national states and private capitals devised managed schemes of international labour migration to combat labour shortages. There were pockets of long-term unemployment created and remaining within some of these regions, but these pockets were on a sufficiently limited scale to enable the state to manage them through public sector housing construction and allocation policies (see, for example, Benwell Community Development Project, 1978b; Byrne and Parson, 1983; Hudson, 1988a; Taylor P.J., 1979).

After 1975, however, registered unemployment grew rapidly and also changed in character: a growing proportion of these unemployed had ceased to be part of a reserve army of labour, and had instead become part of a burgeoning surplus population (see subsection 5.2.1). The forms of state management of the unemployed – through selective public sector housing allocation – that had developed in the United Kingdom in the 1960s were themselves rapidly rendered redundant. The scale of unemployment required in its management new forms of state involvement that would address the issue of employment more directly, and in the 'old' industrial nations of northwest Europe there was an emergent emphasis on government schemes which create temporary jobs, retraining schemes (but for what?), and so on. In localities, such as Consett, within these OIRs the main form of employment became government temporary jobs, with state agencies such as the Manpower Services Commission dominating the local labour market by means of such schemes. And as Osterland (1986, 8) points out, work on these government temporary job and training schemes often mirrors the less desirable characteristics of 'informal' activities, such as an absence of or reduction in social insurance cover. Even so, such government schemes do absorb large numbers of people who would otherwise be counted among the registered unemployed. Osterland (1987, 16–18) demonstrates how such job creation programmes in Bremen reduced registered unemployment by 16 per cent in 1984, and others have carried out similar analyses elsewhere (for example, see Robinson and Sadler, 1984). Osterland (1987, 18) goes on to make the point that in circumstances of mass unemployment such state policy initiatives to alleviate labour market conditions quickly reach 'the limits set by political and economic conditions'. As he puts it (p. 21) 'a government policy strengthening demand only has limited scope of action and chances of success in view of the unfavourable requirements on the company side'. Or, as Aneurin Bevan put it more bluntly in 1944, in the long run, profits and full employment are incompatible within a capitalist economy.

5.3   From Fordism to flexible accumulation in the old industrial regions?

Harvey and Scott (1987) argued, *inter alia*, that it is necessary to theorize, and not merely describe eclectically, the bewildering diversity of expressions of labour market change that emerged in the advanced capitalist economies in the 1970s and 1980s. It is difficult to dispute this argument. They went on to suggest that these changes are understood most appropriately in terms of a transition from a Fordist to a flexible regime of accumulation; that is, a change in the characteristic form of social organization from standardized production and mass consumption patterns (orchestrated via extensive state involvement in economic and social life) to a new, more diverse pattern associated with 'the new regime . . . distinguished by a remarkable fluidity of production arrangements, labour markets, financial organization and consumption' (pp. 2–3).

As a broad macroscopic statement of contrast between the 1960s and 1980s, there is some validity in this position. Even so, it tends to use the concept of regime of accumulation at a rather more general level than does the French regulation school of thought.[11] Lipietz (1986, 26), for example, identified 'two historically and theoretically linked but relatively distinct phenomena' that combine together to define 'Fordism' as a regime of accumulation (other regimes could be defined in a parallel fashion). These phenomena are 'Fordism as a mode of [intensive] capital accumulation', and 'Fordism as a mode of regulation of continued adaptation of mass consumption to productivity gains . . .'. As he put it, 'all this supposed a modification of the role of the State and of the forms of money management, including the substitution of credit money for gold-based currency'. Now, the significance of this is that capitalist states are constituted on a territorial, usually national, basis, and to a degree are in competition to make their own territories centres of accumulation. Although there are structural limits to a particular mode of capital accumulation, national states also possess some room for manoeuvre regarding their choice of combinations of policies which might sustain a particular regime of accumulation. This room for manoeuvre, and the resultant national variations around the central themes of a given regime of accumulation, must be acknowledged. This is not all that must be acknowledged, however. For even at national level, among the advanced capitalist states, pursuit of a Fordist regime of accumulation was not a unanimously accepted goal within the decisive centres of state power; nor was a Fordist regime of accumulation established equally at national level for each of the major capitalist states. For example, 'Britain, because of the resistant strength of its working class and the weight of its finance capital, which is too internationalized to be given over to this internal revolution [of intensive accumulation], has partially missed the boat of Fordism . . .' (Lipietz, 1986, 30).[12] Thus, even at national level within the advanced capitalist world there are problems with the characterization of regimes of accumulation, in the post-1945 period of 'a long wave with an undertone of expansion', as being unambiguously 'Fordist'. Uneven development at national level must be recognized, and if this conclusion holds at national level then it will hold, *a fortiori*, between regions within these national territories. Acceptance of these conclusions raises general questions about the conceptualization of change in terms of national regimes of accumulation, and, more importantly for the present argument, raises considerable doubt as to whether a transition from 'Fordism' to 'flexible accumulation' provides a suitable framework for understanding the changes in the conditions of labour markets, production and consumption in the 'old' industrial regions. As much of the evidence presented in the previous section suggests, it is by no means obvious that the starting point for these changes is one that can be characterized as 'Fordist'. Neither is it obvious that all the various forms of labour market and labour process changes are in the direction of greater flexibility (the systematization of subcontracting indicates greater fixity in relations between capitals and between capital and labour, for example), but in other cases it is clear that 'flexibility' amounts to capital reasserting its power over labour by means of managerial strategies to reorganize production, so as to preserve pre-Fordist and Fordist systems of production. It is certainly not the case that these changes are associated with emergent 'flexible' systems of production, as defined by Storper and Scott (1989), for example (see below). This raises questions as to the circumstances in which the concept of 'flexible mode of accumulation' is useful. Its use may, for example, conflate different tendencies in capitalist restructuring strategies.

These various points give rise to some important issues. Although one must acknowledge the dangers of an over-obsession with description of the myriad changes, one must be equally wary of an over-simplified theoretical interpretation of them. Thus the contrast between Fordism and flexible accumulation may in some circumstances be an illuminating one, with the following provisos.

(1) It is acknowledged that flexible accumulation requires much tighter specification of, for example, the branches in which post-Fordist strategies have been constructed, and of the strategies that capitals have used to reorganize

production along post-Fordist lines (often drawing upon pre-Fordist forms of organizing production and of managing the labour process; that is, there is no general, simple and unidirectional historical sequence of modes of capital accumulation). Storper and Scott (1989) define 'flexible production systems' as 'forms of production characterized by a well developed ability to shift promptly from one process and/or product configuration to another and to adjust . . . output rapidly up or down without any strongly deleterious effects on levels of efficiency', with the latter defined presumably in terms of the interests of capital rather than of labour. They refer (p. 2) to 'ensembles of flexible production sectors such as (a) selected high technology industries, (b) revitalized craft speciality industries and (c) producer financial services'. If we leave aside for the moment the heterogeneous character of these often loosely specified sectors, the crucial point is that they are notably absent from the OIRs (and it remains a matter for debate as to how generalizable such 'flexible' approaches are in terms of their applicability to many sectors of commodity production).

(2) It is acknowledged that Fordism, as a mode of accumulation, did not penetrate evenly and equally to all branches of the economy and localities; put another way, even within what, from some points of view, may be legitimately characterized at national level as a Fordist regime of accumulation, quite a lot of labour processes, even in manufacturing, remained and remain *of necessity* organized on a non-Fordist basis. Moreover, this was especially so in relation to some of the 'old' industries of the 'old' industrial regions. Certainly over a broad swathe of basic chemicals production (and to a lesser degree, bulk iron and steel production) these was an increase in automation, but hardly one that would be regarded as typical of assembly-line Fordist production.[13] Indeed, in some senses this was post-Fordist at a time of the very peak of the success of the Fordist regime. In many other branches of manufacturing, such as special steels production, shipbuilding (despite attempts at standardized designs and products) and heavy engineering, the penetration of Fordist methods was of necessity much more limited, and in many cases negligible, because of the sorts of commodities they produced and/or the technical conditions of production in them. This was *a fortiori* true of the coal mining industry, and the sorts of changes that are currently occurring in the United Kingdom in terms of attempts at automating deep-mined production or switching to more or less automated opencast production (as happened in the USA in the 1970s), represent an ongoing project to impose a quasi-Fordist organization of production in the 1980s. Thus there may well be a situation in which Fordism did have an impact on labour processes and on the organization of production in industries in which it could not itself be established, but this is very different to the 'Fordization' of all branches of production. Indeed, the really significant questions that arise from this are concerned with the relations between production in branches organized on non-Fordist and on Fordist lines, and between both of these and the overall process of accumulation.

(3) It is acknowledged that the impacts of a Fordist mode of regulation were unevenly distributed in terms of variations in life-styles and living conditions, between localities and regions as well as between classes, ethnic groups and gender groups. The precise form of these uneven distribution patterns reflected political priorities and the differing degrees of organization, power and influence of different social groups and classes. In particular, the transformation to mass consumption life-styles in the OIRs was a very partial and uneven one. Certainly state transfer payments, as part of a national welfare state system, provided a floor to consumption levels for many people who were unable to sell their labour power on the market. Undoubtedly, simple standardized mass-produced commodities, such as packets of cornflakes or blue jeans, became widely available from retailing establishments in these OIRs. But more sophisticated consumer goods such as cars, consumer durables, 'white goods', etc., often associated with growing home ownership and privatized life-styles, remained both restricted in scope and uneven in intraregional distribution. Not least this reflected the absence of a high productivity and of a high wage economy to provide effective mass demand for such commodities.

In summary, the transformation of economic and social life in the 'old' industrial regions during the 1960s was at best a pale shadow of what Fordism was meant to be about; and in some regions the shadow was very pale indeed. In fact, this failure to establish patterns of production and of consumption, characteristic of a Fordist regime of accumulation, reflects the failure of national state projects to link regional modernization reciprocally with faster national growth (for further details, see Hudson, 1989a). This failure of national states to implement successfully these policies had implications both at national and at regional levels – it may be stated (at the risk of oversimplification and, maybe, overdramatization) that it ensured a crisis in, though certainly not the demise of, Fordist systems of production and in the Fordist regime of accumulation. At the same time, in the late 1970s and early 1980s, it established some of the preconditions necessary for the OIRs to become locations in which capitals could seek a solution in Fordist terms to this crisis.

This raises important questions about the capacity of national states to manage the trajectory of economic and social change in their territories, questions which are central to a regulation approach but which cannot be explored further here. A couple of simple points can be made in connection with this, however. This failure of state policies to generalize the conditions of a Fordist regime evenly over national territories reflected both the past trajectory of uneven capitalist development within these states, and the growing internationalization of capitalist production in the 1960s. Capitalist development in these national states (again at the risk of some overgeneralization) had previously occurred in two distinctive phases, centred on two distinctive ensembles of industries, in two distinctive sets of regions, and in two distinctive though interrelated modes of accumulation.[14] The first phase centred upon extensive accumulation around 'old' industries in 'old' industrial regions, and the second upon intensive accumulation around a different set of 'new' industries and 'new' regions. During the tight labour-market conditions of the post-1945 'long wave with an undertone of expansion', there was some deconcentration of branch plants from the second set of regions to the first set, as well as into formerly unindustrialized rural areas, as big capitals reorganized production into new intranational spatial divisions of labour (see Hudson, 1987). Precisely which regions became recipients of the new branch plants depended upon cultural, economic and political variations between them, and also upon the political strategies of national states. These new branch plants often brought Fordist production methods to these regions for the first time. Though the extent of this innovation was limited, its potential longer-term significance, in terms of politics and culture within these regions, was considerable. But in the short term the amount of employment provided in the new branch plants was usually much less than that lost from the old industries, and, moreover, capitals establishing these branch plants were often enticed by the availability of cheap, pliant and 'green' labour power (see subsection 5.2.3.1). These sorts of labour market change emphatically did not provide the basis for a high wage or high productivity regional economy that could sustain a generalized transformation in life-styles around the consumption norms of a Fordist mode of regulation. Furthermore, many of these new branch plant jobs quickly disappeared from the OIRs as capitals were able to reorganize production on an international scale in an increasing range of industries; production was moved to locations of greater profitability both in the newly industrialized countries and, often more importantly, elsewhere in the advanced capitalist countries. In these circumstances of increasingly internationalized production, national states were less and less able to manage the intranational location of economic activities within their national territories, further undermining their attempts to generalize the conditions of a Fordist regime of accumulation evenly over these territories.

Even so, the failures of state policies in the 1960s and early 1970s to integrate these OIRs into the mainstream of Fordist production and consumption had important effects in the late 1970s and the early 1980s. As a result of renewed high levels of unemployment in these regions, persistent political and social problems were posed. As national states draw back from interventionist policies (because of a desire to cut public expenditure as part of the fight against inflation) and put more reliance upon the market as a steering mechanism for resource allocation, the OIRs were not favoured as places for the flowering of a new enterprise culture. Sengenberger and Loveman (1987, 101) suggest that:

> Social class structure associated with property relations, income sources and social divisions may provide reasons why it has proven so difficult to revitalize the old, entrenched industrial regions in Europe . . . all of these regions have in common a century or more of extensive industrial production based on industries like coal or ore mining, iron and steel, or shipbuilding that produced a highly dependent working class. When the industrial base began to shrink and employment declined new employment was not brought in at a rate to compensate for the dislocation. The old class of big industrialists did not do enough to reinvest, transform and

modernize the economic base of these regions, and the working class people were too far divorced from entrepreneurship to do it themselves.

Because of a past history of extensive proletarianization, the basis for an economic and social transformation around the expansion of an 'old' or the emergence of a 'new' middle class was strictly limited; indeed at best there have been very modest increases in self-employment, and those often at the margins of the black economy. Under these circumstances it seemed to some people in these regions that the only way to find waged employment was to leave, and to many others it appeared that the only way to retain some waged employment was to do one or both of two things. First, to agree to new, more flexible forms of work in existing industries, although in fact these 'new' forms often involved a reversion to older ones as the improvements in working conditions that trades unions had gained in earlier periods were eroded in the new political-economic climate. Second, trades unions had to accept new flexible conditions of employment in order to attract inward investment in the form of branch plants. Indeed, it appeared that they must go beyond this and collaborate actively in the competition for such investment, as companies sought a solution to their problems within the frame of reference of Fordist production by selectively relocating into areas where the bargain between capital and labour could be redefined more in favour of the former.

So in summary, there is no doubt that there is a great diversity in the form of labour market changes occurring in the old industrial regions. However, in the OIRs, do these changes in working conditions and in employment practices constitute a transition to a new post-Fordist flexible regime of accumulation, and to new post-Fordist flexible production systems? Is this the most appropriate interpretation to put upon them? Although it is perhaps dangerous to draw definitive conclusions, because these processes of change are still being worked through, the answer to these questions at this stage (in 1988) must be 'no'. There are certainly changes in the capacity of management to recruit selectively, to compose particular types of workforce, to redefine relations between core and peripheral workers through increased use of subcontracting, and to impose new forms of employment contract and work conditions at the point of production. Undoubtedly, the climate created by the New Right political economy (that has reasserted the market as the pre-eminent economic steering mechanism, and weakened organized labour by a variety of legislative changes and above all by creating massive unemployment through national economic policies) has been crucial in providing an arena in which such changes could be pushed through. As a result of this, and parallel drops in the levels of welfare provision, the capacity of many people in the OIRs to participate in the mass consumption life-style of a Fordist regime is further weakened; even more so the capacity to participate in the new post-Fordist life-style centred around the consumption of individualized designer commodities is eroded. Even the financial ability to purchase simple mass-produced commodities in many of these regions has been decreased. Widespread poverty has re-emerged as national governments have, to varying degrees, redefined the relation between the state and the market as an allocative mechanism. In this sense the OIRs are even further from the consumption norms of a Fordist regime of accumulation, let alone a post-Fordist one, than they were two decades ago.

Therefore, rather than constituting the emergence of a new flexible regime of accumulation in these regions, the related changes in production and in consumption are most appropriately interpreted as being part of strategies by capital to preserve old modes of accumulation in a political climate very different from the welfare state Keynesianism of the 1960s. This is the case in two senses. First, they involve changes in the 'traditional' pre-Fordist industries of these regions. These changes increase labour productivity and competitiveness within big plants and production complexes by the cutting of employment and the redefining of conditions of employment for those remaining in work within them. Second, they involve corporate strategies to preserve Fordist mass production of standardized commodities – by the careful location of new capacity in OIRs and by the raising of labour productivity through selective recruitment and the intensification of work pace. This is most clearly exemplified in the archetypically Fordist car industry, and in the location of plants such as Daimler-Benz (Bremen) and Nissan (Sunderland). Such regions of high unemployment offer managers the scope to recruit selectively and to dictate the terms on which a wage is offered, and this helps to reinforce quasi-Fordist high volume production rather than facilitate its replacement by post-Fordist flexible production systems. In both cases, then – the old 'traditional' industries and the branch plants of the 'new' ones – the common linking theme is the reassertion of managerial control over labour and over the labour process. Changed labour market conditions, and a backdrop of very high unemployment in these OIRs, allow the relationship between capital and labour to be redefined, and allow management to recover or, for the first time, impose authority and control over labour in a variety of ways that depend upon particular technical conditions of production in an industry, and upon the specificities of local labour markets (for example, in terms of culture and politics, as well as the skill composition of the workforce, and so on). But this does not constitute the emergence of a new flexible regime of accumulation in the OIRs, but rather a partial reworking of old production strategies, that is designed to reproduce, through a selective transformation, the complex of existing and new modes of accumulation in such areas, and in a political-economic climate in which the role of the national state has been reduced *vis-à-vis* the market in regulating relations between production and consumption. In so far as decisive branches of production remain in these areas and organized in these ways, it suggests that (to date at least) 'flexible' production remains fairly marginal within the overall sweep of capitalist production; claims that a 'new hegemonic model of industrialization, urbanization and regional development has been making its historical appearance in the US and western Europe' (Storper and Scott, 1989, 1–2) are, at best, both overstated and premature.

Certainly there is some evidence of the emergence of 'flexible production systems' in particular activities and locations. But the evidence of changes in conditions, terms and levels of employment as the strength to bargain of management and workers has altered in the OIRs, means that different strategies for responding to crisis in other activities are being worked through in these regions. Recognition of this offers one way of meeting another important criticism that emerges in Harvey and Scott's paper (1987). This concerns the significance, for theory, of locality with regard to uneven development within capitalism. In a very literal sense, the 'old' industrial regions were constructed, materially and socially, as an integral part of the initial birth and subsequent growth of capitalism as a global system of commodity production and exchange. Crucial commodities were produced and traded both in and through these regions, and economic and social life within them developed a sort of structured coherence, rooted in place, within the limits imposed by the anarchy inherent in capitalism. Harvey and Scott (1987, 8) rightly point to the danger of 'a fixation on the specificity of the local as opposed to a continuing concern for elucidating the generality of capitalism in its totality'. But a consideration of the historical evolution of working, learning and living in the 'old' industrial regions, without relating this to the generality of capitalism, is possible but rather difficult. One cannot understand what has happened over a long period in these regions without relating these occurrences to the way the contradictions of capitalism have developed on a wider scale. Conversely, the development of the 'generality' of capitalism cannot be understood without reference to the initial emergence of capitalist production in these now 'old' industrial regions and to the subsequent changes within them. So although I would concur with Harvey and Scott's insistence on the need to relate the specificity of locality to the generality of capitalism, it seems to me necessary to remember these specificities. In this context, it becomes crucial to recognize that in the post-war period from the mid-1950s to the mid-1970s Fordism was not established as the dominant regime of accumulation, and so did not decisively shape the patterns of production and consumption in the OIRs. This recognition is necessary in order to grasp the theoretical significance of the sorts of changes, in production organization and in labour markets, that have been occurring over the last decade or so since the mid-1970s in what have increasingly become relatively more peripheral (although by no means insignificant) regions rather than core regions of global capitalist production. Not least, these changes have partly involved projects to preserve Fordist production in sectors such as vehicles and consumer electronics, and to retain or restore the competitive viability of production in world markets in sectors which cannot be organized on Fordist (or post-Fordist) lines. The mosaic of interrelated localities, production systems and industries has taken a still more complex turn, and attempts to unravel this complicated web of relationships present a major theoretical and empirical challenge.

5.4 Some concluding comments: the strategies of labour and the trades unions, and the future for learning, living and working in the old industrial regions

In the previous section I touched briefly upon some of the theoretical issues that arise from the bewildering variety of changes that are observable in the labour markets of the 'old' industrial regions. It is clearly very important whether one interprets these changes as evidence of a transition to a new flexible regime of accumulation, or as a reworking of existing accumulation strategies as capital takes advantage of its greatly enhanced strength *vis-à-vis* labour on the market. It seems clear that, intentionally or not, capitalist restructuring involves redefining and dividing the working class, but it would be wrong to see members of that class merely as passive objects in this process. In this section, therefore, I want to end by making some comments on the role of trades unions and workers (and their communities) in helping to shape these various changes in the 'old' regions; also, I would like to comment on how they interpret them, and on what the future might hold for such regions.

At various points I have alluded to or touched upon some of the political implications of the ongoing redefinition of work and employment. In some regions, job losses from the 'old' industries were bitterly contested for a while, but more generally there has been a reluctant acceptance of these losses as inevitable; and in much the same way, the transition to more flexible working conditions has been seen by those who remained in work as 'inevitable' (see, for example, Clark, 1987). Now clearly there was no absolute 'inevitability' about such changes, but in a political-economic climate dominated by the New Right it certainly looked that way to many residents of the OIRs. Indeed the hegemony that such views have come to hold is itself a testimony both to the weakness of the trades unions in such a climate, and to the weakness of any sort of socialist alternative (in terms of its ability to attract a broad base of political support, a lack which, to a degree, has to be related to an equation of socialism with the failures of Keynesianism and the political economy of the social-democratic welfare state).

What sort of challenges are posed as a result of this new situation of mass unemployment and of increasingly diverse forms of waged labour? One challenge is clearly related to the generalized expansion of permanent long-term unemployment, for this raises related theoretical and political questions. In particular, is a growing proportion of the long-term unemployed most appropriately characterized as part of a surplus population rather than as part of a reserve army of labour? If they have become 'clients of the Welfare State' (Pugliese, 1985), how are their class position and its political implications to be understood? This question is especially relevant to people in regions where working-class politics grew up around social-democratic reformism based upon (male) workplaces, particularly in the OIRs. For those who have become part of the surplus population, it implies, very strongly, that a workplace-based politics is and will remain more or less irrelevant. Can an alternative – and socialist – working-class politics be constructed in these regions around the demands that can be made of the state, around issues such as living conditions and life-styles, ecological concern for the environment, and so on? How might it relate to a more productionist working-class politics in other regions? For the moment these are questions that remain unanswered. A second issue relates to trades unions attitudes towards the growth of subcontracting, of new flexible work practices, and of part-time work, especially when these are associated with 'off the cards' work. In the United Kingdom at least, there is a considerable ambivalence that is detectable in trades union positions on these issues.[15] Some advocates of the 'new realism', notably the Electrical, Electronic, Telecommunication and Plumbers' Union (EETPU), but also the AEU, have eagerly embraced one-union deals that are linked to new flexible working conditions, whereas others (such as the ISTC) have come to accept them more reluctantly, but accept them nonetheless. However, still others, such as the General, Municipal and Boilermakers' Union (GMBU), remain active in their opposition to them. In other cases trades unions have begun to recognize that the problems that, for instance, growing part-time work poses are problems that they may not like but cannot ignore, and unions such as the Transport and General Workers have begun to consider strategies specifically to unionize part-time and temporary workers. Other unions such as the Association of Professional, Executive, Clerical and Computer Staff (APEX), a white-collar union, have decided against making any special effort to organize part-time workers. Although such differences in trades union positions are partly explicable in terms of the sorts of jobs that their members undertake, this diversity of stances towards new work practices helps to create the space in which these new practices can be sustained. Undoubtedly, this space is greatest in the OIRs, blighted by mass unemployment. As recent events in the coal industry in the United Kingdom have shown, if there is opposition within a union to flexible working, British Coal can hold out the threat of rejecting the National Union of Mineworkers (NUM) and dealing with the Union of Democratic Mineworkers (UDM) in its new superpits. This in turn is driving a wedge between different areas of the NUM. And so it goes on. And for those unions that want to oppose the trend to 'flexible working', these fragmentations and divisions within one national territory represent formidable barriers to any attempt to organize against it; rewritten on an international scale, the magnitude of these barriers becomes even more formidable.

Given this sort of scenario, what is the future for living, learning and working in the 'old' industrial regions? In some localities, community groups have developed which reject reindustrialization strategies and further state involvement, turning to an inward-looking idealism, hoping for solutions to emerge from within the community (see, for example, Hudson and Sadler, 1986c). On the one hand, maybe such idealism provides one way of coping with a desperate situation, as it becomes clear that neither the attraction of big branch plants nor the encouragement of new small firms is making any serious impact on unemployment levels there. On the other hand, it is not at all clear that there is any coherent basis for making the sorts of political demands for resources that would allow the survival of such communities which are without the incomes that waged labour supplies; nor is it clear, were such a basis to emerge, that governments would respond to it. Osterland (1987, 21–25) has recently addressed this question from a more practical perspective, and he begins by pointing out that 'conventional' state policy responses, aimed at creating new jobs through either regional policies or government job creation and training schemes, can at best delay the process of structural change in the productive structures and labour markets of the OIRs. Although this may seem to imply that such structural change is therefore 'inevitable' – which of course it is not – this is nonetheless a reasonable judgement, given the prevailing political-economic climate. The alternative of a state-led reconstruction of the productive base that integrates national sectoral strategies with local and regional development programmes sensitive to the specificities of their areas, is simply not on the political agenda at the moment; nor does it seem likely to be for the foreseeable future. Even so, within 15–20 years from now (1988), new possibilities may open up as the interplay of labour demand and supply, especially the changing demographic situation, brings demand and supply more into balance, so that the chance of 'full employment' in the OIRs could re-emerge. But because competition between areas for jobs will remain intense, these chances will not be translated into reduced unemployment unless there is a great emphasis placed upon ensuring that appropriately qualified and skilled labour is then available in the OIRs. As Osterland puts it (1987, 20), 'In the long term (as of the year 2000), therefore, the qualifications of the workers will be of increasing significance. Municipal strategies must promote the qualifications of today's young unemployed on a large scale via secondary labour markets, alternative forms of business (cooperatives), training and further education, informal work within the scope of communal and household production. In order to improve the situation of the private household as a productive unit, the space, money, and workers available to it would have to be increased.' Furthermore, 'the town as a place where people live and work would have to be strengthened'. This seems to be a political strategy which accepts that there can be a sufficient cultural transformation in the OIRs for them to be incorporated into the emergent flexible production systems of the economy of the twenty-first century. But this perhaps overestimates the extent to which existing production systems can be displaced over a wide range of commodity production. For Daimler-Benz at Bremen, for example, 'skill' is seemingly used as a criterion to recruit suitably socialized workers, rather than an expression of demand for their specific technical skills. In many ways, then, these approaches are highly provocative, a sort of radicalized reformism, born of and tailored to the 'new' labour-market conditions of the final two decades of the twentieth century. It may not be socialism (for example, the old tensions between attachment to place and class remain apparent), and even in its own terms it may not succeed, but is it the best that might realistically be hoped for in the OIRs, given their present condition and given the likely future macropolitical-economic environment? If so, it doesn't look like much of a future for many.

Notes

1 This is, of course, a considerable simplification of the productive structure of many of these regions where, for example, textiles production formed a very important part of the industrial structure, and led to a much greater incorporation of women into the waged labour force. Indeed, it was textiles production that

often formed the birthplace of industrial capitalism in such regions. Even so, there are advantages to be gained from focusing upon Department I (means of production) rather than Department II (consumer goods) production in these regions in terms of the argument I want to develop.

2 In terms of empirical examples in what follows, I will draw quite heavily on research I have been involved with in the regions of the UK, France and Germany, as well as on the work of others in these countries and in the USA. It is important to emphasize that evidence of many aspects of changes in labour markets and in labour processes is unobtainable from official public sources. It must, of necessity, be collected through interviews and discussions with those involved in and/or observation of – say – the organization of work in a coal mine, steelworks or car plant (which can mean an involvement with a particular plant over a period of time). In this sense, the sorts of data that are drawn on here make no claims as to being statistically representative, and there are often problems in deciding how much importance should be attached to particular tendencies. On the other hand, the tendencies identified here are representative of the sorts of processes through which labour markets and labour processes in the OIRs have been and are being altered.

3 In practice, many of the changes discussed here occur in combination, but they are analytically separated here for clarity and convenience. Likewise, the distinction between 'new' and 'old' industries is, to a degree, arbitrary, but it allows some important points to be drawn out.

4 Though Clark (personal communication, 1987) has pointed out that in practice this has not led to significant wage cuts.

5 The information about the Bochum plant was obtained in interviews with managers and workers there in the early 1980s.

6 For instance, cumulative Japanese investment in the mid-west of the USA by 1987 exceeded $5000 million (personal communication, Martin Kenney, 1987).

7 For example, in the USA Honda is constructing engine and transmission plants at its Ohio plant.

8 Information on the Daimler-Benz Bremen plant was obtained through interviews with managers and workers there in 1987 and 1988.

9 It would seem, however, that there may be important national variations in the respect. For example, Kenney (personal communication, 1987) suggests that companies in the USA do not display this degree of selectivity in recruitment.

10 The extent to which similar involvement in the international migrant-labour systems is evident in other OIRs remains, for the moment, an open question.

11 The French regulation school refers to a group of authors, including Aglietta (1979, 1982) and Lipietz (1979, 1984, 1986), much of whose work is summarized and reviewed by de Vroey (1984). De Vroey (p. 45) points out that while 'these authors draw on a similar set of concepts, like "regimes of accumulation" or "forms of regulation" . . . it would be incorrect to regard them as one homogeneous school of thought . . .'. It is not my purpose here to provide directly a detailed critique or review of the strengths and weaknesses of a 'regulation' approach *per se*, but rather to examine the extent to which changes in the OIRs may be understood in terms of a transition from Fordism to a flexible regime of accumulation and, in particular, to flexible systems of production. This does, however, raise some questions about a 'regulation' approach.

12 In so far as Lipietz is correct in its judgement about the UK – and there are others who make the same point in a different way (see, for example, Nairn, 1977;

Ingham, 1982; Rowthorne, 1983) – it raises doubts as to the validity of attempts to interpret the political economy of the UK in the post-1945 period, in terms of a transition to an intensive regime of accumulation (see, for example, Dunford and Perrons, 1986).

13 I would take issue here with Storper and Scott (1989, 6) when they claim that 'In its classical guise, Fordism was underpinned by large and highly capitalized units of production consisting of either (a) continuous flow processes, as in the case of petrochemicals or steel production, or (b) assembly line processes (and deep technical division of labour), as in the cases of cars, electrical appliance or machinery.' Assuming that by 'steel' they mean bulk steel production (because a wide range of special steels have always been produced on a small batch basis), this seemingly reduces *all* mass production to Fordism, and in the process of doing this loses the specific features of the labour process and managerial control over labour at the immediate point of production that characterize Fordism. The differences in labour process between what they refer to as 'continuous flow processes' and 'assembly line processes' are too significant to be summarily swept away in this manner.

14 It is important to stress that in both these main phases, forms of economic organization and labour processes, with their origins in earlier periods, lived on; and of necessity pre-Fordist forms continued to coexist with Fordist forms of production throughout the golden age of Fordism, in the postwar period (correctly stressed by Storper and Scott, 1989, 5–6).

15 This account draws on information gathered primarily from the 'Labour News' section of the *Financial Times*, over the first six months of 1987. As virtually every issue carried something on these topics, I have not attempted to give references to particular issues.

Chapter 6

New production concepts, new production geographies? Reflections on changes in the automobile industry*

6.1 Introduction

During the last decade or so, there has been considerable debate as to the existence and significance of different forms of organization of industrial production within capitalism and the geographies of production associated with them (see, for example, Hudson, 1988b, 1992c). This debate has been predicated upon a recognition that there is no simple linear sequence whereby one method of production automatically and non-problematically succeeds another, within either a company, an industry or the economy more generally. In sharp contrast, capitalist production is conceptualized as necessarily taking place in and through a variety of different organizational forms, linked together through complex geographies that are themselves an integral part of the organization of production.

In part, this debate was stimulated by the burgeoning crisis in the capitalist world economy from the second half of the 1970s. This led some commentators to a recognition that 'Fordism', both in its 'narrow' sense of a particular model of mass production and in its 'broad' sense of a specific macro-scale model of capitalist organization and growth, had reached its limits and was in crisis. For some, Fordism in its narrow sense of a production methodology grounded in Taylorist control of the labour process and strongly vertically integrated within individual companies, an approach widely regarded as the pivotal mass production method of the long post-war boom, had reached or was in danger of reaching its limits (see, for example, Piore and Sabel, 1984). Recognition of such an emerging crisis took on a wider significance, however, for Fordism had also been endowed with a broader meaning by the regulation school (see, for example, Aglietta, 1979; Lipietz, 1986). One of the strengths of regulationist approaches is that they emphasize the social and political character of production and the constitution of the economy. Such an emphasis directs attention towards the relationships between the main social actors involved in these processes, not least of which is the national state which has a crucial (though within regulationist approaches, somewhat undertheorized:

* First published 1994 in *Transactions of the Institute of British Geographers*, **19**: 331–345, Royal Geographical Society, London

see Mayer, 1992) role to play in establishing a viable regulatory framework within which production can occur and economy and society can be successfully reproduced (for fuller consideration of regulationist approaches, see Dunford, 1990).

In a regulationist context, then, Fordism is defined as a specific mode of social and economic development that revolves around and is characterized by a particular combination of mass production, mass consumption and state regulation to ensure that production and consumption remain broadly in balance, thereby containing the crisis tendencies that are inherent within a capitalist economy. There is, however, an evident risk that these crisis tendencies may be transposed from the economy into the operations of the state itself or into civil society, thereby endangering or destabilizing the mode of regulation (see Habermas, 1976). The severe economic and political crises of the 1970s and 1980s, therefore, led some to claim that the era of Fordism was over and that *a* post-Fordist successor was in the process of emerging. This was expressed in a variety of ways: some, for example, emphasized the central role of flexibly specialized small firms in crossing the second Industrial Divide (see, for example, Piore and Sabel, 1984); for others disorganized capitalism had succeeded organized capitalism (see Lash and Urry, 1987) while for still others, flexible accumulation had become the dominant tendency in the new post-Fordist era (see Harvey, 1989).

It is impossible fully to review the often polemical debates concerning the alleged transition from a Fordist to a post-Fordist model of capitalist development within the constraints of this chapter, but it is necessary to insist on one point that is extremely relevant to what follows here. There are grave dangers in attempting to reduce complex social relationships and processes of social change to simple dichotomies such as Fordist versus post-Fordist (see Sayer, 1989). This is especially so as there is abundant evidence that even in the Fordist – in the broad sense – era, much industrial production and economic life more generally of necessity remained organized around pre-Fordist and non-Fordist principles, with different organizational forms linked into and through complex geographies (see Hudson, 1989b; Pollert, 1988).

Within this chapter, I want to focus on a small segment of this wider debate. In particular, I wish to examine the extent to which, and the locations in which, 'Fordism' in the narrow sense remains a viable model of mass production and the ways in which it relates to other organizational forms of mass production that have emerged in recent years, forming a complex and changing pattern of geographies of production. One implication of recognizing this complexity in the organization of production is that it is untenable to argue that Fordism's successor, in either the broad or the narrow sense, is readily identifiable, or even that Fordism has *a* successor (a point that I will return to in the conclusion to the chapter).

Recognizing that different industries are characterized by different approaches to production organization, the focus of this chapter will be the car industry. This is because it has, historically, played a pivotal and revolutionary role in pioneering changes to the social organization, technology and geographies of production. Indeed, it came to be seen as emblematic of the emergence of 'just-in-case' strategies for mass production with the development of 'Fordism' (in the narrow sense).[1] More generally, its influence was perceived as being so pervasive that this term came to characterize a particular mode of capitalist development. It would be difficult, therefore, to overestimate its significance, both in relation to the trajectory of capitalist development and to our understanding of that process.

Moreover, it remains at the forefront of the contemporary search for viable alternative approaches *of* mass production rather than for alternatives *to* mass production. As Fordism, in both the broad and narrow senses, slumped into crisis, companies involved in car production were at the forefront of the search for new mass production strategies.[2] This involved seeking changes in – *inter alia* – the regulatory regimes within which production takes place, the social relations of production (both those between capital and labour and those between capitals), the technologies of production, and the geographies of production, though these have been emphasized to varying degrees within different corporate strategies. It has therefore involved seeking changes on a wide variety of related dimensions, including the following.

- A search for and experimentation with new concepts and models of high volume production such as lean production, dynamic flexibility and mass customization, seeking to combine the advantages of economies of scale with those of economies of scope.
- A search for new forms of capital–labour relations as a crucial part of seeking out viable new production strategies.
- A search for new capital–capital relations, involving three main elements: first, the construction of new relations of competitive cooperation between the major car assembly companies through the creation of strategic alliances and joint ventures; second, the restructuring of relationships between the assembly companies and their first-tier component suppliers, away from antagonistic relations based on price competition and towards cooperative relations based on trust and the forging of long-term links, stretching back to research, development and component design; third, a parallel restructuring of links between first-tier suppliers and at least some of their suppliers further down the production *filière*.
- A search for new regulatory frameworks within which these capital–labour and capital–capital relations could be sustainably refashioned; in short, a search to discover both micro-regulatory and macro-regulatory frameworks appropriate to the new models of production and accumulation.
- As an integral part of these processes of change, a search for new geographies of production, involving decentralization, reconcentration and a subtle blend of both within particular corporate strategies.

Clearly these diverse but interrelated changes are being worked through on a global stage, as part of an ongoing redefinition of relations between processes of globalization and localization (see Amin and Thrift, 1994; Humbert, 1993; Hudson and Schamp, 1995). As the crisis of the car industry deepened, however, the competitive struggle between the major companies involved in it both became more severe and came to focus increasingly on Europe, and in particular (as 1993 and the Single European Market approached) on the European Community, the single largest market area within the world market (see Dicken, 1992b; Sadler, 1991b). This intensifying struggle was given further impetus by the inward investment strategies of the major Japanese companies as they responded to pressures to restrict imports by rapidly building up productive capacity in the United Kingdom. This concentration of investment within the United Kingdom reflected a combination of relatively low wages and moderately high labour productivity in the context of a national labour market regulatory regime that was less restrictive than those in the other major countries of the Community. Moreover, the United Kingdom had a Conservative government that positively encouraged and welcomed such investment, not just as a way of revitalizing car production within its national territory but also as a means of radically restructuring industrial relations and working practices throughout manufacturing. Thus the United Kingdom became a base from which Japanese producers were to launch an assault on the European Community market from within its protective tariff wall. This role was perhaps most evocatively and powerfully symbolized by Jacques Calvet, Chief Executive of Peugeot, when he referred to the United Kingdom as 'Japan's fifth largest island' and as 'a Japanese aircraft carrier off the coast of Europe', an offshore base from which its companies could launch a fatal attack on mainland Europe's car market. Had he developed the analogy further, he might have gone on to add that northeast England, the location of Nissan's major assembly plant, was the initial flight deck (see Hudson, 1992a, for fuller details). The strident militarism of these images of battle and hostility between countries was far from accidental.

In the remainder of this chapter, therefore, the focus will be on changes in the car industry within Europe, though with reference to other parts of the world where this is relevant, in terms of the dimensions of change outlined above. The next three sections explore various aspects of relationships between companies within the overall car production *filière*. The first of these examines the dialectic of competition and cooperation in the strategies of the major car production and assembly companies within Europe, especially in relation to the threat posed by inward investment from Japan. The relationships between car assembly companies and their component suppliers are then analysed, leading to consideration of competition and cooperation between the major component companies themselves. There is no doubt that such capital–capital relationships are important in understanding the organization and geography of production. Much of the recent literature has emphasized inter-corporate relationships (see, for example, Sayer and Walker, 1992), though sometimes in a rather partial and one-sided way (see, for example, Scott, 1988b). A corollary of this emphasis (though by no means a necessary one) has been a tendency to play down the significance of the capital–labour relationships which are central to capitalist production, leading to an impoverished and partial understanding of the social process of production for profit. Section 6.5 therefore switches the focus to the relationships between capital and labour and the organization of the labour process and of work. Recognizing that production is built around antagonistic structural relationships, which could erupt into crisis and undermine the basis of production itself, the next section considers the regulation of production by the state, at various levels and spatial scales. Section 6.7 then moves on to examine the implications for local and regional development of recent changes in the organization and geography of the car industry within Europe. The concluding section then seeks to draw together some of the main conclusions and implications of the chapter.

6.2 Competition and cooperation between assembly companies

During the 1980s the market for new cars within Europe became increasingly competitive and the major car companies strove to maintain or enhance their position in this competitive struggle. One of the main indicators of the relative competitiveness of different companies and factories is their level of labour productivity. By the start of the 1990s there were very marked differences in labour productivity and quality levels between indigenous European and US-based automobile companies within Europe. There were also similar differences between plants within the same company, often producing identical models. For example, in 1990 the labour time required to produce a Fiesta in Ford's Cologne plant was 29.9 hours, whereas in its Dagenham works it took 52.2 hours to produce an identical car; the labour time needed to produce an Escort in Ford's Saarlouis plant was 33.9 hours, compared to 63.8 hours at Halewood. Such differences were not seen as a consequence of different ensembles of fixed capital in these plants but were, above all, perceived to be a result of the climates of industrial relations and the forms of organization and regulation of the labour process. As Mr Lindsey Halstead, Chairman of Ford Europe, put it: 'It isn't that the facilities are different, there is not a damn thing wrong with the Halewood facility. It is the way labour is organized and the way the labour functions . . .' (for further details, see Hudson, 1992a). As a result of such productivity differences, unit production costs were often lower in higher-waged countries within Europe.

Such productivity and cost differences were dwarfed, however, by the sharp dichotomy that was emerging between productivity and quality levels associated with lean production in the new UK plants of Honda, Nissan and Toyota and the factories of all other non-Japanese assembly companies within Europe. In the first half of 1991, for example, Nissan was achieving productivity levels that were more than 50 per cent higher than its main rivals within the United Kingdom. It was then producing at an annual rate of 43.4 cars per employee, compared to 26.0 at Vauxhall, 12.8 at Rover and 9.0 at Ford (excluding Jaguar).[3] Moreover, Toyota was forecasting a comparable productivity level of 58.8 cars per employee at the factory that it was then constructing at Burnaston, in Derbyshire (see also Womack *et al.*, 1990, for more general data on productivity differentials between Japanese and non-Japanese car producers). The rising productivity norms are intimately linked to fierce competition among the Japanese producers themselves for a growing share of the European Community market. One indication of this was Nissan's decision in 1993 to reduce employment at and output from its flagship factory at Washington, against a background of rising production at Honda and Toyota and stagnation in that market.

A combination of slow market growth and increasingly fierce competition from the Japanese led to a frantic search for strategies to close the gap between the Japanese and non-Japanese plants. Broadly speaking, this involved the non-Japanese producers responding in one or more of the following five ways.

(1) Firstly, seeking to adopt or adapt lean-production strategies, predicated above all on intensification of work on the production line (and so raising questions to do with labour market conditions, recruitment and industrial relations: see below).

(2) Secondly, seeking to discover other non-Fordist high volume production strategies such as those of dynamic flexibility or mass customization. In investigating such approaches, companies were searching for ways to combine elements of economies of scope, to allow some degree of flexibility of output in response to shifts in consumer choice and demand, with the pre-eminent concern to realize economies of scale. Since such strategies imply very heavy fixed capital investment in automated production technologies, they are more in evidence in areas of labour shortage such as Japan[4] than in areas of labour surplus, such as can be found widely and easily in Europe. Such locations offer possibilities for other routes to increased labour productivity, enhanced product quality and a degree of flexibility in product mix, through intensification of the labour process and the introduction of new forms of work organization within relatively low technology plants.[5]

(3) Thirdly, non-Japanese assemblers have responded by entering strategic alliances with Japanese assemblers as a way of acquiring access to their production and technical expertise in exchange for ready market access and distribution networks: for example, the alliance between Honda and Rover gives Honda access to European Community markets whilst it provides Rover with access to Japanese production technology and know-how (for further details, see Hudson, 1992a). The most significant of the remaining such alliances is that between Mitsubishi and Volvo in the Netherlands, a collaborative venture that thereby also involves Renault (see below).

(4) A fourth response is to forge strategic alliances between European *and/or* US-based companies to share R&D costs, spread the risks of new product development, share knowledge about best practice in production, etc.: for example, the collaboration between Ford and VW over the development of a new 'people carrier' in Portugal (see Ferrao and Vale, 1993; 1995) or that between Renault and Volvo (see Savary, 1993; Malmberg, 1991, 1992).

(5) Finally, companies have responded to the threat of competition by re-organizing their geographies of production. This has involved *either* closing plants in low productivity *and/or* high cost locations *and/or* seeking to

preserve existing production strategies by seeking out new areas with suitable labour market conditions (as in southern Europe and parts of eastern Europe: see Sadler *et al.*, 1993; Swain, 1992) *and/or* relocating production as a way of introducing new experimental approaches to production in new satellite branch plants, as with companies such as General Motors at Eisenach in eastern Germany: see Schamp, 1992).

The clear implication of the persistence of the first and second of these three spatial strategies is that it would be premature to pronounce the death of Fordist approaches, even within the car industry. Although there is clear evidence of experimentation with and adoption of non-Fordist high volume production approaches, the potential of Fordism is far from exhausted. Whereas in the past intra-national spatial differentiation offered a temporary solution to the problems of Fordist production, the search for a new spatial fix at global scale continues, both at the level of strategies to develop new models (such as the Mondeo, Ford's latest world car launched in 1993, designed by a single global team) and in seeking to source particular components on a global basis.

6.3 Cooperation between component suppliers and assembly companies

One of the characteristics of the social division of labour within the car industry is that many of the components that are assembled into finished products for the final consumer market are manufactured by component companies which are linked to one another and to the assembly companies within the overall structure of the production *filière*. The patterns of relationships between assembly and component companies are, however, variable through time and over space. Historically, relationships between component suppliers and assembly companies in Europe were regulated by price competition, with the component sector characterized by a plethora of fiercely competitive small firms and quality typically suffering as a consequence. In contrast, Japanese producers evolved a very different structure of links with their suppliers, which involved buying in a much higher proportion of components from a pyramidal supplier network structured in terms of a privileged first tier of suppliers, in turn fed by second-, third-, etc. tier suppliers within the structures of *keiretsu*. Relationships between the assemblers and first-tier suppliers (which were increasingly supplying sub-assemblies rather than individual components) were constructed as long-term relations of trust and cooperation rather than short-term ones of antagonistic competition, with reliability and quality of supplies the pre-eminent considerations. A corollary of this is that component prices increasingly became negotiated on the basis of 'open book' discussions between assembler and component producer rather than created via transactions in a competitive market. It is, however, important to stress that the continuing renewal of such relationships was and is by no means unconditional but dependent upon the attainment of agreed targets for increases in labour productivity and quality.

While this structure of relationships was very different from that in the USA and Europe, there were also important differences in this respect between the USA and Europe. In particular, in the USA a much higher proportion of components was produced in-house by the major assembly companies themselves or by their own subsidiaries (Rubinstein, 1992). The more limited extent of intra-company vertical integration in Europe led to the initiation of a process of restructuring of these assembler–component supplier linkages to something approximating the pattern found in Japan *prior* to the establishment of Japanese assembly plants in the UK (and this had important implications in terms of local and regional development, which are discussed below). This change reflected a growing awareness of the need to respond to the Japanese challenge on the world market. In turn, it led to the beginnings of a transition to a structure in which there were a few dominant first-tier suppliers, companies such as Bosch and Valeo which were of global significance, and others such as GKN, Lucas, and T and N which were more confined to Europe. Each of these companies had rationalized its product range so that it specialized in a particular market segment or limited set of products, often serving the entire European market from one or two factories. This restructuring of the component sector greatly facilitated the insertion of the Japanese assemblers into production within Europe (see Amin and Smith, 1990, 1991; Sadler, 1991a; Amin and Sadler, 1992).

The process of change was, however, greatly accelerated by the arrival of the Japanese assembly plants, not least because they were constrained by political pressures to source components locally as the issue of 'local content' (with 'local' meaning manufactured within the European Community) became of increasing salience in the arguments over the costs and benefits of inward investment from Japan. The Japanese assembly companies took great care in the process of screening potential suppliers and in choosing those which eventually were awarded contracts to supply them. These were governed by very rigorous criteria as regards timing of deliveries and quality, with the component companies expected to take on an increasing share of the costs of R&D as part of the price for long-term (but never unconditional) contracts. This is very clear from consideration of the ways in which Honda, Nissan and Toyota came to choose their suppliers in Europe. Toyota, for example, over a period of 18 months, systematically reduced a list of 2000 potential suppliers to 250 actual suppliers via a rigorous assessment of their capabilities in relation to four key criteria: management attitude and capability; manufacturing production facilities and levels of investment in new technologies; philosophy and systems of quality control; and research and development capabilities. Only when firms had been considered successful in relation to these criteria did price become a relevant factor, although the subsequent insistence by the Japanese that efficiency gains in component production be passed on to them as price reductions was often a contentious issue (see Hudson, 1992a; also National Economic Development Office, 1991, for further detail regarding Nissan's approach). This simultaneous pressure for both quality improvement and price reductions created tensions between the assembly companies and their component suppliers which suggested a strong thread of continuity with at least some of the earlier patterns of buyer–supplier relationships.

Such a combination of a search for lower prices along with greatly increased attention and care to ensure component quality was an example which non-Japanese producers had no choice but to try to emulate. Consequently, companies sought to reduce the number of their suppliers. Between 1988 and 1992, for example, Ford cut its list of suppliers in Europe by 15 per cent, to 900. It intended to reduce it further to 600 by 1995 (though this would still leave it with many more suppliers than the Japanese in Europe). There were varying emphases between companies in terms of speed of response to the new challenges and in the relative weight placed upon increasing quality and decreasing prices. General Motors, for example, placed considerable emphasis on cost reduction through insisting on radically altered contracts with its suppliers in the 1980s and early 1990s – once again the strands of continuity with the earlier emphasis on price competitiveness are clearly visible. Belatedly, in 1993, even Volkswagen, which until then had remained relatively unresponsive to the Japanese challenge, was acknowledging the need for radical change in its component sourcing policies if it was to respond effectively to the growing pressures of competition within Europe. This was symbolized most dramatically in its controversial acquisition of Ignacio Lopez from General Motors, with a brief radically to cut Volkswagen's costs in the same way as he had previously done for General Motors.

6.4 Competition and cooperation between component companies

An important implication of the changing relationships between assembly and component companies was to generate pressures for a further restructuring of the production *filière* in terms of relationships between the component companies themselves. This involved a restructuring of the components sector in Europe into a first tier of major suppliers, themselves major multinationals (and often diversified conglomerates which produced car components as one of many lines of activities), with privileged relations to the assemblers, and below these second-, third-, etc. tier suppliers. This in turn implied a redefinition of the relationships between and within the different tiers of component producers. Increasingly, the first-tier suppliers sought the kind of relationships with *their* suppliers that they had negotiated with, or which had been imposed upon them by, the assembly companies. This is not to suggest that price competition disappeared as a regulatory mechanism, especially (in Europe as in Japan) as one descended the supply chain, but the balance between competition and cooperation altered significantly in structuring relationships between component producers.

Increasingly, the first-tier suppliers concentrated on specific products and segments of the product market, in part motivated by a hunt for technological and/or monopolistic rents. For example, GKN increasingly concentrated on the manufacture of higher value-added products deriving from its innovative constant velocity joint technology, while T and N shifted its emphasis away from friction materials and towards high precision engine components (see Sadler, 1991b, for fuller details). As part of this process, they reorganized production so that particular components were produced in one plant (or at most a small number of plants) to supply the whole European market and allow the successful realization of crucial economies of scale. For example, Bosch established a new plant at Cardiff to supply the entire European market with particular types of alternator. This spatial dispersion was the very antithesis of a spatial reconcentration of production into component complexes within close proximity to major assembly plants.

Indeed, decentralization of production within Europe was simply one element in a global reorganization of the production strategies of the major multinational component companies. Especially for those components with very labour-intensive production processes, new plants have often been set up in cheap labour areas. Such plants are located either within Europe in the cheaper labour areas of southern and eastern Europe or in selected locations in parts of the Third World, where labour costs are only a fraction of those in Europe. As Marcus Bierich, chairman of Bosch, pointed out (in a comment that puts Bosch's investment at Cardiff in an interesting light), 'If we want to stay competitive with our products we must work in low-pay countries where wages are up to 90 per cent lower than in Germany' (see Hudson, 1992a). From such locations, the multinational component companies supply their customers scattered around the world.

Compared to the USA, the influx of Japanese component companies into Europe has been much more muted, reflecting institutional differences in the structure of the industry in the two continents and dashing many hopes of local and regional development centred around car production complexes (see below). By 1990 there were over 50 Japanese component companies operating in Europe, with 30 of these involved in joint ventures or licensing arrangements. There have been a few takeovers of European companies but very few new greenfield factories (see Hudson, 1992a). Nonetheless, there is evidence that they could in future provide fierce competition because of much greater labour productivity and efficiency in production, especially as the scale of Japanese car production within the European Community increases the incentive to locate there and as the Japanese component companies seek to globalize their operations in order to remain competitive with their non-Japanese rivals (see Boston Consulting Group and PRS Consultancy International, 1991). On the other hand, the established first-tier producers within Europe have already taken strategic decisions to restructure their own operations and there are strong political constraints which push the Japanese assemblers towards components produced in Europe. As a result, the threat of an influx of Japanese plants could be 'largely discounted' (Economist Intelligence Unit, 1992). Indeed, such political pressures have been influential in the decision by many inwardly investing Japanese component companies to take over, merge with, or otherwise establish collaborative relations with existing European component producers, even in those relatively rare cases in which they have set up greenfield factories.

6.5 Capital–labour relations

One of the main ways in which companies compete with one another is in their structuring of labour relations and their control over the labour process and the organization of work. Thus the character of capital–labour relations is an integral part of the struggle between capitals. A central feature of the Japanese model of lean production is a particular type of industrial relations regime, a particular culture of work, and a specific form of control of the labour process. This took on great ideological significance in so far as it was presented by its supporters as empowering production workers, offering an enriching form of work with a job for life. In sharp contrast, it was criticized as disempowering and disabling by its critics. For example, Nissan projects an image of work at its Washington factory as taking place in an empowering environment, built around the themes of flexibility, quality and teamwork. Others see it as a disempowering regime of subordination, characterized by control, exploitation and surveillance, bound together through team working (see Garrahan and Stewart, 1992, 109–111).

It is important to recall that the Japanese model of industrial relations is not some timeless cultural characteristic of Japanese society but emerged from bitter class struggle, in specific circumstances in the 1950s. This resulted in the defeat of vigorous left-wing unions, following a series of strikes, culminating in the decisive strike and lockout at Nissan in 1953 (Cusumano, 1985; Kenney and Florida, 1988). As a result of this, it subsequently proved relatively easy to socialize workers into the dominant managerial conception of production (aided by promises of employment for life in deeply segmented labour markets characterized by precarious employment for many). Establishing such a factory regime in locations such as the United Kingdom, with very different histories of trades unionism and industrial relations (not least in the car industry itself: see Beynon, 1984), was always going to be more problematic. Hence the Japanese companies exercised great care in selecting locations with appropriate labour market conditions and histories (in essence, high current unemployment and an industrial past that excluded any previous experience of car plants).

All three companies have, from the outset, displayed equal care in selecting and composing their workforces. Consider, for instance, the selection process used by Nissan in the recruitment of its first 240 production workers at Washington in 1986. It began with a 'paper screening', by means of completion of a comprehensive questionnaire relating to past work experience, reasons for wanting to work at Nissan, and attitudes to overtime and shift-work. Those who successfully passed this stage were then given a series of tests: aptitude tests, with a minimum level of competence required in Mathematics, Physics and English; informal discussion with groups of applicants and supervisors;

practical tests, to check dexterity and hand-to-eye coordination. 'After this', according to the Personnel Manager, 'we interview them – that's why we call this a comprehensive selection procedure' (for further details, see Hudson, 1992a). It has not subsequently become any less comprehensive, at Nissan, Honda or Toyota. At all three Japanese plants in the United Kingdom, appropriate attitudes of commitment to the company were deemed much more important than technical skills as criteria for recruitment.

Nissan and Toyota were equally fastidious in identifying the union that they chose to represent their workers in single-union deals. In contrast, Honda has refused to recognize any union at its Swindon plant, which remains non-unionized. While the Amalgamated Engineering and Electrical Union continues to press for the recognition there that it has at the other two companies, as yet (1993) Honda has shown no indication that it intends to accede to this request. There is no doubt that the new Japanese plants either have achieved or will achieve very high levels of productivity and quality, posing a major challenge to all other car assemblers within Europe (as well as competing fiercely between themselves for shares of the crucial European Community market).[6]

This has led to a frantic attempt by many – both other assemblers and increasingly component suppliers – to imitate the Japanese approach to the regulation of capital–labour relations. In fact, this is only one of a range of identifiable non-Fordist capital–labour relationships (see Leborgne and Lipietz, 1988). In broad terms echoing Friedman's (1977) distinction between direct control and responsible autonomy as strategies for regulation of the labour process, these involve a spectrum of combinations of positions on two axes, which together structure capital–labour relations: one running from coercion and direct control of labour to negotiated involvement and cooperation, the other extending from inflexible to flexible labour market conditions. Feasible options to Fordism include Neo-Taylorism (seen as characteristic of the United Kingdom, the USA and Spain), the Californian model, Toyotism (seen as characteristic of Japan), the West German model, and Kalmarism (seen as characteristic of Sweden). However, at least in a European context, there is evidence to suggest that working in Japanese car plants bears more than a passing resemblance to the supposedly discredited and discarded practices of Fordism (see Garrahan and Stewart, 1992, 55–56 and 125), and Nissan itself has been at pains to stress that it is not simply slavishly copying working practices from Japan but is seeking to develop its own distinctive style, based on blending elements of 'best practice' from wherever it finds them (see Hudson, 1992a).

It may therefore be wise to avoid any simplistic stereotyping of different models of industrial relations, working practices and strategies for control of the labour process, identifying particular places with particular approaches. Nevertheless, whether in its Japanese or European variants, the Japanese version of cooperative coercion is proving much more viable than the rigidities of Fordist line management. There is also evidence that it is proving more potent than European cooperative alternatives such as Kalmarism and the German collaborative approach. Only time will tell if Volkswagen has, in the words of the Russian proverb, been 'too late in running from the bear' in formulating its strategic response to mounting competition in the European market. It is important to note, however, that the recent closure of the innovatory plants at Kalmar and Uddevalla in Sweden may have had more to do with product mix, overall overcapacity and managerial conservatism in responding to this by concentrating production in the larger Gothenberg plant than it did with lower productivity in the plants that were closed.

6.6 Searching for new regulatory regimes

There are unavoidable tensions – indeed class structural contradictions – in the relationships between the main social actors involved in the process of capitalist production. Yet for production to take place successfully, involving the creation and realization of surplus value as profit, these contradictions must routinely be kept in check within tolerable limits. Whilst it is important not to reduce the processes of regulation through which this containment occurs simply to the activities of national states, there is no doubt that the latter have played a key role in constructing regulatory regimes. These must be created and reproduced, at a variety of scales ranging from the individual workplace to the international market. Such regimes are constructed within the relationships of civil society, as well as within and between those of the state and companies themselves.

Clearly the Japanese automobile companies within the United Kingdom have been able, more or less successfully, to create the micro-scale regulatory regimes for the control of production and the labour process that they consider as necessary within their own factories. This has been greatly facilitated by national government politics and policies in the 1980s which eagerly encouraged inward investment from Japan. This was seen both as a way of reviving car production within the United Kingdom and as a means of restructuring industrial relations and working practices throughout manufacturing against a background of high unemployment. The success of the Japanese translated into pressures on both other assembly companies and component suppliers, both within and outside the United Kingdom, either to imitate these regimes or to construct different but equally competitive ones.

This search for supportive regulatory regimes extends beyond the factory to the surrounding localities and regions. Locations with established car plants are anxious to hang on to them as competitive pressures mount; places with no history of car production wish to compete in the place market for a share of the investment in new plants as a way of tackling localized unemployment and reviving flagging local economies. Nissan, for example, was lobbied by no fewer than 53 different locations in the United Kingdom alone before it finally chose to establish its operations at Washington (see, for example, Hudson, 1992a, on the specific case of Nissan; more generally, see Hudson, 1992b). In both types of case, local, regional and often national governmental authorities (as well as trades unions) are eager to give the car companies as much assistance as they can and create a supportive regulatory environment in their areas around them. The construction of such supportive regimes is not just a question of appropriate attitudes and practices within the state but extends into the fabric of civil society within these areas, often with a wide spectrum of community groups aligning themselves with pro-growth coalitions, in support of inward investment by Japanese car companies.

Beyond this, however, for those located within the boundaries of the European Community, there is a further important actor involved in setting the conditions of the regulatory framework (see Sadler, 1991a). The Community sets limits and conditions on the types and levels of support that can be given at national and sub-national levels. Beyond this, and more significantly, it also sets limits on the extent to which Japanese producers can penetrate the Community market via imports produced outside its borders, as well as influencing the volume of Japanese production within the Common External Tariff wall. Its trade policies are a crucial part of the supra-national state regulatory regime, while the definition of 'local content' and the level at which this is to be set became focal points in the arguments concerning the costs and benefits of inward investment by the Japanese car companies. The United Kingdom government's willingness to defend the interests of the Japanese companies in the debates over the definition of acceptable levels of 'local content' with other governments and companies in Europe was important in influencing the destination of Japanese investment. The attraction of the United Kingdom was reinforced by its government's insistence on maintaining a national regulatory regime that diverged in important respects on issues such as labour market conditions from European Community norms. As Jacques Calvet's comments (cited above) make clear, there is a continuing fierce political struggle, involving various representatives of sectorally and spatially (at local, national and Community levels) defined interests, that seeks to determine the volume and location of such inward investment by influencing the shape and content of this regulatory regime.

6.7 Local and regional economic development implications: just-in-time and in one place?

During the 1980s there was growing concern as to the existence of feasible local and regional development strategies in and for Europe's older industrial regions. In many of these, an economy constructed around the carboniferous capitalism of coal mines, steel works and associated engineering industries had been in decline for two or three decades. A dominant policy response in those years had been central and local government attempts to attract mobile branch plant investment. In the wider context of a changing international division of labour in the 1970s and 1980s, however, these were often revealed to be vulnerable global outposts within international but intra-corporate geographies of production. They both lacked linkages to the surrounding regional economies and were increasingly liable to close completely. At the same time, the cultural and labour market histories of such regions did not make them fertile ground for the newly emergent policies that emphasized entrepreneurship and small-firm growth in a putative enterprise culture (see Hudson, 1992b). This raised serious questions as to the future of such areas and led to a frantic search for alternative bases of economic regeneration there.

Against this background, the possibility that more embedded branch plants, and in particular new production complexes based around the automobile industry, might provide a basis for renewed growth began to attract serious attention. The new geographies of production and the spatial relationships between new Japanese assembly and component plants in the USA seemed to point to a reconcentration of production within regional economies, raising initially great expectations as to the local and regional developmental potential of the restructuring of the car industry within Europe. This was one reason why the competition between places to try to attract investment in the assembly plants was initially so fierce. For while such changes would pose a major threat to regions that were already locations of car production, it raised hopes in many areas that had not been involved with this industry. Since their labour forces lacked experience of working conditions and norms within automobile plants, they were much more attractive as locations to companies seeking to introduce radically new recruitment and working practices and industrial relations policies. There is, however, evidence neither of the relocation of existing component plants in response to pressures for just-in-time production in existing non-Japanese plants, nor of the majority of such new Japanese component plants as have been attracted to Europe clustering in close proximity to the new Japanese (or indeed the old non-Japanese) assembly plants there. In practice, then, there has been a sharp difference in their locational behaviour in the two continents.

This largely reflects significant differences in the structure of the car production *filière* in the USA compared to Europe. Relative to Europe, in the USA a much higher proportion of components was produced in-house by the major assembly companies themselves or produced for them by their own subsidiaries. This made it much more difficult for Japanese companies to plug into the existing component supply network in the USA. Consequently, of necessity, the new Japanese assembly plants were followed by component producers from Japan. Not surprisingly, the latter often then located in relatively close proximity to the new assembly plants, leading some commentators to elide 'just-in-time' production with the spatial reagglomeration of production and the reconstruction of regional economies around integrated car production complexes (see, for example, Mair, 1991). In this way, necessary and contingent features of the emergent geography of production became confused.

Certainly, a few plants, generally making bulky, low value-added components for which there was regular demand, did locate near to the new Japanese assembly plants, but this could be easily explained by reference to old arguments about transport costs rather than new ones about the relevance of 'just-in-time and in one place'. One consequence of this is that hopes of regenerating depressed regional economies around new integrated car production complexes are fading fast. Although the possibilities of inward investment in the form of branch plants without any particular articulation to the regional economy remains, this represents the latest form of an old and heavily criticized regional development model rather than the emergence of a new and better one (for fuller discussion, see Hudson, 1992a, 1992b).

There *is* some evidence, nevertheless, of what initially *appears* as a sort of pseudo-'just-in-time', which involves component companies constructing new warehouses near assembly plants in order to guarantee continuity of supply (see, for example, Schamp, 1991, 1992). This is especially the case where there are difficulties over transportation (as in eastern Germany, for example: see Swain, 1992). This is not so much an implementation of the principles of 'just-in-time', however, as a new variant of 'just-in-case', with again extremely limited local and regional economic developmental potential.

6.8 Conclusions

In view of the considerable uncertainty surrounding developmental tendencies in the car industry, as well as the ongoing theoretical debates as to the most appropriate interpretation of them, it may well be thought premature to try to draw any definitive conclusions. Nevertheless, there are some important points to be made precisely because of this state of flux and the competing interpretations that have been put upon it. Moreover, there is sometimes a dangerous slippage between description of what is happening in some companies and plants and prescription as to what ought to – indeed will have to – happen in all companies and all plants, not least in some of the more messianic accounts of lean production (for example, Womack *et al.*, 1990). It is by no means obvious that a single model of production must emerge, or is emerging, as universally dominant. It is equally by no means certain that Fordist production strategies are merely part of the historical geography of the car industry or of capitalist production in general. In 1993, Ford launched its latest 'world car' – the Mondeo – on to the market, the most recent of its attempts to preserve, albeit in modified form, its traditional approach to production in the face of rapidly changing global industry.

There undoubtedly have been experiments with a variety of high volume production strategies, involving different forms of capital–capital and capital–labour relations. Some of these have clearly been more successful than others, but it would be both premature and dangerous to assume that a single Japanese lean production strategy had become dominant, let alone hegemonic. Not least, it is clear that there are significant differences in the strategies of different Japanese, as well as other, producers and in those of a single producer in different places at the same time. Moreover, the Japanese producers themselves are engaged in an ongoing search to retain competitiveness which is involving radically new relations between the major assembly companies and between them and their component suppliers within Japan itself. In 1993 Nissan announced that it was taking the unprecedented step of buying parts from a supplier within Japan affiliated with its arch-rival Toyota. This not only points to new relationships between the two major Japanese assembly companies but is indicative of potentially drastic restructuring of the established pattern of linkages between assemblers and component suppliers, with hierarchical networks of component suppliers each uniquely attached to one assembly company. This will involve reducing the number of component suppliers, including some in the privileged first tier, thereby rupturing long-established inter-corporate linkages. Furthermore, there is evidence that the notion of lifetime employment, so central to the structuring of capital–labour relations, is beginning to be eroded, or at least amended, with workers aged over 50 years increasingly encouraged to retire early (and so forego part of their anticipated peak earnings late in their career). The Japanese model itself could well be on the verge of radical and potentially destabilizing change, further suggesting the need for caution in advancing claims as to its apparent hegemonic status.

It is also clear that there are no simple deterministic relationships between particular forms of organization of production and their geographies. Different organizational principles can take different spatial forms in different institutional, cultural and political contexts. This point was sharply reinforced by the different spatial patterns associated with the introduction of 'just-in-time' approaches as a direct and indirect consequence of inward investment by Japanese car assemblers in the USA and in the European Community. Within the USA, there was certainly some evidence of a spatial clustering of Japanese component companies around the new assembly plants. This raised expectations as to the local and regional development possibilities of such inward investment in Europe. In practice, these expectations have not been realized. Within the different institutional and political context of the European Community, there has been much more limited inward investment by Japanese component companies, while the new Japanese assemblers have generally inserted themselves into the pre-existing production *filière* and sourced from already established component companies. There has certainly been a profound restructuring of the components sector in response to the quality control and product development demands of 'just-in-time' and lean production. This has not, however, involved any significant reagglomeration and the reconstruction of regional economies around 'just-in-time' production complexes. In fact, even *within* the component sector, there are divergent spatial trends. Some new component plants, making bulky and/or relatively low value items such as seats and trims, for which there is a regular but varying demand, have indeed been established in close proximity to some of the new assembly plants. In this sense, one can validly talk of some limited establishment of synchronous production but it is also clear that there are strict limits as to the generalizability of organizing production in this way. More generally, however, the dominant spatial tendency has been one of decentralization, especially for those components with labour-intensive production processes, with the establishment of plants in peripheral locations within (especially southern and now eastern) Europe and increasingly within Newly Industrializing Countries as part of a more extensive process of global shift (Dicken, 1992a).

More significantly, the point to stress is that of the co-existence of different production strategies, conjoined in various ways, in a complex geography of the car production *filière*. An integral part of these processes of change is an ongoing and continuous search for new geographies of production, involving decentralization, reconcentration and a subtle blend of both within particular corporate strategies. This is true even within Europe; when Europe is placed into its broader global context, the point is simply reinforced even more strongly.

There is a further implication of this pattern of change that relates to regulation approaches and claims as to the emergence of a post-Fordist successor to Fordism. If it is the case that even within the paradigmatic car industry the picture is one of the persistence of Fordist 'just-in-case' approaches to mass production, alongside and in relation to other approaches to high volume production based on the different organizational principles of 'just-in-time' and lean production or those of flexible automation and mass customization, then it is premature to herald the death of Fordism (in the narrow sense) even within this industry. This conclusion is simply reinforced by recognition that, even at the height of the Fordist era, only a minority of activities were actually organized around Fordist principles. This also throws considerable doubt on the validity of those claims that a fundamental qualitative transition has taken place from Fordism to some identifiable post-Fordist (in the broader sense of a regime of accumulation) successor. There is no doubt that there have been profound changes in the political economy of capitalism over recent years, but these are more appropriately interpreted as evidence of crisis, and perhaps terminal decline in the old regime of accumulation (see, for example, Gordon, 1988), rather than as evidence of the emergence of a successor to it. This is particularly so because while there is evidence of experimentation with new production concepts – even new models of accumulation – there is no evidence of the emergence of a stable mode of regulation that would be appropriate to the establishment of some new regime of accumulation (see Tickell and Peck, 1992). If Fordism does have a successor, it is far too early to specify what it (and its geography) will look like.

Notes

1 Sayer (1986; see Sayer and Walker, 1992) argues that it is misleading to elide 'just-in-case' with 'Fordism' in the narrow sense, pointing out, correctly, that there were variations from a very early stage in the mass production strategies of the dominant US-based automobile companies. In particular, General Motors followed a rather different approach from that of Ford.

2 This is not to deny that one element in the responses to the crisis of Fordist mass production in other industries was a tendency to switch production into smaller firms and/or plants (though it is important not to confuse small firms with small plants). One result of this was growing attention in intellectual and political debates to competitive small firms as part of an emergent 'enterprise culture' and to flexibly specialized, small batch production via small firms, linked into industrial districts by relationships of cooperation and trust which yield significant economies of scope (see Amin, 1989). A revival of interest in flexible production by small

firms was linked to the growing significance of product market flexibility in response to rapidly changing demands for certain types of consumer goods. Equally, it was clear that the generalizability of such a model of production organization was strictly limited by the material and social demands of commodity production – though this is not to say that a concern to introduce a greater degree of flexibility in consumer choice was absent from mass production industries such as cars (see, for example, Coriat, 1991; Veltz, 1991). Any elision between production within small firms and greater consumer choice ought, however, to be resisted.

3 Such data overstate the magnitude of the productivity gap between Ford and Rover (which are involved in full manufacture and assembly operations in the United Kingdom) and the other companies (which only assemble cars there). Nevertheless, even allowing for this, the productivity gap between the Japanese and non-Japanese is a formidable one (for fuller details, see Hudson, 1992a; see Williams *et al.*, 1992, for a comprehensive review of the problems of equating productivity measures).

4 Nevertheless, it is important not to overgeneralize about the Japanese experience. For example, Nissan, facing far more difficult labour market conditions in the localities around its production facilities in Japan than did Toyota, became involved with strategies for automation of parts of the production process from a much earlier stage (see Cusumano, 1985).

5 Whilst it may be the case that 'small is beautiful' in some areas of commodity production, car production is not one of them and economies of size and scale remain pre-eminent. But it is also clear that a dynamic flexibility (Coriat, 1991) or customized mass production (Veltz, 1991) strategy requires very considerable fixed capital investment in automated production technologies, which implies not only high levels of product demand but also shortages of suitable labour – for in situations of abundant labour there may be other and cheaper ways of introducing the required levels of labour productivity and flexibility on the line via intensification of work and multi-tasking (a point that Coriat and Veltz tend to ignore).

6 Since the new Honda and Toyota plants are only beginning production on a limited scale in 1993, it is now (in 1994) difficult to give definitive data on actual – as opposed to projected – productivity levels in these new factories. It seems certain, however, that they will aim for at least parity with Nissan's factory in northeast England.

Chapter 7

The end of mass production and of the mass collective worker? Experimenting with production and employment*

7.1 Introduction

There has been growing recognition of the changing character of work in the advanced capitalist countries (Beynon, 1995), and of the decline of industrial employment there as it expands in newly industrializing countries (Dicken, 1992a). These changes in the volume, character and spatial distribution of employment are one facet of profound changes in the character of contemporary capitalism, manifestations of 'the crisis of Fordism' and the search for post-Fordist successors to it. This is the case both at the micro-level of a particular form of organizing production within the workplace and at the macro-level of a model of societal, political and economic development. The meaning and status of the changes continue to be hotly debated. Broadly speaking, there are competing alternative interpretations, with different implications for labour and the spatial organization of the economy. For some, a shift of epochal significance has occurred. Fordism's successor is already known, although the terminology used to describe it varies. One implication of this is that the mass collective worker, characteristic of the big urban factories of Fordism, is a thing of the past, disempowered in and by the new economy of decentralized small flexible firms (Murray, 1983). Others contest this, arguing that the death of Fordism may have been prematurely announced, based on a partial reading of the evidence. Seen from this perspective, large-scale production in big factories is far from being a thing of the past. The search for Fordism's putative successor(s) remains an ongoing process and may or may not represent some epochal shift. Claims to the contrary are premature and potentially damaging, theoretically and politically. The shape of the future remains to be determined but its developmental trajectory will be more complex than a neat, clean break from a Fordist past to a post-Fordist future. This is so for two different sorts of reasons. First, even at the high point of Fordism as a macro-scale development model, only a minority of labour processes were organized on Fordist lines (Pollert, 1988). 'Fordism' was constituted as an uneven mosaic of places, production and labour processes. Secondly, as Fordism reached its limits in the core countries of capitalism, companies began to search

---

* First published 1997 in R Lee and J Wills (eds) *Geographies of Economies*, Arnold, London

elsewhere for locations in which Fordist production would remain economically viable. This intersected with the desires of governments in peripheral countries (initially those which became characterized as the first generation of NICs and more latterly others such as China) to promote industrialization as the route to development. The crisis of Fordist production in the core of advanced capitalism thus became the proximate cause of changes in its location. It continues, but nowadays in new locations in the peripheries of the global economy. Consequently, the mass collective worker has not necessarily simply become a subject of history.

As these are ongoing changes in the character of a *capitalist* economy, understanding them requires a political-economy approach that acknowledges this and also recognizes that there are limits to capital (Harvey, 1982). The ways in which, and the perspectives from which, we represent the world are certainly important. These are, however, competing accounts of a material world, socially produced according to 'rules of the game' in which people make historical geographies of employment and production. The economy remains subject to the structural class relations and boundaries that define capitalism. Consequently class relations, especially those between capital and labour in the labour market, in the wage relation and at the point of production, remain of central significance. Class relations in the workplace can be reconfigured but they cannot be erased. It is for this reason, and because such relations are treated in a one-sided and idealized fashion which seeks to deny their class character in much of the literature on new post-Fordist forms of work, that they are the focus of attention here. Companies must be able successfully to purchase labour power and then organize and deploy it to ensure the production of surplus value and the possibility of profitable production. This emphasis on the central importance of the class relation between capital and labour, of value analysis in emphasizing the asymmetrical but mutually defining power relations between capital and labour, may smack to some of essentialism (Barnes, 1996). But insistence upon the pivotal significance of capital–labour relations in a capitalist economy does not deny the significance of other processes and relationships which are deeply involved in the reproduction of this class relationship; nor does it reduce them to mere reflections of it (*cf.* Massey, 1995). Capitalist societies are not *simply* divided along class lines. There are also divisions along dimensions such as gender and ethnicity, which are intertwined with those of class in a variety of ways. Attempts to understand the ongoing restructuring of production and work that fail to acknowledge the class basis of these changes are, thus, at best partial, theoretically impoverished and politically dangerous.

Capitalist societies and the conditions that allow production to take place within them are not, however, automatically reproduced. The constitution and representation of the social relations of capitalism remain contested and a focus of struggle. Accounts which deny this, or gloss over its implications, do not do so innocently. It is, therefore, important to acknowledge the processes and institutions through which the class basis of production and wage labour is reproduced and regulated and through which contradictory interests are kept within limits that permit profitable production. While regulation involves more than the activities of states, these are integral to the institutions and processes that, in different ways, help make production possible. Social relationships within households and the institutions of civil societies are, however, also central to the reproduction of labour power and are bound up with the (re)formation of class relations as the spheres of production and social reproduction are reciprocally, but not necessarily equally, determining. State policies help define the shifting boundaries and articulations between commodified and non-commodified social relationships. One implication of this shifting articulation, as well as of the different forms that class relations can take, is that the broader social relations of capitalism can be cast in varied moulds within these structural limits.

Given the continued salience of class relations in production, the focus in this chapter is therefore upon changing forms of work and competing interpretations of the implications for labour of capital's attempts to experiment with and find new methods *of* high volume production (HVP) in manufacturing. Other aspects of contemporary economic restructuring and employment change such as the search for alternatives *to* mass production,[1] the impacts of human resource management practices on employment in parts of the services sector, or indeed the Taylorization of large swathes of service sector employment (Beynon, 1995), will be considered only in passing. This is not because they are unimportant but rather because of the continuing significance of large-scale industrial production and the leading-edge role that this has occupied in debates and discourses about the redefinition of work. The focus is micro- rather than macro-scale, although some consideration will be given to those macro-scale conditions that make micro-scale changes both necessary and possible. The remainder of the chapter is organized as follows. First, the broad lineaments of HVP will be outlined. Secondly, the implication of HVP for the character of work and geographies of employment will be discussed. Thirdly, the reconfiguration of the collective worker and the implications of new forms of work and industrial relations practices for labour will be explored. Finally some conclusions will be drawn.

7.2   Experimenting with new models of high volume production

The term 'high volume production' denotes approaches to production that seek to combine the benefits of economies of scope and greater flexibility in responding to consumer demand which are characteristic of small batch production with those of economies of scale characteristic of mass production. They include approaches designated as 'just-in-time' (Sayer, 1986), 'lean' production (Womack *et al.*, 1990), dynamic flexibility (Coriat, 1991), flexible automation (Veltz, 1991) and mass customization (Pine, 1993), which, with its ambition of batch sizes of one – uniquely customized commodities assembled from mass produced components – epitomizes the goal of HVP. While typically presented as distinctive, all these different HVP approaches emphasize the emergence of new structures of relations between companies and of new forms of work, work organization and industrial relations practices – in part because they typically refer to the same set of exemplar companies and industries. Since one of their shared defining features is a blurring of the distinction between mass and craft methods of production, HVP approaches also show considerable continuities with, as well as differences from, existing approaches such as 'just-in-case' Fordist production and small batch production. In practice they combine elements of 'old' methods of production with new production concepts and practices, both within and between firms. These innovations in production methods have often been associated with changes in corporate anatomy. Growing company size as a result of acquisitions and mergers and an increasing prevalence of strategic alliances among these bigger companies in search of economies of scale (in R&D, product development and so on) and scope has often been a prelude to experiments with HVP, with the how and where of production (Rainnie, 1993).

These new approaches have typically involved a reorganization of production in new factories which achieve enhanced labour productivity via some combination of new fixed capital investment in more automated production technologies, more intensive ways of organizing work and the labour process (so that employment declines), and more efficient ways of processing material inputs. These *are*, however, forms of *high volume* production, incorporating variations around the basic mass production theme. Consequently, there are strict limits, in terms of the material and social requirements of commodity production, that define the range of industries and products in which HVP approaches *can* be applied. Companies such as Dell and Motorola may provide commodities such as pagers, PCs and workstations on a mass-customized basis (Pine, 1993). Other major companies, such as Ford, are seeking to move more towards mass customization as an automobile production strategy. It is difficult to see how soap powder or screws could be profitably produced in this way, however, and they will continue to be mass produced. Conversely other commodities, major items of fixed capital equipment such as power stations, will continue to be produced on a 'one-off' basis while exclusive fashion goods will continue to be produced in small quantities in small production units. Consequently, there is not *necessarily* a complete divide between small batch production, mass production and various forms of HVP.

Equally, while these new HVP approaches share characteristics in common, there are also significant differences between them. Companies thus face a choice in deciding the what, how and where of production. This has definite implications for geographies of employment, the amounts and types of waged work on offer, and forms of organization of the labour process. Furthermore, the economic viability of such approaches assumes that certain labour and product market conditions will be fulfilled. The choice of a particular HVP approach involves seeking a balance between responding to more differentiated product markets through economies of scope and the need to retain scale economies. For example, especially in those approaches, such as flexible automation, that encompass very highly automated production processes, there are (usually tacit) assumptions as to very high levels of demand and of capacity utilization to allow fixed capital investment to be depreciated sufficiently quickly. In addition, the introduction of such automated approaches has sometimes been in response to tight labour market conditions (such as prevailed in the Japanese automobile industry in the 1980s) in circumstances in which relocating in order to maintain the competitiveness of existing production technologies was infeasible. As a consequence, the introduction of robots to replace human labour in automated production processes has been very uneven, sectorally (concentrated in automobiles and also electronics) and spatially (with the greatest concentrations per worker in Japan, Singapore, Sweden, Italy and Germany (United Nations Economic Commission for Europe and International Robotics Federation, 1994). Conversely, there is evidence of companies switching to rather different automated HVP technologies in the 1990s, with robots as an adjunct to rather than a replacement for human labour at the point of production. These offer greater scope for combining flexibility with profitability as aggregate levels of demand have declined and/or as labour market conditions have allowed the requirements of such HVP approaches for very specific types of worker to be satisfied (see Hudson and Schamp, 1995).

7.3 Work, workers and HVP and its geographies

The literature on HVP has paid considerable attention to new forms of relations between companies (Crewe and Davenport, 1992; Hudson, 1994a). Restructuring the links between companies impacts upon relations between workers, upon those between workers and managers, and upon forms and conditions of work. These changing forms of work and of capital–labour relations reveal both breaks from and continuities with existing forms of work and the ways in which it is organized.

From one point of view HVP approaches are represented as abolishing the mass collective worker and class conflict at the point of production, incorporating more satisfying, individualized forms of work. From another perspective, they are seen as transforming the forms of, rather than abolishing, class conflict at the point of production and reshaping rather than removing the collective worker. These competing claims about the character of employment in HVP approaches can be illuminated by comparing it with the character of employment under Fordism. Fordism is characterized by a deep technical division of labour, a sharp distinction between occupations requiring mental and manual labour, informed by Taylorist views of scientific management within companies. It is marked by a strong vertical hierarchy of control; individual production line workers are confined to single, specialized – often deskilled – tasks within a finely disaggregated detail division of labour (Braverman, 1974). Management relies upon increasing line speed and intensifying the labour process in *this* way as the route to greater labour productivity. This leads to extremely monotonous jobs, which fail to capture the knowledge that workers develop through doing the job. This can result in problems – for capital – of alienation, lack of motivation, and resistance from workers, dis-rupting production via strikes and other forms of industrial action, resulting in decreased productivity and profitability (see, for example, Beynon, 1984). This is especially so in large factories, locationally concentrated in major urban areas, in labour market conditions of 'full employment' – in short, in circum-stances in which the capacities of the mass collective worker spontaneously to organize and resist the demands of capital are favourable. Such conditions are not, however, automatically realized and capital deploys a variety of strategies, in terms of the how and where of production, precisely to prevent them from being brought into existence. The invention of new methods of HVP and associated ways of working constitutes one element in this repertoire of tactics.

Alternative forms of HVP require workers to perform a wider range of tasks than on the Taylorist line; in that sense the technical division of labour is not so deeply inscribed. This is partly because of the incorporation of principles of 'just-in-time' production into HVP approaches but there are other reasons associated with claims about producing better and more satisfying forms of work. There is considerable debate as to the implications of this change for those actually carrying out the tasks of production. There are those who see workers as multi-skilled and much more creatively involved in the process of production. Companies (and their intellectual supporters, such as proponents of HRM) emphasize that these new forms of work provide, from the perspect-ive of workers, better jobs in an empowering environment, built around the themes of flexibility, quality and teamwork (see Wickens, 1986). There are strong claims, which allude to an alleged golden age of craft production, about the re-emergence of multi-skilled polyvalent workers, employed in jobs which recombine the mental and the manual which Taylorism had torn asunder. Florida (1995, 168), for example, writes approvingly of the emergence of 'high-performance manufacturing', in sectors such as automobiles and con-sumer electronics in the USA. He associates this with a shift to a more knowledge-intensive economy in which the keys to success are harnessing the ideas of all workers from the R&D laboratory to the factory floor to create the high-quality, state-of-the art products that the world's consumers want to buy.[2] Under this new form of organization, the factory itself is said to be becoming more like a laboratory, with knowledge workers, advanced high-technology equipment and cleanroom conditions free of dirt and grime. This does indeed powerfully suggest that the Taylorist distinctions between manual and mental labour are being swept away to the advantage of all workers.

Others dispute this. They stress not the qualitative differences from the old Taylorist model but rather the continuities between the old and the new. In part, this is because of the inconsistencies between the claimed advantages of the new forms of work and the realities of the labour market and the labour process (Blyton and Turnbull, 1992; Peck, 1994; Pollard, 1995). Critics con-tend that what is involved is not multi-skilling but multi-tasking, part of a search for new ways of intensifying the labour process. The production line keeps running all the time but not *necessarily* at its maximum possible overall speed. Indeed, the aim is not to maximize line speed but to minimize the number of workers needed for a given speed, as dictated by the implementation of just-in-time and 'lean' production principles. The emphasis in lean production on 'halving the human effort' has drastically reduced the number of jobs available. In *this* way is the labour process intensified and, in terms of its direct inputs of labour power, production becomes 'lean'.

There are therefore serious doubts as to whether the jobs on offer in these factories are actually any better than those on Taylorized mass production lines. In contrast, critics claim that the new jobs display very clear parallels with the discredited employment forms of Fordism and indeed can be thought of as an intensified extension of them (see, for example, Garrahan and Stewart, 1992, 125). From the point of view of labour, therefore, HVP approaches involve greater intensification of work and greater stress than before (see Okamura and Kawahito, 1990). There are fewer jobs on offer and greater competition for them. The notion of 'lifetime employment' has ceased to be meaningful for the majority of employees, not least because intensification of work leads to a physical incapacity to meet productivity norms with increasing age and the average age of workforces has fallen sharply. As there are fewer jobs and so greater competition for them, especially in locations blighted by high unemployment, firms can be extremely selective about who they recruit, and about the terms and conditions on which they offer employment. Against a background of high unemployment, recruitment is typically based at least as much on 'appropriate attitudes' and expressed commitment to the company, and age, personal and family circumstances and physical fitness, as it is on requisite technical knowledge and skills. As well as changing the terms and conditions on which full-time jobs are offered, there is a tendency to replace full-time with part-time and casualized jobs, as a further stage in customizing labour supply in relation to fluctuating demands in product markets. If the new forms of work and HRM approaches within HVP are based upon fostering loyalty to the company, then this is a loyalty born of fear of unemployment in circumstances where individual jobs are no longer guaranteed. This suggests a commitment that may be shallowly based and very much dependent upon the context of labour market conditions.

Enhanced selectivity in recruitment in turn allows radical and regressive changes in the organization of the labour process. Work is organized within a disempowering regime of subordination, characterized by control, exploitation and surveillance, bound together through team working (Garrahan and Stewart, 1992, 109–111). There is, however, a tension between an emphasis on team work and increased individual competition for jobs, promotion and pay. There are strong pressures for workers to regard themselves as competitive individuals, dealing with the company individually rather than collectively over wages and terms and conditions of employment. At the same time, through the rhetoric of teamwork, workers discipline themselves and their colleagues (identified as their 'customers' further up, or 'suppliers' further down, the line). This in itself both increases stress and changes the nature of the mode of regulation of the labour process. No longer is it 'us' versus 'them'; 'they' are now part of 'us'. Considerable ambiguities and uncertainties follow from this change of identities, not least in terms of forms of organization and representation of workers' interests. In so far as there is collective representation of workers' interests, there are pressures for this to be via works councils rather than trades unions – although the latter may still be permitted. The net result is that jobs within HVP approaches are actually worse jobs than those on offer within Taylorized mass production approaches.

Thus, to summarize the argument so far, these new forms of HVP must be understood both as a response to a profound crisis of mass production and as enabled by it. As part of an attempt to restore profitability, they have to respect and so reproduce the defining structural limits and parameters of a capitalist economy. This implies structuring work and the labour process so that profitability criteria are met. In the process they redefine the social relations of production. Whatever the rhetoric about empowerment of workers, production has to be organized in such a way that it is sufficiently profitable. In order to introduce new ways of working, there have been corresponding changes involving greater selectivity in recruitment, putting more emphasis on appropriate attitudes, making sure companies hire the 'right' people on the labour market, workers who will accept and adapt to new ways of working. At this point links to broader macro-economic and geographical contexts become crucial. High unemployment, spatially concentrated, both in deindustrialized old industrial areas in which the power of the mass collective worker can be broken, and in non-industrialized areas in which it never existed, provides the context in which the texts about employment practices can be rewritten.

A crucial precondition for companies being able to introduce drastic changes in the organization of the labour process therefore has been either to find locations untainted by an earlier history of capitalist industrialization or to break the power of existing forms of workers' organizations in areas with such a history. The historical geography of Fordism clearly demonstrates these twin tendencies, as mass production of consumer goods was established in new locations, and companies subsequently sought a series of spatial fixes to find viable new locations for such production once it became problematic there. Similar points can be made about other forms of production organization. The defeat of the militant Japanese trades unions in the 1950s allowed the introduction of innovatory new forms of HVP by automobile producers there, for example. The selective location of experimental new HVP plants in parts of the European periphery in the 1980s and 1990s, often directly or indirectly associated with inward investment from southeast Asia, is another example (Hudson, 1994a; Hudson and Schamp, 1995). The creation of locationally concentrated mass unemployment has often been a crucial precondition for attracting such investment by eroding the power of established trades unions. This again emphasizes the significance of relationships between state policies, macro-economic conditions and feasible factory regimes of production (Burawoy, 1985). In particular, it highlights the importance of the replacement of Keynesian welfarism by neo-liberal workfarism in establishing labour market conditions which made the experiments with new forms of HVP possible over much of the advanced capitalist world.

The geographies of employment associated with the introduction of HVP show that spatial variation in labour market conditions is clearly a critical consideration in decisions as to whether, and where, to introduce new forms of HVP. Localized high unemployment has been equally important in permitting established firms with more 'traditional' forms of work and labour organization to restructure and introduce new ways of working, informed by the precepts of HRM, often in response to the competitive challenge of HVP. The literature on geographies of HVP tends to emphasize the importance of new forms of relations between companies rather than forms of capital–labour relations, however. It initially seemed that just-in-time approaches and close inter-firm linkages would lead to a regional reconcentration of production, offering growth opportunities and a considerable increase of new sorts of industrial jobs in at least some peripheral regions. This could be seen as heralding an opportunity for the resurgence of the mass collective worker. While acknowledging the emergence of new forms of relations between companies, such an interpretation underestimates the significance of locationally concentrated high unemployment in attracting such clusters of factories in the first place. Their production strategies are predicated on the ability selectively to recruit particular types of employee and prevent the reconstruction of a militant mass collective worker.

There is no 'obvious' geography to other new HVP methods such as lean production, flexible automation or mass customization in the way that there initially seemed to be with just-in-time. The relationship between location and labour market conditions is indeterminate and variable, though it may well involve a sophisticated use of spatial differentiation. As the strategies of companies such as Motorola and Dell reveal, there is considerable flexibility over choice of production location, enabled by advances in information technologies in communication and production (Pine, 1993). As a concern with economies of scale remains central in many industries, lean production almost certainly means fewer factories. There will be intensified place market competition for them, though *where* these factories might be located remains an open question. From one point of view, one might expect heavily automated assembly plants, requiring only relatively small inputs of living labour, to be drawn to 'core' regions, near the main markets of contemporary capitalism. The greater weight attached to the role of R&D and the increasing emphasis placed upon post-sales services as part of the product could add to the attractions of core locations for mass customized production. On the other hand, the ready availability of labour in peripheral locations, facilitating continuous shift working and new working practices, plus the availability of substantial state financial subsidies, make *these* attractive locations (Conti and Enrietti, 1995; Schamp, 1995). Furthermore, labour-intensive production, organized on classic Taylorist principles, will continue to find the cheap labour peripheries an attractive destination.

7.4   So is this the end of the mass collective worker?

Fordism was created as a way of breaking the power of the craft worker to control the production process in the 'traditional' heartlands of industrial capitalism. This it did, at least in certain industries, times and places, exploiting existing spatial variations to create its own distinctive geographies of employment. At the same time, however, in concentrating production in massive factories in major conurbations, Fordism created the conditions for the emergence of a militant mass collective worker that in due course threatened its viability as a production strategy in what had become *its* 'traditional' heartlands in the advanced capitalist world. Consequently, from the moment that the mass collective worker emerges as a threat to it, capital has striven to achieve '... *the destruction of the spontaneous organisation* of the mass worker on a collective basis' (Murray, 1983, 93 – emphasis added). Capital remains, of necessity, committed to this objective of destroying challenges to its power to organize the production process on its terms. Capital is not, however, necessarily opposed to the existence of the collective worker *per se*. Many companies continue to require large numbers of workers organized into individual workplaces and there are limits to the extent to which they can deal with them and organize their work individually, as opposed to on some sort of collective basis. There are tendencies to underestimate the extent to which production must remain a collective enterprise, both in the HRM literature and in some of the critiques of it. One implication of the continuing collective character of production is that spatial variations in labour market conditions remain of great significance to companies in their search to produce a compliant collective worker.

One expression of capital's continuing requirements for large amounts of malleable labour is the growing shift of routine mass production to branch plants in parts of the peripheries of the First World and into some parts of the Third World. There has been a succession of such fixes, first intra-nationally within the advanced capitalist world, then internationally, as companies have sought to contain crises of profitability via decanting routine production into peripheral branch plants (Hudson, 1988b). Capital's latest search for a spatial fix to preserve the viability of mass production and recreate the mass collective worker involves investment in locations such as the Special Economic Zones of China. With labour costs in industries such as clothing and textiles in China a mere 2 per cent of those in Germany, and those in other parts of the Far East less than 4 per cent of German levels (according to Coats Viyella: see Rich, 1996), there is considerable scope to preserve mass production strategies on the basis of the availability of very large masses of very cheap labour. While such shifts in production challenged the powers of trades unions both in their 'traditional' manufacturing heartlands and in those peripheral locations in which earlier branch plants were located, they also open up potential opportunities, at least for a while, in new locations. Such potential has often remained latent, however, as a consequence of broader macro-economic and political conditions in the new destination areas.

At the same time, within the First World, different strategies are being followed. The emergence of new forms of HVP in the First World involves the reconfiguration and reformation, in slimmed-down form, rather than the death, of the collective worker. While the current 'round' of HVP factories require less labour than their Fordist predecessors of 30 or more years ago, many still require a lot of labour. While companies can be much more selective as to who they recruit and retain as compared to the decades of 'full employment', workers still need to be organized in ways that conform to capital's needs for profitable production. The rhetoric may emphasize individually more satisfying work, but producing profitably necessarily remains a collective and social process that has to be regulated to conform to the disciplines of commodity production. Companies therefore still need workers with *particular* collective forms of labour organization and activity that align the interests of workers with rather than against those of the company. Therefore companies searching for viable models of HVP of necessity have sought out new ways of disorganizing and then reorganizing labour, typically with the assistance of national states and their restructuring of regulatory regimes. Consequently, HVP approaches both require and permit the shattering of old forms of trades unionism and the institutions of labour and recasting them in new no-union or one-union moulds. There is a much greater emphasis on plant-level or local rather than national-level bargaining, in part linked to competition between places for jobs as well as between unions for members. Trades unions have often been willing to trade off sole bargaining rights for various 'sweetheart' deals as one way of combating their own falling memberships. Furthermore, 'selling places' often involves emphasizing the passivity, flexibility and malleability of their workers as compelling attractions to mobile capital. These new deals are thus grounded in a very different conception of capital–labour relations from the previously dominant one as the already asymmetrical power relations between capital and labour have swung sharply in favour of the former but in a context in which production nonetheless remains a collective and social process.

This situation is not, therefore, without dangers for capital and potential opportunities for labour. On the one hand, the new models of production are undeniably predicated on there being *no* return to 'full employment'. Such labour market conditions would at a stroke destroy capital's capacity to be so selective in recruiting labour and deciding which individuals will become incorporated as part of the collective worker. Former trades union strategies based on organizing workers in large industrial plants on a national basis in a 'full employment' economy have undeniably been weakened. Companies are nevertheless acutely vulnerable to interruptions to production in 'lean' production strategies built around vertical disintegration, just-in-time principles, sole supplier deals and minimal stock levels. These characteristics can lead to production being compromised by rapid labour turnover and to companies being particularly susceptible to the effects of industrial action. This was sharply demonstrated, for example, by strikes at Ford's UK factories in 1988, which soon disrupted production in Belgium and Germany, and at Renault's Cleon plant in the autumn of 1991. The Cleon plant supplied a very high proportion of Renault's engines and gearboxes and Renault had recently adopted just-in-time production principles. Potentially, these new forms of HVP could strengthen the position of labour more generally, subject to two important caveats. First, that aggregate labour market conditions become more favourable, possibly as a direct result of alternatives to neo-liberal regulatory regimes being put in place. Other regulatory regimes, embedded in a different conception of economic development policy, which sought to lower aggregate unemployment, reduce spatial variability in labour market conditions, improve working conditions, and create better and more humane working environments, would clearly generate a more favourable terrain for organized labour. Such regulatory regimes would need to be constructed internationally – the Social Chapter of the European Union's Maastricht Treaty and European legislation on minimum wage levels and working conditions represent small but nonetheless invaluable steps in this direction. At a minimum, they serve as a reminder that there are alternatives to the free market rhetoric of neo-liberalism. Secondly, trades unions need to devise new forms of international organization that recognize the increasingly sophisticated ways in which companies use spatial differentiation in labour market conditions. This could, for example, involve trades unions in strategies of collaboration rather than competition for transnational investment, possibly seeking to build new forms of combines between plants that are dispersed globally but linked within the production networks of companies, or seeking to build alliances across sectors and industries and national boundaries. Workers in places are increasingly bound in dense webs of interdependencies woven through corporate global strategies that offer the potential for cooperation. Trades unions could thus seek to use the economic linkages between companies as the basis for their own forms of industrial and political organization and strategy. There is evidence that they are beginning to do so. Organizations such as the International Conference of Free Trades Unions are increasingly sensitive to the challenges posed by globalization, for example, and are seeking to develop new ways of cooperating across national boundaries in response to them. Such strategies will, however, also need to be sensitive to the politics of place, the varying conditions of local labour markets with different industrial histories and traditions of labour organization, and the extent to which production is necessarily embedded in particular places. These variations in the geographies of labour markets and production will influence the possibilities for the sort of strategies that trades unions will be able to pursue in particular places within a framework which recognizes the collective but spatially segmented character of production.

7.5 Conclusions and reflections

The crisis of mass production and the introduction of new methods of HVP have led to a sharp reduction in industrial employment, profound changes in the character of work and continuing changes in the geography of industrial employment. HVP has not, however, replaced mass production, any more than mass production replaced craft production, but has come to co-exist alongside it. It is therefore unlikely that HVP will provide a general long-term solution to capital's problems of profitability, any more than did the once-revolutionary innovations of Taylorism and Fordism. This is because, as with mass production methods, there are strict limits on the conditions under which HVP strategies can be profitably deployed. Some companies will undoubtedly successfully restructure. Some will do so because of the continuing potential to discover spatial fixes to preserve mass production. Others will do so in part precisely because of the persistence of conditions within a global labour market that allows them to introduce new models of HVP. The enhanced organizational and technological capacities of major companies to exploit differences within an increasingly differentiated global production space have led to new geographies of employment in a global labour market that is simultaneously more deeply integrated and more sharply segmented. There are increasingly marked differences in labour market conditions between and within countries, regions and cities – *a fortiori*, within the so-called world cities (Sassen, 1991). There are enormous qualitative and quantitative locational differences in the terms and conditions on which companies can purchase and deploy labour power as a part of their routine repertoire of tactics in search of profits. In many areas companies find no difficulty in recruiting workforces that will accept the terms and conditions which the new models of production necessitate – or that allow old production technologies to be preserved (depending on whether companies are pursuing strategies of 'strong' or 'weak' competition: Storper and Walker, 1989).

Labour's position within this emergent new order would seem, from this point of view, to be necessarily worse in terms of number and types of available jobs. The restructuring of employment is disabling and dissecting, and then reshaping and reforming, the collective worker as part of a more general process of restructuring of capitalism. This process is unfolding on terms that are increasingly disadvantageous to labour and which outflank its 'traditional' forms of sectoral and territorial organization. The capacity of the mass worker to organize spontaneously and collectively to challenge the imperatives of capital has been seriously eroded. Corporate restructuring and inter-place competition will continue to pose acute problems for community organizations and trades unions seeking to come to terms with a shifting labour market terrain, not least because they have been active, albeit unwilling, subjects in producing these changes. The mass collective worker may remain in the peripheries of the global economy but in a weakened position. In the heartlands of industrial production, however, for the foreseeable future at least, the collective worker will exist in emaciated form, no longer possessing the capacity for spontaneous autonomous action that challenges the imperatives of capital.

On the other hand, the introduction of HVP approaches enhances the possible vulnerability of the production process to disruption by industrial action, and in important ways this potentially enhances the power of organized labour, subject to some important broader conditions being met. For realizing this potential requires that trades unions evolve new forms of cooperation between companies and across national borders so that they can take positive advantage of the new forms of relations between companies and between companies and their workers. In short, a renewed recognition of the collective character of production must become much more prominent in framing trades union strategies. Secondly, and most importantly, it requires the construction of new regulatory regimes centred around greater social and economic equality, perhaps linked to notions of greater environmental sustainability (Weaver and Hudson, 1995). This presupposes, at a minimum, that the dominance of neo-liberal discourse in economic policy formation is broken. There are clear signs that it is creaking under the weight of its own internal contradictions and that the necessary is becoming impossible, the impossible necessary, in reproducing this particular mode of regulation. There is no doubt that the neoliberal regulatory regimes and their representation of the character, constraints and opportunities of contemporary capitalism, which have dominated over the last decade or so, have had enormous influence. Not least, they suggested that a pattern of changes which was inimical to the interests of workers, their families and communities was desirable, necessary and unavoidable. It was none of these things. The trajectory of changes that they have set in motion is not, however, sustainable, economically, socially or environmentally (Hudson, 1995d). It is highly improbable that a new stable macro-scale model of growth and mode of regulation will be discovered that does not break sharply with its legacy. There are undoubtedly alternative approaches to production and regulation that place more priority on the interests of workers, their families and communities and which offer greater possibilities for the future. The future for labour is not therefore necessarily one of a continuation of the debilitating trends of the last two decades but remains to be fought and struggled over as part of a process of redefining the political and regulatory terrain of contemporary capitalism.

Notes

1 There is now a very extensive literature dealing with these alternatives *to* high volume production, notably the proliferation of small firms producing in small quantities and, more specifically, flexibly specialized production within networks of small firms organized into industrial districts (see, for example, Asheim, 1996).

2 Thus such a view attributes considerable weight to the notion of consumer sovereignty. It ignores the extent to which companies shape consumer tastes and structure markets via their advertising and marketing strategies.

Part 4

Territorial politics and policies

Introduction

The focus of the four chapters in this part of the book is territorial politics, state territorial policies, and the links between the two. These issues have been alluded to in previous chapters but in this part of the book become the focus of concern. Capitalist development is an inherently uneven process, with growth taking place along a crisis-prone trajectory. The conditions necessary for successful capitalist production and smooth accumulation are not automatically guaranteed, as Marx demonstrated forcefully in Volume II of *Capital* in setting out the conditions that would need to be met for long-term expanded accumulation of capital. One aspect of this unevenness is territorial, with capitalist geographies constituted as a complex mosaic of places of growth, of stagnation and of decline, at varying spatial scales. Managing territorially uneven development so that it remains within 'acceptable' limits has become a key task for capitalist states. It is, however, also a very problematic task as crisis tendencies from the economy are transposed into the state and emerge in various forms as crises of the state and its mode of operation – as rationality or legitimation crises (Habermas, 1976) or as fiscal crises (O'Connor, 1973), for example. Part of the reason as to why managing territorial inequality is problematic for capitalist states is that different social classes and groups seek to shape the geographies of capitalism in ways that will favour their interests (within the structural limits that define capitalist economies as such). Constituting geographies of capitalism thus becomes an arena of struggle. National states form part of the arena, as social groups struggle to shape the content and form of territorial development policies. At the same time, however, such states are also key actors in the struggle as they seek to deal with issues of uneven development as part of their own policy agendas. State involvement can become particularly complex in situations in which some state policies (for example industrial policies or policies towards nationalized or publicly owned industries) become the proximate cause of territorial development problems which other state policies then seek to rectify. As a result of recognizing the often central role of state policies in shaping geographies of industries and territorial development trajectories, economic geographers have become increasingly concerned with issues of politics, state policies, regulation and governance. Equally, the ways in which they have sought to conceptualize state involvement have evolved and changed.

Initially, in the 1970s, economic geographers were heavily influenced by Marxian approaches to the state. The State Derivation debate among German state theorists had a marked impact (for example, see Clark and Dear, 1984; Holloway and Piciotto, 1978). The focus of this debate was upon explaining why the state took the form that it did, constituted in a political sphere that was formally separate from the economy, and as a result had a variable degree of relative autonomy in its scope for action in policy formation and implementation. As this debate developed, it became clear that to seek to derive a theory of *the* capitalist state was a flawed project and the focus of concern became the national state and the relationships between state policies, uneven development and geographies of economies. State policies are structurally constrained by the parameters defined by capitalist social relationships but within these parameters are open to influence from a range of social forces and interests. Although there is a clear tendency for the interests of capital to dominate, they do not necessarily do so on all occasions and in any case competition between companies and various capitalist concerns renders any simple notion of the state non-problematically 'meeting the interests of capital' problematic.

Chapter 8 reflects such a Marxian approach. It seeks to analyse the development of Washington New Town as one element in a regional modernization programme for northeast England in the 1960s (which is also discussed in Chapters 2 and 3). The development of Washington is interpreted as a way of enticing capital to the northeast as a result of the state meeting part of the costs of production. The state did so through a series of infrastructure investments to help provide general conditions of production, grants and loans to specific companies, and, especially, by the assembly of a pool of labour there via public sector housing policies. At the same time, development of the town helped meet social democratic objectives of improving environmental and housing conditions and increasing living standards. The increased incorporation of married women into the wage labour force as a way of raising household incomes became a critical link between attracting new investment, economic restructuring and enhancing material living standards. While it seemed for a while that state policies could simultaneously meet this variety of objectives, it soon became clear that these forms of state involvement themselves became crisis-prone and problematic.

In part the need for new employment in Washington reflected the rundown of the nationalized coal industry in the town, so that the state was involved in complicated and paradoxical ways in the restructuring of the local economy and labour market. This is also a prominent theme in Chapter 9 which explores the links between region, class and the politics of steel closures in the European Community. The steel industry in western Europe was enmeshed in a deep crisis in the second half of the 1970s, as effective demand for steel fell sharply, corporate profits slumped and companies sought to cut capacity and jobs to try and restore profitability. Responses to crisis often encompassed complex patterns of state involvement. The industry in the UK had been (re)nationalized in 1967, and a major expansion programme was announced in 1973. The subsequent abandonment of this programme, and its replacement with savage cuts in capacity and employment, met with little resistance from the steel industry trades unions and communities. In part, this reflected divisions within and between the steel unions, deeply sedimented as a consequence of the historical geography of the industry, and in part it reflected more immediate pressures flowing from the political strategy of the Thatcher government. Such resistance as there was remained firmly within peaceful and democratic channels. Although there was a national steel strike in 1980, this was for higher wages rather than preserving communities and jobs and was easily defeated. The case of Consett is used to exemplify the way in which trades unions and steel communities argued for generous redundancy provisions for those who lost their jobs and programmes of economic regeneration and employment creation rather than vigorously resisting and contesting closures. In Lorraine and the Nord in France, in contrast, a powerful campaign was mounted by left-wing trades unions against closures in what was initially a privately owned steel industry. Protests spilled over beyond 'normal' democratic channels to violence and direct action on the streets. Once the Conservative government of Barre was replaced by the socialist government of Mitterand, however, the emphasis moved from resisting closures to arguing for generous redundancy and regeneration packages. These, however, put a grievous strain on the finances of the French state, threatening to provoke a fiscal crisis. On the other hand, abandoning the steel regions ran the risk of provoking a legitimation crisis. The problems of managing rundown and regeneration proved much more problematic for the French state than they did for the UK state.

Chapter 10 reflects the influence of the institutional turn in economic geography. Much of the literature that explores the significance of regional institutions emphasizes their positive role in underpinning economically successful regions (see, for example, Dunford and Hudson, 1996). In contrast, the emphasis in this chapter is upon the ways in which territorial development strategies can become imprisoned in particular trajectories by a combination of cognitive and institutional 'lock-in'. Institutional forms and ways of viewing the world that at one time were extremely relevant to a particular economic structure become a problem in seeking to adapt and move to a new structure and developmental path as the old economic rationale is destroyed. As a result, in such economically problematic cities and regions, there are typically attempts to create new sorts of institutional structures that will facilitate the transition to more propitious trajectories. This often involves seeking to learn from the institutional structures of successful regions, sometimes slavishly imitating them. This is a project fraught with dangers and difficulties as the causal powers inherent in institutional structures cannot be transplanted in these ways with any firm expectation that they will have their intended effects (and only these). Moreover, there is a clear relationship between the sorts of urban and regional development strategies that are put in place, the types of institutions that are created to try to facilitate the successful emergence of new economic activities, and the type of national state regulatory regime that is in place. Rather than the sub-national being an alternative to the national, the latter exercises a strong constraint upon the room for manoeuvre of the former.

As national states cut back on the scope of territorial development policies, more and more responsibility for these issues was pushed back onto local and regional political authorities and communities. From the point of view of national states, this was a way of reducing public expenditure and easing fiscal pressures. From the point of view of local politicians and communities, this was often a move to be welcomed. It was seen as devolving power and the scope to tailor policies specifically to local and regional needs to local and regional levels (though often responsibility rather than resources were devolved, and this was especially true in the UK). Chapter 11 focuses upon the former steel town of Consett and examines the sorts of regeneration strategies put in place in and for the town in the wake of the steelworks closure (discussed in Chapter 9). It contrasts the formal top-down state regeneration strategy (discussed briefly in Chapter 10) with an alternative, locally based strategy that sought genuinely to build a 'bottom-up' approach to regeneration, informed by locally expressed needs. This centred around music as a substantive focus and a co-operative as an alternative organizational form to the orthodoxies of entrepreneurship and wage labour in predominantly small firms and/or plants as the dominant social relationships of production. It discusses the relative merits of these two approaches. It also illustrates the way in which the radical alternative of co-operative development around making music, in seeking to avoid reliance upon the market and the orthodoxies of capitalist social relations of production, became entangled in webs of dependency upon state funding. This is a point of some significance in the context of subsequent debates as to the possibilities of a 'Third Way' in constructing economic development strategies (see also Hudson and Weaver, 1997).

Chapter 8

Accumulation, spatial policies, and the production of regional labour reserves: a study of Washington New Town*

8.1 Introduction

It has been argued by Damette (1980) that regional labour reserves have come to occupy a central role in the process of capital accumulation by providing a source of surplus profits as other sources of these began to disappear. Furthermore, it has been argued (see also Hudson, 1979a) that the state has increasingly become involved in producing such reserves through its spatial policies, often with unintentional results in relation to their stated aims, and these ideas are explored further in this chapter. Section 8.2 considers more fully the general theoretical arguments concerning the importance of labour reserves in the accumulation process and the role of state policies in creating such reserves. Following on from this, the remainder of this chapter focuses upon one particular set of interventions by the state which served to create such a labour reserve through the development of Washington New Town, WNT (see Figure 3.1).[1] First, a brief outline is given of the context in which the decision to develop WNT was taken, the stated objectives of its development, and the ways in which the creation of labour reserves in WNT was justified and legitimated as socially progressive. The mechanisms by which they were created are then more fully considered, as well as the extent to which the specified policy goals have been attained.

8.2 Capital accumulation, regional labour reserves, and state policies: some key concepts

Accumulation, the driving force of the capitalist mode of production, is inherently crisis-prone. It depends upon an expansion of not only the mass but also the rate of profit, and this mode of production is characterized both by a tendency for the rate of profit to fall and by counter-tendencies set in motion temporarily to overcome or to avoid the actualization of crisis by reducing the value of the component parts of capital. Both the onset of crisis and this restructuring at the level of capital in general are brought about and made possible by competition between individual capitals. These deploy various

---

* First published 1982 in *Environment and Planning A*, **14**: 665–680, Pion Ltd, London

strategies to try to guarantee or boost their profits. One way to achieve this objective and so avoid a crisis of profitability is to seek surplus (that is, above average for a branch of production) profits. A variety of routes is available to capitals striving to attain this goal (see Mandel, 1975, 77–78).

One way is to discover and exploit fresh labour reserves. Indeed, Damette (1980, 77–78) has gone so far as to suggest that such reserves currently constitute the main source of surplus profits for monopoly capital. Although Damette perhaps overstates and oversimplifies a strong case, it is undoubtedly true that for many capitals, both big and small, the discovery of new labour reserves is of decisive importance in avoiding or overcoming profitability crises (see, for example, Borzaga and Goglio, 1979; Friis, 1980; Hudson, 1981a).

Now in a more general sense the capitalist state has also become centrally involved in crisis management and this led it to pursue a variety of policies in an attempt to guarantee private profitability. These are intended either generally to reduce the value of elements of one or more of the component parts of capital or specifically to take over part or all of their cost for some capitals but without necessarily lowering their value – rather the state takes on responsibility for an increasing share of their costs of production and reproduction. One important way in which the state intervenes to attain such objectives is through spatial policies, and through their implementation it has become intimately involved in the production and reproduction of labour reserves (see, for example, Damette, 1980; Breathnach, 1982; Toft Jensen, 1982). However, this can also raise problems for the state in terms of justifying and legitimating such policies.

8.3   Legitimating the development of labour reserves in Washington New Town: intra-regional uneven development as the route to social progress

Washington was designated in 1964 under the 1946 New Towns legislation with the expressed intention of reducing male unemployment in the area around it (Llewelyn-Davies et al., 1966). This formed part of a wider programme of modernization in the region, set in motion in the wake of the proposals of the 'Hailsham' White Paper (Board of Trade, 1963) and intended to transform social conditions in the region. This objective was to be achieved through a temporarily greatly expanded programme of public expenditure which would help establish some of the preconditions needed to bring about a fresh wave of private investment, which would in turn diversify and modernize the region's industrial structure.

In fact, this particular set of proposals must be located historically within an ideology which originated in the response to crisis put forward by the region's bourgeoisie in the 1930s in an attempt to guarantee their own interests (see Carney and Hudson, 1978) and which subsequently came to be generally regarded as the best – indeed only – way to abolish the northeast's status as a 'problem region'. For, crucially, such policies came to be accepted as a legitimate solution by the Labour Party and working-class organizations within the region: modernization policies became the basis of a consensus politics, a class alliance as to how best to solve the region's problems. This was possible because of the political hegemony enjoyed by the Labour Party (and, more generally, Labourism) within the northeast, which enabled it to justify, within a tradition of reformism, the implementation of modernization policies as being socially progressive (as indeed they often were, leading to improvements in housing and living conditions for parts of the working class, for example). At the same time, such policies objectively met the requirements of some capitals.

However, it has consistently been recognized that the promotion of fresh industrial investment could not proceed evenly throughout the region. Consequently, a persistent theme in the ideology developed to legitimate implementation of such state policies is that modernization must of necessity proceed unevenly within the region. Crude areal classifications have been and continue to be proposed in which various areas and places (growth zones, growth points, etc.) that are already experiencing growth are classified as locations on which modernization policies and public expenditure will be focused. In turn, the concentration of public investment there (for example, in housing and industrial estates) would create attractive environments for private investment. In particular, it would set in motion desired inter-regional, but more particularly intra-regional, migration patterns and stem out-migration of young people, developments seen as necessary in order to assemble pools of labour which would prove attractive to capital. Conversely, other places not so designated are relatively starved of public investment and so of private capital. However, it is asserted that such locations and, consequently, the whole region will receive the favourable impress of change as beneficial effects diffuse outwards from the selected nodes of modernization in the region (for a critique of such policies, see Carney and Hudson, 1974).

WNT is such a node of modernization, representing in some ways the epitome of this policy of locationally concentrating public sector investment. The development of the town as part of a deliberate policy of intra-regional uneven development was justified in terms of this being necessary in order to reduce social and economic differentials between the northeast and other regions and to promote social progress within the region, but had as a deliberate consequence the assembly of considerable labour reserves there. It is to a more detailed consideration of the mechanisms involved in this that we now turn.

8.4 Reducing the cost of variable capital and the reconstruction of a labour reserve in and around Washington New Town, 1964–1978

Washington was designated with the intention of helping to reintegrate northeast England into the mainstream of the accumulation process by creating there a focus of fresh economic growth – the reasons why this was seen as politically and socially necessary were briefly discussed in the preceding section. However, to attain this objective necessitated both considerable public investment in infrastructure to create a new built environment and the reconstruction of a reserve of labour in and around the town: both the expansion of labour supply and a restructuring of the labour market (sectorally and spatially). In general, we will argue that rather than assembling specific workforces for individual firms, a general pool of labour has been assembled in WNT and in the surrounding area through a combination of the effects of corporate strategies and state policies, in particular those of Washington Development Corporation (WDC) and the National Coal Board (NCB), as well as those of central government, although more specific provisions have been made in the case of a limited number of key workers.

In one sense, then, the pattern of expenditure by the WDC can be seen as part of a deliberate policy to assemble workforces – both by providing housing within the town and (in collaboration with other parts of the state apparatus) by improving road links between the town and its surrounding subregion (see also section 8.5). In subsection 8.4.2, issues of housing provision, migration, and journey-to-work patterns are considered in relation to this issue, but first the impacts of NCB policies are briefly examined.

8.4.1   National Coal Board policies

Some 2400 mining jobs were lost in WNT between 1964 and 1973, associated with the closure of Harraton Colliery in May 1965, Washington F Colliery in June 1968, Glebe Colliery in 1972, and Usworth Colliery at the end of 1973. Moreover, these figures underestimate total job loss caused by the run-down of workforces prior to closure. The Master Plan for WNT had assumed that mining would continue in the designated area until 1981 (Llewelyn-Davies *et al.*, 1966). As well as job losses because of colliery closures in WNT and the surrounding region, mining employment also declined because of increasing mechanization in those pits that did remain open (for a fuller account, see Regional Policy Research Unit, 1979, part 4).

8.4.2   Washington Development Corporation policies

8.4.2.1   Housing provision in WNT

All housing completions up to 1967 in WNT resulted from decisions taken prior to designation, but with the passage of time the WDC came to take an increasingly important role in housing provision: by 31 March 1978, over £41 million – over 50 per cent of net WDC capital investment (Table 8.1, overleaf) – had been allocated to house building and 6774 houses built by the WDC for rent. Some 4850 houses had been added to the housing stock of WNT by 31 December 1972, the end of the first phase of the town's growth – but this nevertheless fell below the target set in the Master Plan of 6200 houses. Despite this shortfall, the resultant growth of population and of resident

Table 8.1 Washington Development Corporation: net capital expenditure in each category, 1966–78, as a percentage of the total from the date of designation to 31 March of the year indicated

Year	Percentage net capital expenditure[a] by expenditure heads[b]											Total (£)
	1	2	3[c]	4[c]	5	6	7	8	9	10	11	
1966	48.84	0.43	10.08	nil	nil	9.27	0.44	3.88	nil	nil	27.05	550 612
1967	53.25	5.11	6.36	1.65	0.44	10.07	5.26	1.96	0.17	0.99	14.51	2 319 085
1968	36.03	15.36	4.32	0.95	0.08	16.64	10.32	1.42	1.22	4.67	8.31	4 455 298
1969	22.14	16.39	2.45	0.47	1.15	26.11	11.38	0.91	2.86	11.25	4.83	7 925 866
1970	15.11	17.85	1.51	0.29	1.11	34.52	10.66	0.74	3.96	10.81	3.45	12 646 882
1971	10.65	17.36	1.04	0.18	1.64	43.34	10.08	0.51	5.07	8.10	2.01	20 201 251
1972	9.22	18.81	0.84	0.15	2.22	45.87	9.29	0.43	4.66	6.93	1.58	24 432 135
1973	8.14	19.19	0.72	0.13	3.22	45.90	9.79	0.42	4.86	6.66	1.00	28 570 500
1974	8.32	17.99	n.d.	n.d.	2.47	45.20	10.98	0.70	5.19	6.36	4.66	35 700 064
1975	8.50	16.16	n.d.	n.d.	2.10	48.41	10.08	0.78	5.01	4.81	2.79	46 327 902
1976	10.33	16.52	n.d.	n.d.	1.83	52.53	10.65	0.74	5.52	0.29	1.59	58 408 752
1977	10.13	17.25	n.d.	n.d.	1.82	51.59	9.89	0.68	6.22	0.47	1.95	69 220 631
1978	9.87	17.09	n.d.	n.d.	2.68	50.98	9.00	0.69	7.22	0.69	1.78	81 399 815

[a] Gross expenditure minus transfers, disposals and depreciation
[b] Expenditure heads:

1 land
2 site development works
3 buildings: administrative
4 buildings: agricultural
5 buildings: commercial
6 buildings: housing
7 buildings: industrial
8 furniture, plant and equipment

9 main (primary) roads
10 sewers and sewerage works
11 other

[c] n.d.: no data; from 1974 subsumed under 'other'

Source: WDC, 1964–1979

Table 8.2 The build-up of housing in Washington New Town, 1964–78: cumulative completions as of 31 December of the year indicated

Year	Rent			Sale		
	LA	WDC	Total	WDC	Private	Total
1964	22		22		34	34
1965	70		70		130	130
1966	70		70		218	218
1967	492	22	514		226	226
1968	743	171	914	4	234	238
1969	743	516	1 259	12	304	316
1970	743	1 305	2 048	54	477	531
1971	833	2 524	3 357	145	728	873
1972	919	2 845	3 764	287	1 034	1 321
1973	927	3 131	4 058	331	1 763	2 094
1974	1 019	4 102	5 121	331	1 959	2 290
1975	1 093	5 137	6 230	350	2 182	2 532
1976	1 189[a]	5 620	6 809	331	2 210	2 541
1977	1 317[a]	6 374	7 691	331	2 375	2 706
1978	1 417[a]	6 774	8 191	373	2 518	2 891

[a] Includes completions by Housing Association as follows: 1976/21; 1977/33; 1978/101
Source: WDC, 1964–1979

labour forces was initially on a much greater scale than job creation; for example, by the end of 1970 some 2579 houses had been completed since designation, but only 1285 jobs were available on the town's industrial estates. However, given the assembly of this pool of labour, employers have subsequently been attracted to the town and differences between housing completions and employment creation narrowed, at least for a time (see Table 8.2 and Table 8.3, overleaf).

Of particular importance in assembling labour in WNT are state policies of locational concentration of new housing (for a review of these, see Hudson, 1976a, volume 1, 80–88 and 187–194). Consideration of public sector completions in WNT in relation to those in its immediate subregion throws the impact of designation into sharp relief (see Table 8.4, page 189). Between 1964 and 1966, less than 1 per cent of public sector completions in that subregion were in WNT. This rose to 10 per cent in 1967, and despite fluctuations in total completions, did not fall below 30 per cent between 1970 and 1973, the peak being 47.3 per cent in 1971 (the year with the largest absolute number of completions, 3052). Even defining the surrounding subregion to include all of Tyne-Wear County, it is clear that in the 1970s WNT continued to account for a high, though declining, share of public sector completions. Within WNT itself the private sector has come to play an increasingly important role in housing provision, the target percentage of private sector to total completions having been raised from the 30–38 per cent set in the Master

Table 8.3   Employment growth in Washington New Town, 1964–79[a]

Sector	Employment in WNT					
	1973			1976	1977	1978
	Male	Female	Total	total	total	total
Manufacturing	5 025	2 640	7 665	6 266	6 537	7 042
Services	1 015	945	1 960	2 346	2 347	2 595
Shops	210	880	1 090	888	900	1 400
Education	165	190	355	n.d.[a]	n.d.	n.d.
Offices	560	380	940	1 500	2 850	3 390
Construction	2 060	15	2 075	n.d.	n.d.	n.d.
Total	9 035	5 050	14 085	11 050	12 684	14 436

	Employment (as at June) on industrial estates only[b]						
	1970	1973	1975	1976	1977	1978	1979
Male	n.d.	3 300	4 788	4 754	4 886	5 225	5 366
Female	n.d.	1 700	2 279	2 334	2 685	2 887	3 434
Total	1 285	5 000	7 067	7 088	7 571	8 112	8 800

[a] n.d.: no data
[b] This does *not* correspond to manufacturing employment
Source: WDC, 1964–1979; and unpublished data made available to the author by the WDC

Plan to 50 per cent in 1967 by the Labour government, although private sector completions have been much less locationally concentrated than those in the public sector. There has, however, been a more marked locational concentration of private sector completions in WNT since 1971 (see Table 8.4, opposite). This reflects two types of factor. First, the availability of residential land, coupled with the site preparation activities and more general capital investments of the WDC. Second, WNT represented a favourable investment opportunity for speculative house building, the activities of the WDC and other state organizations helping to guarantee a favourable rate of return on capital advanced. This increased private sector involvement has also led, of necessity, to the WDC having to provide a higher proportion of smaller dwellings, which are more expensive per unit area to build and do not offer a sufficiently high rate of return to attract private capital.

This relative concentration of new housing stock into WNT has meant that housing opportunities, particularly for those limited to the public sector, are severely constrained. Local people wishing to move house, especially households seeking a house for the first time, are channelled towards WNT for this reason. The relative concentration of private sector completions in WNT may

Table 8.4 House completions in Washington New Town in relation to those in eight Local Authority areas[a]

Year	Public sector		Private sector		Total	
	All areas	WNT area (%)	All areas	WNT area (%)	All areas	WNT area (%)
1964	1 909	1.2	1 523	8.4	3 432	4.4
1965	3 051	1.6	1 171	8.2	4 222	3.4
1966	2 744	0.0	1 395	6.3	4 139	2.1
1967	3 002	10.0	1 169	2.7	4 171	8.0
1968	2 683	28.8	1 426	0.6	4 109	13.8
1969	2 849	12.4	1 076	6.5	3 925	10.8
1970	2 484	30.1	878	13.2	3 362	25.7
1971	3 052	47.3	1 158	23.1	4 210	40.6
1972	1 524	35.8	1 243	24.4	2 767	30.6
1973	860	39.0	2 213	28.6	3 073	31.5
1971–74	16 492	18.5	11 321	12.9	28 263	16.2
1975–78	22 073	14.0	8 077	8.7	30 150	12.6

[a] For the 1964–73 period this is defined as the eight Local Authority areas of Gateshead, Felling, Boldon and Sunderland, Houghton-le-Spring, Chester-le-Street RD and UD, Washington UD. Prior to the boundary change in 1967, numbers for Sunderland CB and RD have been combined. These are *gross* completions (that is, they do not allow for demolitions). However, demolitions are likely to have been more concentrated outside WNT: by 31 March 1976 some 1037 dwellings within WNT had been demolished or disposed of – the number of each category is unknown (WDC, 1964–1979, *Eleventh Annual Report*). For the periods 1971–74 and 1973–78 this is defined as Tyne-Wear County, together with Chester-le-Street District
Sources: DoE, 1964–1973; Northern Region Joint Monitoring Team, 1980, Table 8.4

be exerting a parallel effect among owner-occupiers, although they face a qualitatively different set of constraints to those facing people restricted to public sector housing policies; this is discussed in subsections 8.4.2.2 and 8.4.2.3.

For the moment, we consider an issue closely related to these, that of the management and tenant selection policies of the WDC and Local Authorities in WNT in the period up to 1978.[2] By and large, Local Authority (LA) housing in WNT is reserved for existing residents of the town. Although a certain amount of WDC housing has been reserved to rehouse people from dwellings demolished in the town, it is much less tied in this way than is LA housing. Between 1968 and 1972, 28 per cent of WDC lettings were to existing town residents – although these included people moving from one WDC tenancy to another – but, more commonly, they are to in-migrant households, whose heads are increasingly aged 45 years or less (as are those in owner-occupied housing in the town). Some considerable priority is given where necessary to people from outside the town with key jobs in Washington and, although this is not necessarily the sole criterion for access to WDC

Plate 8.1   New Town housing development in the 1970s: Blackfell, Washington New Town.

housing, it nevertheless can be important in relation to attracting new industry and is seen in this way by the WDC. A WDC document of 12 September 1974 talks of the:

> . . . strict adherence of the Housing Department to an allocation policy directly related to employment within the designated area. Key workers living outside the area were given top priority, followed by workers already resident in Washington. In cases where there was no key worker in the household, priority was given to families who were already resident in the town.

In general, however, housing allocations are not made to satisfy *specific* labour-power requirements of *specific* firms – the one exception being in the case of key workers, a small number of managerial or specialist technical staff with skills that cannot be supplied through the local labour markets who are brought in by particular firms: only 2 per cent of economically active residents of WNT moved there because of a transfer by their employer (as revealed in the 1978 WDC/Building Research Establishment Household Survey).[3] Indeed, it is unnecessary that such precise arrangements be generally made and there have been no wholesale transfers of workers to WNT with incoming firms. Rather, because of changes in labour processes and the emergence of a new spatial division of labour, companies often simply seek locations with large masses of relatively unskilled labour-power – a point amplified in section 8.5. These can be assembled by the locational concentration of new public sector housing or by in-commuting (for new employment has been even more severely concentrated in WNT than has new housing: see section 8.5).

There are also other relationships within WNT between housing provision and capital accumulation. The quality of rented housing provided by the

Table 8.5   Washington Development Corporation: total housing subsidies,[a] 1966–78, for years ending 31 March

Year	Subsidy (£)	Year	Subsidy (£)	Year	Subsidy (£)
1966	—	1971	297 438	1975	1 736 623
1967	560	1972	519 800	1976	2 869 330
1968	3 146	1973	1 342 615	1977	3 639 365
1969	49 334	1974	1 234 736	1978	4 084 850
1970	94 834				

[a] Includes sums in lieu of Rate Fund contributions (1967); Statutory Contributions (as a category: 1967); Exchequer Subsidies (1968–1972); Housing Subsidies Act 1967 (1968–1972); New Towns Act 1965, Section 42(2) (1968–1975); Home Improvement Grants (1973); other (1973)
Source: WDC, 1964–1979

WDC tends to be higher than much of that generally available within WNT and the surrounding area (see Hudson, 1976a, volume 1, 194–196). The costs of housing for WDC tenants are high, often being on a par with mortgage repayments. Wage levels in those industries that have located in the town are often relatively low, both for men and for women. The existence of potentially cheap labour-power was a strong incentive for many capitals to locate there. At the same time, the conjunction of low wages and the costs of housing served to push married women into wage-labour to help meet these and associated living costs, as higher rent or mortgage levels were not translated into higher male wages (see Hudson, 1980a, 1980b). Thus, although female wages are generally very low, they tend to be sufficient to boost household incomes to the level necessary to sustain the required level of consumption, as well as corresponding to the production requirements of the employing firms; equally, although skill levels are generally low, this reflects the labour-power requirements of employers.

At the same time, however, the WDC acknowledge a discrepancy between the labour-power requirements of incoming firms and their own rent levels and statutory requirements to balance their housing revenue account. Recognizing that rent levels were high in relation to wages in the town, in 1970 the Board of WDC decided that no further rent increases could be imposed in the near future. This effectively constituted a direct subsidy to some capitals, the WDC taking on part of the costs of reproduction of labour-power. The WDC has also attempted to cut housing costs in other ways to keep pressure off wages. Sites for low-priced housing were sold to private developers without recouping the cost of servicing or general development charges, in response to a 1967 Ministry Circular that the greater part of New Town housing for sale should be aimed at the poorer wage earner whose income was about £25 per week. More generally, there have been substantial transfer payments in the form of housing subsidies to the WDC from the Treasury (Table 8.5); indeed, it is only these which give the appearance of a positive net rental income from

WDC housing, serving to conceal the real loss-making character of WDC housing investments (see Hudson, 1979a).

8.4.2.2 Migration and the assembly of labour

The concentration of investment, particularly in housing, in WNT has led to a rapid growth by in-migration of both the total population and resident labour force. Both resident male and female labour forces in WNT have expanded: between 1966 and 1972 the former increased from 6630 to 8450, the latter from 6010 to 8468. At the same time the average age of the resident male labour force was also being reduced by in-migration; for example, between 1968 and 1972, over 50 per cent of male in-migrant heads of households were 34 years of age or less. Thus, as well as expanding quantitatively, the nature of the population was changing qualitatively. Increasing proportions of the younger, mobile elements of the regional population were being drawn there via migration (for details, see Hudson, 1979a).

The reasons for this migration, the geographical origins of migrants and the distance moved are closely related. The concentration of new housing into WNT has had marked impacts on migration patterns, channelling people to the town and resulting in a preponderance of short-distance moves to find housing there (see Table 8.6). Information as to the reasons why people moved to WNT was obtained from the 1970, 1974, 1976 and 1978 WDC/ Building Research Establishment (BRE) censuses (see Table 8.7, overleaf for the 1972 results; those for the other years are virtually identical). Slightly less than half of respondents gave 'housing' as their main reason for moving, whereas rather more than half gave it as a subsidiary reason. Housing availability was seen as particularly important to those 34 years of age or less and also to those aged 65 or more. Smaller proportions of migrants moved to WNT from outside the Northern Region – about 20 per cent in 1972 – and/ or for reasons directly connected with employment – about 30 per cent. Some of these people were key workers. As the flow of new firms to WNT decelerated

Table 8.6 Place of origin of household heads who migrated to Washington New Town one year prior to survey, as percentages

Origin	Year of survey			
	1968	**1969**	**1970**	**1971**
Durham County	68	59	62	64
Northumberland County	9	16	15	15
Rest of Northern Region	2	8	3	3
Elsewhere	21	17	20	18
Total	100	100	100	100
Number	442	351	649	1 551

Source: WDC/BRE *Household Surveys for 1968, 1969, 1970* (see note 3, page 200)

Table 8.7   Reasons for moving to Washington New Town, 1962–72, as percentages

Item	First reason	Subsidiary reason
Housing	47.7	58.6
Job	28.5	10.0
Personal/Social	17.8	18.0
Amenities/Environment	2.8	9.7
Other	3.2	3.7
Total	100.0	100.0
Number	3 919	2 564

Source: WDC/BRE *Household Survey for 1972* (see note 3, page 200)

in the latter part of the 1970s, so did interregional migration of such key workers: by 1978 only 6.3 per cent of heads of households resident in WNT had moved there from outside the Northern Region, compared to some 20 per cent in 1972. The joint availability of jobs and housing was also important. Many people citing employment as their main reason gave housing as their second reason. This points both to the operation of WDC policies linking housing provision to employment and to the concentration of new jobs and houses in WNT (see above). Of those resident for less than 10 years in WNT in 1978, over 10 per cent gave their main reason for moving as 'to be nearer their job', suggesting a change of residence after finding a job in WNT; in contrast, only 2 per cent moved before getting a job in the belief that one would be available for them.

Thus the availability of jobs, together with that of housing, came to be increasingly important in attracting migrants to WNT from 1969 but in the sense of a general pool of available jobs there and in the surrounding area rather than particular employers specifically assembling workforces to meet their own labour-power requirements. Despite the growth of employment within WNT, however, many migrants to the town work outside the Designated Area. In part this reflects the slower build-up of employment relative to population, particularly in the 1960s and early 1970s, but, more generally, this out-commuting by WNT residents suggests that the reconstruction of a labour reserve was a rather more complex process than simply concentrating population into WNT itself.

8.4.2.3   Journey-to-work patterns and the spatial restructuring of the labour market

The combination of aggregate growth and sectoral changes in demand for labour in WNT has found expression in changing journey-to-work patterns. Increased provision of housing and employment there has been accompanied by increased commuting: in 1961, 72 per cent of employed residents of Washington worked in the town, but by the early 1970s this had declined to about 50 per cent (see Hudson, 1979a). This increase in commuting was proposed in the WNT Master Plan – but, in any case, the tendency in the area prior to designation was in this direction (see Hudson, 1976a, volume 1, 157–162).

Increased commuting reflects not only a slower expansion in aggregate labour demand than in supply in WNT because of net in-migration and employment loss within the town, especially in coal mining, but also differences between the type of jobs created and the 'skills' of those already living in or moving into WNT. The mix of housing provided in WNT has led to an influx of migrants who have been unable to find suitable employment there, particularly in the case of white-collar workers living in owner-occupied housing (see Hudson, 1979a).

However, trends in male and female commuting out of WNT have differed. Whereas there has been an absolute and relative increase in male commuting out to work, total female out-commuting has shown a relative decline, although many female residents continued to work outside WNT – in 1966 12 per cent of female employed residents worked in WNT, but by 1972 this had risen to 15 per cent. At the same time as out-commuting by WNT residents was increasing, the number of those resident outside WNT who commuted in for work likewise rose, trebling between 1961 and 1972. In a sense, these in-commuters fill the void resulting from the discrepancy between the job requirements of WNT residents and the number and types of jobs available in the town.

The increased volume of commuting, its spatial pattern (especially that of commuting into WNT), and the differences between male and female commuting patterns are partly related to and indeed contingent upon changes in patterns of private and public transport. The planning of WNT was based on assumptions of almost 100 per cent household car-ownership in the town by 1976 and on car-ownership rates of 0.45 per person by 1980, assumptions which came also to be applied to the northeast more generally. Consequently, considerable road investments were made both in WNT and in the region. Although car-ownership levels have risen, they have not approached those assumed in the Master Plan, and this increase has been uneven within the region (see Hudson, 1976a, volume 1, 160–161 and 253–261). A corollary of considerable road investment and emphasis on the private car for transport was a neglect of public transport provision, which in turn had led to pressures to purchase cars in order to be able to make journeys to work, as well as shops, schools, etc. (see also Mandel, 1975, 393–394).

Thus those without access to a car are subjected to objective pressures to obtain one and constrained in their employment choices. This constraint is reflected in the differences between male and female out-commuting from WNT and in the areas of origin of those who commute into WNT. On the one hand, women in one-car households resident in WNT are often deprived of the household car because it is used by the husband or partner for work. As two-car households are comparatively rare, the effect of this is to restrict the employment opportunities of women in these households who wish to work. Given the poor provision of public transport in WNT and its expense (it is seen in this light by a large majority of WNT dwellers; see Hudson, 1976a, volume 1, 274–281), many of these women are effectively restricted to working within WNT or even to certain industrial estates in WNT – a restriction that dovetails with the labour-power requirements of many of those firms expanding their productive capacity in WNT. However, it is important to stress that transport availability is neither the only nor necessarily the most important reason constraining women to work within WNT. Given the prevailing division of labour between men and women within the family and the absence of child-care facilities such as nurseries within WNT (see Hudson, 1980b), the generally greater involvement of women in the process of social reproduction also acts as a major constraint on where – indeed whether – they become involved in wage-labour.

8.5 State expenditure, policy intentions and outcomes

A crucial element – indeed, *the* crucial element – in the growth of WNT has been state expenditure, particularly that made through the WDC. The state has taken on revenue expenditures and capital investments that in themselves are unattractive to private capital, yielding either losses or an insufficiently high rate of return, yet which are necessary for sufficiently high rates of return to be possible to attract private capital to the town. A major part of state expenditures is in those spheres of collective consumption (such as roads and transportation, public sector house construction and housing site preparation, sewerage, etc., and education) which provide necessary material preconditions for the assembly of labour reserves and the reproduction of labour-power. In addition state expenditures are made in respect of industrial and commercial land assembly, site preparation, and factory building. Direct subsidies to individual capitals which reduce the price of elements of fixed capital result from central government grants for plant and machinery, free depreciation and free rents (for fuller details, see Hudson, 1979a).

Although such state policies have made WNT an attractive location from the perspective of some capitals, from the point of view of the state, however, these investments are loss-making. Implementation of these policies has resulted in an increasing burden of indebtedness specifically for the WDC (Table 8.8, overleaf) and more generally for the state (on this general thesis, see O'Connor, 1973). Deficits from successive years are simply rolled forward and allowed to grow. Moreover, their real magnitude is disguised by the fact that much of the income of the WDC is in the form of transfer payments – notably housing subsidies (Table 8.5, page 191) – received from other parts of the state. Although this deficit has been exacerbated by the rapid rise in interest rates, especially in the 1970s at a time when many rents on WDC property have been fixed, this merely strengthened an already existing tendency (for details, see Hudson, 1979a). In order to attract private capital, promote accumulation and so provide new employment, WDC investments were deliberately loss-making – and increasingly so – over the period 1964–1978.

Nevertheless, despite this state investment and consequent private investment, there remains a considerable gap between the stated intentions for the

Table 8.8   Washington New Town: General Revenue Account, 1966–78 (£)

Year[a]	Expenditure	Income[b]	Deficit	Year[a]	Expenditure	Income[b]	Deficit
1966	24 678	1 450	23 228	1973	2 366 795	2 135 698	231 097
1967	53 556	11 847	41 709	1974[c]	1 107 989	552 409	555 580
1968	199 144	52 109	147 035	1975[c]	1 643 449	725 736	917 713
1969	416 953	152 892	264 061	1976[c]	1 582 379	240 287	1 342 092
1970	727 139	311 168	415 971	1977[c]	2 310 269	448 930	1 861 339
1971	1 270 030	642 097	627 933	1978[c]	2 844 882	66 706	2 778 176
1972	1 945 795	1 129 807	815 988				

[a] The figures are for year ending 31 March
[b] In the General Revenue Account as published, certain sums are written off as 'amount capitalized as other development expenditure' and treated as income. These are *not* so treated here, but categorized as part of the deficit. The sums involved are for 1966 £22,218, and for 1967 £39,087. In the Account for 1968 onwards this convention is adopted and so the 1966 and 1967 data are now consistent with those for 1968 onwards
[c] For these years, the Housing Property Revenue Account is separated from the General Revenue Account, the balance of income plus subsidies less outgoings from the Housing Account being transferred. This leads to much smaller amounts in the total expenditure and income categories
Source: WDC, 1966–1979

Table 8.9   The expansion of unemployment, as a percentage of the previous year's total, in and around Washington New Town,[a] 1968–77

Period	Male	Female	Total	Period	Male	Female	Total
1967–68	+17.7	+22.3	+18.4	1972–73	−21.7	−31.3	−23.3
1968–69	+8.6	−37.2	+1.3	1973–74	−3.8	−16.0	−5.6
1969–70	+2.9	−2.4	0.0	1974–75	+30.4	+64.1	+35.2
1970–71	+20.3	+63.9	+24.6	1975–76	+28.5	+71.5	+35.7
1971–72	+7.9	+51.9	+13.9	1976–77	+3.0	+34.5	+9.6

[a] Defined as the following Employment Exchange Areas: Birtley, Chester-le-Street, East Boldon, Felling, Gateshead, Houghton-le-Spring, Jarrow and Hebburn, South Shields, Sunderland (including Pallion), Washington
Source: Author's calculations from data made available by Department of Employment, Watford

development of the town and actual outcomes of its growth, especially in relation to employment and unemployment. Although a considerable number of capitals have established themselves in WNT (for example, no fewer than 12 per cent of the 307 in-migrant firms to the Northern Region between 1960 and 1974 located in WNT: Hudson, 1976a, volume 1, 139–140), male unemployment has in fact risen in WNT, in its surrounding subregion, and indeed, more generally in the Northern Region, in part because of other state policies as mediated through the nationalized industries, in contrast to the intended outcomes of developing WNT (see Table 8.9).

For example, within the designated area of WNT itself, there was a considerable loss of employment in coal mining between 1964 and 1973. Yet redundant miners have not been incorporated on any substantial scale into the workforces of incoming firms. Rather one of four things has happened to them: some have migrated from the town (which few seem to have done); others have continued to live there. A few of the latter group obtained temporary employment as unskilled labourers in the building industry (whose number in WNT expanded by 75 per cent between 1966 and 1970 as construction expanded in the town), but access to such employment was limited because of the nature and organization of workforces in the construction sector. Others commuted out to work in other pits, although again the scale of this was restricted as even in pits that were expanding output over the period, workforces were often being run down in response to increasing mechanization. Thus, finally, and most commonly, they became and remained unemployed.

In part, this rising unemployment reflected the fact that, in aggregate, the number of jobs lost in the period up to 1973 exceeded that of those created for men in the town. In part, it reflected the fact that ex-miners tended to be trained in skills specific to mining and to be older, less willing, and/or capable (at comparable costs) than younger men of being retrained in those modern skills required by incoming firms. Perhaps more crucially, as the sorts of skills demanded by many firms required minimal training, is the fact that, on average, the productivity of older workers was seen as likely to be less than that of younger workers.

Unemployment in WNT was and to some extent still is particularly associated with those men 50 years of age or over made redundant by colliery closures. In 1970 it was reported that over 50 per cent of males registered as unemployed in WNT were in this age group (*Washington Chronicle*, 4 September 1970). Data from the joint WDC/BRE Survey for the same year confirm this: 60 per cent of those males recorded as unemployed were 50 years of age or more, and these data may well be a more accurate estimate of unemployment than those of males registered as unemployed at their local Employment Exchange. In contrast, males 50 years of age or more formed only 20 per cent of the resident male labour force. Successive WDC/BRE censuses from 1968 to 1976 have shown a progressive increase in the proportion of unemployed males who are between 50 and 64 years of age. Thus there is considerable evidence to suggest that this particular group of redundant miners 50 years of age or more constitutes 'a pool of surplus labour, much of which, in that particular situation, was unemployable' (Hole et al., 1979, 29). Rather than being incorporated into the workforces of new companies, this group has become part of the stagnant reserve army, permanently excluded from the labour force: in this respect state policies demonstrably failed to have their intended impact.

At the same time, female employment has expanded rapidly and, in particular, there has been a strong growth in the employment of married women within WNT, largely to help meet the increased living costs associated with a move there (see Hudson, 1980a, 1980b). Indeed, one can suggest that rather than the growth of WNT removing male unemployment, the expansion of male unemployment has in fact been a necessary condition for the attraction of capital to WNT, both directly in creating a pool of surplus male labour and indirectly in pushing married women into the wage-labour force. The reconstruction of a reserve of labour in these ways has enabled companies to hold down wages and enabled surplus profits to be obtained by capitals located in WNT.

To understand why this has occurred in WNT, it is necessary to examine why some capitals chose to locate all or part of their productive capacity there. It is clear that in a general sense the state acts to reduce the production costs of private capital in WNT. Given this, however, the question remains as to why some rather than other capitals chose to expand there. The decisive point to consider in this connection is corporate strategies to boost profits and accumulate capital. For example, it is overwhelmingly the case that the productive capacity located on industrial estates in WNT is associated with consumer goods production. It was precisely such 'diversification' into a wide variety of consumer goods industries that, in the 1960s and early 1970s, was seen as the solution to the region's social and economic problems (see Carney and Hudson, 1974). At the same time, much of this capacity represents fixed capital investment by major British or foreign multinational companies, particularly those controlled from the USA or Scandinavia. It typically takes the form of branch plants. In this respect the changes in production in WNT can be seen not only as typical of those in the region (see Northern Region Strategy Team, 1976a), but also as an expression of more general tendencies in late capitalism and of the emergence of a new international division of labour (see Fröbel *et al.*, 1980).

The key factor in such companies investing in WNT in these particular branches is the changing character of production processes, the increasing separation of the stages of control and administration from those of production *per se*, and the spatial separation of different parts of the production process, in response to the imperatives of accumulation (on this point, see Lipietz, 1980b; Massey, 1978; Perrons, 1979). Much of the manufacturing located there involves production processes which, in general, require relatively large masses of cheap labour-power – such as those assembled in WNT via WDC and other state policies as to capital expenditure on aspects of collective consumption, in particular housing. Thus those parts of the production process that are located in WNT tend to be those based on assembly work and deskilled labour processes which require only a limited amount and some types of skilled labour. Training requirements are often minimal (see Hudson, 1980b) and employees can be switched between jobs and firms relatively easily. Thus for many capitals the assembly of a labour reserve, often specifically cheap, non-unionized female labour, in and around WNT, from which they can assemble their own workforces, together with the reduction in production costs arising from state expenditures and policies to cater for such key workers as are required, provides compelling reasons for expanding capacity in WNT. Furthermore, this pool of female labour has also served to attract service activities, both private sector and state, to WNT so that employment in these has expanded sharply. To some extent this growth simply reflects the increase in the town's population. However, it also reflects restructuring processes within service sector activities and changes in the division of labour due to the impact of technological change on labour processes. Of special importance in this context has been the location within WNT of the Department of Health and Social Security Child Benefit Centre as part of the restructuring of the state apparatus itself: this largely accounted for the very sharp increase in (mainly part-time female) office employment in WNT between 1976 and 1978.

The issue of the nature of the workforces employed in these activities which are new to WNT is of considerable importance. Although in itself the increase in female wage-labour may have a progressive character, it remains the case that the attraction of mainly female-employing activities to WNT can do nothing to alleviate male unemployment (as it was suggested that the attraction of new industries would). Many firms employ either large numbers and/or large proportions of female employees (see Hudson, 1980a, 1980b). In general the availability of female labour in particular – as opposed to labour in general – has been of considerable importance since designation in attracting capital to WNT (on the more general importance of female labour reserves, see Mandel, 1975, 170–171). The continuing growth of female employment in WNT at the same time as female unemployment rose sharply in and around the town after 1974 (see also Mandel, 1978, 16) illustrates both the continuing attraction of WNT for some activities and the increased incorporation of women into the wage-labour force.

8.6   Summary and conclusions

The implementation of state policies has been a proximate cause of the creation of labour reserves in and around WNT. In part, this has reflected the policies of the NCB to restructure coal production, thereby cutting the price of a key component of circulating constant capital that has important implications for the competitive position of British-based manufacturing (see Hudson, 1981a). As a result, high-cost collieries have been closed and other collieries increasingly mechanized so that considerable numbers of ex-miners were released onto the labour market in and around WNT. In addition, the pursuit of avowedly social reformist policies of concentrating public expenditure on new infrastructure into selected locations, justifying such policies in terms of their simultaneously providing better housing, living conditions, and public service provision for parts of the working class while allowing a return to full employment by providing attractive investment environments for private capitals, has served to concentrate population into WNT itself.

However, the resultant magnitude and pattern of labour demand has been at variance with the proclaimed aim of producing full employment. The labour-power requirements of those capitals and parts of the state apparatus expanding their operations in WNT (and more generally in the region) as part of their own processes of restructuring have not corresponded to what were perceived as the social needs for employment, in terms of either the numbers or types of jobs created. The net result has been an insufficient number of jobs, both for women and for men, and those jobs available are relatively unskilled and poorly paid.

Furthermore, this discrepancy between assertions as to intended effects and actual results illustrates the limits to this form of intervention and provides specific empirical evidence to support the general theses developed by, for example, Habermas (1976) on this point. The state cannot overcome the contradictions inherent in the capitalist mode of production between the goal of capital accumulation (the valorization of capital and the appropriation of surplus value by capital) and the means by which this goal is pursued (growth in social productivity and the increasingly social character of production). Rather, state intervention itself becomes inherently crisis-prone as this fundamental contradiction is absorbed into the state apparatus to emerge in fresh forms. Thus, for example, instead of the development of WNT having its intended impacts (and only these), helping to solve the problems of the region, the solutions adopted have exacerbated the very problems that they were supposed to abolish, as well as producing new problems, since state policies have not – and indeed cannot – remove the deeper fundamental contradictions. On the one hand, this can reappear as a fiscal crisis as the burden of debt associated with state investment continues to expand, or, on the other hand, as a rationality crisis as the gap between the intended and actual impacts of state intervention becomes perceived (see Habermas, 1976). It has been suggested that such developments could act to trigger off a challenge to the legitimacy of the state's pattern of intervention (although this is not to imply that this would follow automatically in any simplistic, mechanistic manner); however, in the specific case of the northeast and, indeed, more generally that of Britain, there is little evidence to support such an interpretation. Indeed, the major response has been in terms of the state changing its forms of intervention, in this case by cutting back both the New Towns programme and the scope of regional policy as part of a more general policy of cutting public expenditure, justifying this as a necessary – indeed the only – available policy option, regardless of its wider social consequences. Whether this will generally continue to be regarded as a legitimate policy response remains to be seen.

Notes

1 For fuller discussion of the history of WNT and other new towns within the northeast which fulfilled similar roles, see Hudson (1976a) and the Regional Policy Research Unit (1979).

2 The transfer of the town's housing stock to Sunderland Borough Council from 1 April 1980 will alter housing allocation procedures: WDC, *Fourteenth Annual Report*; see WDC (1964–1979).

3 This is one of a biennial series, carried out jointly by Washington Development Corporation and the Building Research Establishment (BRE). Results from these were kindly made available for the author via Reports on the Censuses prepared by the Development Corporation.

Chapter 9

Region, class, and the politics of steel closures in the European Community*

9.1 Introduction

Following a long and fairly sustained period of expansion, the steel industry in the European Community (EC) slid into a deep and seemingly intractable crisis after 1974, with capacity closures, falling levels of capacity utilization, reduced output, mounting losses or (at best) sharply reduced profits, and rapidly falling employment (for details, see Morgan, 1981, 1983; Sadler, 1982a, 1982b). This decline in the steel industry in the EC has to be seen in the context of the capitalist world economy lurching into what Mandel (1978) has aptly termed 'the second slump' and the associated changes in the global pattern of accumulation and the international division of labour in (particularly bulk) steel production, which has seen a decline in traditional areas such as the EC and expansion in parts of the Third World (for example, see Balassa, 1981; Organization for Economic Cooperation and Development, 1980).

Within the EC, national states have responded in various ways to this situation of crisis; for example, in some cases this has involved *de jure* or *de facto* nationalization, as in Belgium and France; in others, the subsidization and encouragement of private capital's restructuring of steel production, as in Germany. Such responses and policies towards steel production cannot be understood in isolation from those at the supranational EC level, however. For as a result of the provisions of the 1951 Treaty of Paris, which established the European Coal and Steel Community (ECSC) and were confirmed in the 1957 Treaty of Rome, the legal basis of the European Economic Community (see Vaughan, 1976), the Commission of the European Communities (CEC) has formal powers of control over the steel industry within the EC in respect of capital investment, finance and prices. The CEC's initial response to the crisis was the 1977 Davignon Plan (CEC, 1977). This represented an attempt to regulate the EC steel market and to avoid a price war, which had three broad aims: to reduce capacity through plant closures and production quotas, to bring about coordinated price increases, and to abolish subsidies. Implementation of the Plan depended upon voluntary cooperation from private capital and national states: as a result, its effectiveness was partial and uneven. Because of this and the persistence and deepening of the crisis in steel, in

---

* First published 1983 with D. Sadler in *Society and Space*, 1: 405–428, Pion Ltd, London.

November 1980 the Commission, for the first time, used the powers first granted to it in Article 58 of the 1951 Treaty to declare a state of 'manifest crisis' in the EC steel industry. These powers allow it, *inter alia*, to set capacity and production quotas for individual countries, producers and products – by far the most draconian powers available to any part of the EC organization within the sphere of industry and, in principle, a substantial transfer of state power from national to supranational level. Even so, in practice, violations of production quotas have frequently occurred and, more generally, the competence of the EC as an embryonic supranational state to implement policies within its constituent national states essentially continues to depend upon cooperation from the latter (see Morgan, 1983, 177). Consequently, one critical reason for the uneven decline of the steel industry within the EC has been the varying responses of national states to EC-level plans for restructuring, responses that themselves reflect the balance of power between capital, labour, and the states themselves. At the national level, for example, over the period 1974–1981, employment in the steel industry actually rose slightly in Italy, but it declined particularly severely in Britain and, to a lesser degree, in France (see Morgan, 1983, Table 2).

But although uneven at the national level, the locational concentration of decline and its associated impacts have been considerably more marked at regional and local levels. As steel producers have attempted to rationalize and restructure to restore profits or at least stem mounting losses, they have closed capacity and/or drastically cut employment in localities which, because of the historical development of the steel industry, wholly or in large part depended upon steel-making as a source of employment and income. Given the high level of state involvement in the steel industry within the EC – at the level both of the EC itself as an embryonic supranational state and of national states – the concentrated loss of employment in particular localities has often appeared as a transparently 'political' decision rather than one that simply reflects the logic of the market and 'economic' forces. For this (and other) reasons, such closures and job losses have often come to form a focus of protest in the affected localities and regions: for example, in the United Kingdom, northeast England, Scotland and Wales (see Morgan, 1983); in Belgium, Wallonia; in Germany, the Ruhr; and perhaps most notably, in France, Lorraine and the Nord.

What is noteworthy, however, is that, in general, opposition to plant closures and job losses in steel areas has mainly been directed at national states, despite the active involvement of the supranational EC in the restructuring programmes. Indeed, exceptions to this trend, such as the invasion of the EC's Brussels headquarters on 11 February 1981 and subsequent riots and demonstrations on the streets of that city by protesting Wallonian steel workers, a protest which in any case cannot be separated from the location of the EC's headquarters and the long history of regionalist movements in Belgium itself (see Mandel, 1963), have been notable by their absence.

In this chapter we are not concerned with the broad patterns of change in the steel industry and the reasons for them: as we have indicated, these have been discussed elsewhere. Rather we wish to focus upon the different forms of opposition to specific steel closure plans in different regions and localities, on the different forms of expression of attachment to locality, region and class in the face of the destruction of what often was effectively the economic rationale of these areas; in brief, with the politics of region and class in opposing steel closures. These forms of opposition to closure are not to be regarded merely as a response to restructuring decisions taken by private capital and/or national states (see Carney, 1980), or indeed the embryonic supranational state of the EC, but rather as themselves an active and formative element in this process (see also Sadler, 1982a). The two cases that we have selected for further analysis here – northeast England and north and east France[1] – have been chosen because they simultaneously represent very different forms of protest, yet at the same time are linked in various ways.

In focusing upon different degrees and expressions of attachment to locality which steel closures have brought forward, it will be clear that we are touching upon a theme that has long been of interest to human geographers and other social scientists. Indeed, within geography it was an old concern that was given a new twist in the 1960s and 1970s by the emergence of the environmental perception studies (see Pocock and Hudson, 1978) and, more particularly, the humanistic geography approach to sense of and attachment to place (see, for example, Pocock, 1981; Relph, 1976; Tuan, 1977). Although we share this concern for analysing attachment to place, the approach adopted here differs in two important ways from that of the humanistic geography school, and is rather more closely related to that of other social scientists who have analysed attachment to people and place – for example, in the context of the links between work and community life, between the social relations of the workplace and those outside it, and in one-industry settlements such as coal mining villages (see, for example, Bulmer, 1978; Dennis *et al.*, 1956; Douglass and Krieger, 1983; Williamson, 1982). In particular, we seek to locate the emergence of overt expressions of attachment to place in the context of uneven capitalist development, to stress that knowledge of and feelings about place must be related to the material basis of social development: human geographers have seldom explored such issues (see Hudson, 1979b; also see Harvey, 1982; Jensen-Butler, 1981). Furthermore, rather than simply focusing on describing individuals' attachment to place, we seek to relate individuals to the structure of the societies in which they live; to explore the basis and character of shared attachments; and to understand their relation to collective action to prevent closures (or at least to guarantee an alternative economic basis for the affected 'areas of attachments' and, where appropriate, to locate the actions and attitudes of key individual actors within this context). In placing attachment to place in its social context in this way, we seek also to reinforce emergent links between human geography and modern social theory. For although concern with the spatial dimension of social development was relatively neglected within social theory over a long period (though for an important exception, see Mandel, 1963), recent theoretical developments both by human geographers and by social theorists have begun to stress the role of place and space and the spatial aspects of uneven development in comprehending the reproduction of contemporary capitalist societies (see, for example, Giddens, 1979, 1981; Harvey, 1982; Nairn, 1977; Soja, 1980; Urry, 1981; Weaver, 1982).

This growing concern raises important theoretical issues, which can be no more than touched upon here, concerning the relationships between class structure and attachment to place, between class and territory as bases for social organization and political practice. By way of introduction to the empirical analyses that follow, we sketch out some of the theoretical terrain that guided them by means of a few preliminary observations on concepts of class and on the links between territory and class. Within social theory there are various views as to the most appropriate bases for conceptualizing class, class boundaries and location (see Giddens, 1973). Of these, the most useful starting points in terms of comprehending capitalist societies are Marxist ones, which stress the crucial role of relations of production. It is important to emphasize, however, that this does not imply that issues that are seemingly independent of any immediate proximate links to production relations, such as patterns of consumption, cannot form foci of social groupings or organization or indeed of social protests (in fact, there is considerable evidence that they can: for example, in relation to consumption issues as a focus of protest see Pickvance, 1976).

Nevertheless, the character of capitalist production and social reproduction within capitalist societies itself forges links between class position, defined in terms of relations of production, and issues such as levels and patterns of consumption. For at the level of value, the production of commodities as exchange values that have to be consumed as use values for the value embodied in them to be realized, inextricably binds production and consumption within the capitalist mode of production (CMP). At a less abstract level, class position heavily constrains consumption patterns and life-styles, especially for members of the working class. For those who manage successfully to sell their labour power on the market, living conditions depend heavily upon the level of the real wage and upon various forms of social provision by the state, whereas those who, for whatever reason, are unable to sell their capacity to work are more or less wholly reliant upon state welfare provision. Thus conditions and levels of consumption cannot meaningfully be understood independently of relations of production and the varying strengths of the antagonists in the struggle conducted on the labour market over levels of employment and the real wage and in the arena of the state over the social wage, welfare payments, and so on.

In brief, a Marxist conception of class is one that is analytically powerful and useful, capturing essential points about the real nature of social relationships in capitalist societies. Nevertheless, it is important to clarify the relationship between different levels of abstraction in analysis insofar as these affect the conceptualization of class. At the most abstract level, the analysis of the 'pure' CMP in terms of exchange values, the simple dichotomy of class structure into capitalists who own the means of production and the working class who own only their labour power serves a vital analytic function in understanding the inner logic of the CMP and reveals a profound truth about the social relations which underpin capitalist production. But as one moves from the high abstraction of the 'pure' CMP to the concrete realities of actual capitalist societies, social formations under the dominant influence of capitalist social relations, then a more differentiated conception of class and class structure becomes necessary (see, for example, Giddens, 1973; Olin Wright, 1978, 30–110). As well as recognizing the fundamental contradictory relationship and antagonism between capital and labour, it is necessary to take account of competition between capitals in the search for profits, competition between groups within the working class (on the basis of differences by sector of the economy, industry, or occupation of employment, for example), and the possibility of conflict between capital(s) and/or parts of the working class and social groups located either outside capitalist social relations or within them, but organized on issues and lines other than those of production. Clearly, what this suggests is a society riven with conflict and dissension on several planes simultaneously, yet one that nevertheless does hold together in such a way that capitalist social relations are reproduced: just how and on what basis such class groupings form, how the balance of class forces fluctuates, and how capitalist societies are contingently reproduced must ultimately be resolved through concrete analysis and empirical investigation. Nevertheless, in so far as such societies are reproduced as capitalist, with the social relations of capital continuing to be accepted as legitimate by the working class (both in and out of work) and other non-capitalist social groups, though not necessarily without challenge or modification, it is clear that the capitalist state, particularly in the form of the national state, plays a crucial role in mediating between competing classes and interest groups (for a review of theories of the state, see Jessop, 1982). But the modern capitalist state itself is of comparatively recent origin, and one of the potentially critical issues posed by the restructuring of the steel industry within the EC is that it raises the possibility of conflict between national and supranational state interests, competences and powers. This itself may weaken the capacity of national states to fulfil their cohesive role and raises questions as to their legitimacy and authority to regulate class relations – in this way posing a threat to the reproduction of those same relationships.

There is, however, a further dimension to the divisions between and within classes which is recognized in the concept of the national state: this is the role that place, space and territory may play as a basis for the formation of interest groups that are seen as either within or cutting across, obliquely or orthogonally, class lines and hence obscuring the real interests of dominated classes through entering into territorially based alliances with dominant classes. Framed in this way, particularly in the context of the 'national question', one strand of Marxist thought has been to see class and territory as alternative and competitive bases of political organization (see Hobsbawm, 1977). Rather than follow this line of thought, we would argue that, both for practical and for theoretical reasons, a more sophisticated conceptualization is required which recognizes that uneven development over space is an integral part of capitalism (see, for example, Carney *et al.*, 1980; Harvey, 1982, chapters 12 and 13), that space, place and the organization of interest groups around a territorial basis can play a key role in the actual processes of class formation and organization (see also Anderson *et al.*, 1983). This is an approach which recognizes not only capitalism's temporality but also its spatiality (on this concept, see Lipietz, 1980b; Soja, 1980), predicated on the assumption that differentiation on the basis of location in space is not merely something to be added, once a class analysis is completed, but rather is to be recognized as a potentially central element in the identification of class interests and class formation. Location in space, then, is not simply one way of differentiating between groups within a class once the latter has been formed; the point is that class interests, organizations and practices actually are formulated and framed, at least in part, with respect to particular territories; whether they ought to be is, of course, another question.

Thus one dimension of differentiation, and so a possible basis for competition – between capital and labour, between capitals, between groups of wage labourers and between national states – is location in space and attachment to place. In the case of labour, this is acutely so in the case of those one-industry towns and villages where the major source of income – of the means of living – depends upon wages from one factory or mine, one employer: coal mining villages and steel towns typify such places. This is especially so when capitals restructure production so as to combat their own crises or further their own self-expansion. For as Harvey (1982, especially 425–431) has put it, devalorization is place-specific: seen from the point of view of workers and their families in such places, the only feasible solution often appears to be to fight for their mine, their steelworks, their community, accepting that this means the closure of some other mine, some other steelworks, a threat to another group of workers and their community, rather than posing the broader questions as to why restructuring is seen as either necessary or legitimate, given its broader human and social costs. A concern with a more general class solidarity is subordinated to a more immediate concern with work and life in a localized and spatially delimited community. The threat to locally based small capitals (for example, in retailing and services) that such a loss of working-class wages poses may lead to their becoming involved in cross-class alliances to defend life and work in a particular locality or region. It is in part precisely to contain the potential political threat posed by such costs of devalorization and protests against them that national states have devised policies over redundancy payments for those who lose their jobs and other policies directed at the intra-national distribution of jobs and welfare within their respective territories, to try, and to be seen to try, to combat the worst effects of such place-specific devalorizations, even if unintentionally these policies may serve to help reproduce divisions within the working class, both within the affected communities themselves because of the differential level of redundancy payments and more broadly on a territorial basis by offering resources or the promise of resources to some areas and so not to others. Such actions are nonetheless important in promoting the legitimacy of national states, crucially so when, as with steel in the EC, they are visibly and heavily involved (directly or indirectly) in the processes of restructuring that led to devalorization – and hence in underpinning the system of social relations which the capitalist state reflects and represents. Again, however, the often delicate balance of intra-national territorial and class interests is potentially threatened when, as in the case of steel, there is significant involvement by the organs of an embryonic supranational state, possibly posing questions about the competence and legitimacy of both levels of capitalist state form.

To borrow and slightly modify Mandel's (1963) memorable phrase, to begin to understand such issues the focus of attention must shift to the dialectic of territory and class, the relationships between those two potential bases of social organization and action. It is this that we explore in a preliminary manner in the context of two case studies in the rest of the chapter, which falls into three main sections. In the next two sections, a chronology of the politics of closure decisions in Lorraine, the Nord, and northeast England is set out. In the final section, some of the similarities and differences between these two cases and the reasons for them are discussed, while pointing to some of their theoretical and political implications.

9.2   Region, class, and the politics of steel closures: Lorraine and the Nord

During the period 1965–1974, both capacity and output in the French steel industry rose sharply, associated with a major period of sustained fixed capital investment. This was concentrated on the new coastal locations of Dunkerque and Fos-sur-Mer (see Castells and Godard, 1974, and Bleitrach and Chenu, 1982, respectively; also Gauche-Cazalis, 1979), reflecting a growing state concern both with the international competitiveness of French steel production and with the regional dimension of national economic planning in France (see, for example, Ardagh, 1982, 123–205). A corollary of this was that the 'traditional' steelmaking areas of Lorraine and the Nord were starved of fresh investment and suffered a decline relative to the new and expanding coastal production sites (see Damette, 1980).

From 1974, however, the French steel industry slid rapidly into a deep and sustained crisis: total output and output per employee fell from their peak 1974 levels, profits became losses and the indebtedness of the main steel groups (Usinor, Sacilor, Chiers-Châtillon) grew rapidly (Table 9.1, overleaf). The main steel groups attempted to restructure to counter the deteriorating situation and succeeded in negotiating an early retirement scheme with the unions, involving bringing forward the age of retirement to 56 years and eight months at 80 per cent pay and, primarily as a result of this, employment fell by 8000 in Lorraine and by 2500 in the Nord in 1977. Though useful from the point of view of the steel groups, reductions on this scale were in no sense adequate in terms of the depths of the problems facing the steel industry when seen from the perspective of capital and the French state. Thus, in March 1977, because of the perceived central role of steel in the French economy, the

Plate 9.1   Longwy: the site of bitter struggles over steelworks closures.

Table 9.1   The French steel industry, 1973–79

	1973	1974	1975	1976	1977	1978	1979
Output (million tonnes)	25.3	27.0	21.5	23.2	22.1	22.8	23.6
Output per employee (tonnes)	171.8	175.0	137.3	150.0	149.0	186.6	n.a.[b]
Employment (thousands)	151.0	157.6	155.5	153.7	142.7	126.5	n.a.
Turnover[a]	24.3	35.5	28.5	32.6	34.2	n.a.	n.a.
Cash flow[a]	2.5	5.2	−2.6	−2.5	−4.1	n.a.	n.a.
Net profits/losses[a]	0.9	2.3	−3.7	−4.0	−6.1	n.a.	n.a.
Medium and long-term debt[a]	20.5	23.7	28.3	33.9	38.0	n.a.	n.a.

[a] In billions (10^9) of French francs
[b] n.a.: not available
Sources: CEC 1980; *Financial Times*, 1978

French state introduced the first in what was to be a series of measures designed to reduce overcapacity, cut employment and restore profitability. In return for state loans, the three main steel groups agreed not to open any further capacity and began a closure programme directed mainly at old plants in Lorraine and the Nord, preserving the newer capacity in the coastal complexes. Associated with these cuts, the companies were to declare a further 16,200 redundancies over the next two years, concentrated in the north and east, and in addition to the 10,500 jobs lost there in 1977 through early retirement and non-replacement. The threat which these cutbacks posed to the economic basis of Lorraine – and hence to their own interests – was in fact recognized at an early stage by the local bourgeoisie who, in March 1978, formed an organization with 1300 members to protect the region's economy (Noiriel, 1980, 34). This organization, the Avenir du Pays Haut (APH), was to provide an important element in broadening the social base of the campaigns against steel closures, especially in Lorraine itself (at least until February 1979).

In fact, employment was reduced at an even faster rate than called for by the state: 16,000 jobs were cut by September 1978. Yet still Sacilor and Usinor remained in crisis, operating at only two-thirds capacity: Usinor, the largest producer, with an output of 8.3 million tonnes of bulk steel in 1977, had accumulated losses of FFr 4.5 billion[2] over the period 1975–77, and Sacilor, the second largest producer, with an output of 6.4 million tonnes in 1977, had run up losses of FFr 4.3 billion over the same period. More generally, the accumulated medium- and long-term debt of the whole French steel industry had grown to FFr 38 billion by the start of 1977 (see Table 9.1).

Faced with such a potentially explosive situation, the French state acted swiftly and set in motion the *de facto* nationalization of the industry (though vigorously denying it was doing so) via a major financial restructuring. This involved the central government directly taking a 15 per cent stake in the three main steel groups, but, in addition to this, the shareholdings of the major banks and financial institutions effectively gave the government, directly or indirectly, control over some two-thirds of the share capital of the three companies. Essentially, this restructuring centred on a scheme to convert the companies' debts to the government (FFr 9 billion) and major banks, both private- and state-owned (FFr 9.4 billion), into 'participatory loans' – that is, loans on which a nominal 0.1 per cent interest would be paid for five years, in practice converting them from debts to assets to be added to the companies' capital. In return for this, three holding companies were set up to control the production activities of the three groups (though in fact Chiers-Châtillon was merged with Usinor), which to all intents and purposes were under government control.

This rescue from the verge of bankruptcy was not without its price, however. In return, the steel companies had to agree to draw up and implement, rapidly, an enlarged round of employment reductions and plant closures – the costs of which would fall on steelworkers, their families and others in the affected localities dependent upon their expenditure. In anticipation of, and in an attempt to defuse reaction to, these cuts, the French government announced the creation of a Special Industrial Adaption Fund (the FSAI) in September 1978, two weeks prior to the announcement of the intention to accelerate the closure programme. The FSAI was to have a budget of FFr 3 billion to help create alternative employment in areas to be hit by steel closures: from it, companies could obtain grants of 25 per cent of their investment costs plus low-interest loans (3 per cent per year for five years) for another 25 per cent, and the government claimed that it would create 12,000 new jobs. Unimpressed by promises of possible new jobs, but certain job losses, the response of the steel unions in Lorraine was to call a 24-hour strike

Table 9.2   Proposed redundancies under the 1978 plan for the French steel industry, 1979–82

Company	Location	Number of redundancies
Usinor/Chiers-Châtillon	Denain	5 600
	Valenciennes	500
	Sedan	100
	Billemont	95
	Longwy	7 200
	Anzin	400
	Blagny	460
Sacilor	Gandrange-Rombas	2 000
	Hagondange	2 200
	Saint Jacques-Hayange	800
	Joeuf-Homécourt	1 050
	Hayange	170
Sollac	Serémange	1 100
	Ébange	1 100
	Florange	1 100
	Fensch	250

Source: *Le Figaro*, 1979

on 29 September 1978, when the intention to accelerate and expand the closure programme was made public.

It was not until December 1978 that the details of these new closure plans were announced, however. On 10 December, Sacilor announced that a further 8500 jobs would be lost through closing old iron and steel plants between April 1979 and the end of 1980. Shortly afterwards, the Usinor group (including now Chiers-Châtillon) announced that in excess of another 12,000 jobs were to go, concentrated at Denain, Longwy and Valenciennes (see Table 9.2). Another 20,000 jobs were thus to be cut in Lorraine and the Nord over the next two years (see also Gauche-Cazalis, 1979).

These announcements triggered off a swift reaction in the areas most threatened by closure, in part encouraged by the leading local newspaper, *Le Républicain Lorraine* (Ardagh, 1982, 59), which very soon after became actively involved in the protests, sending, on 17 January 1979, a petition with 40,000 signatures to President Giscard d'Estaing, protesting against the closure proposals (Noiriel, 1980, 39). Prior to this, in December 1978, an inter-union coordinating committee, the Intersyndicale, was set up to fight the closures and to link up the actions of individual unions (Durand, 1981, 83; Noiriel, 1980, 32). Government assurances that steelworkers laid off would be guaranteed an income or another job made very little impression on the steel unions; neither did promises of reducing the retirement age to 55 (or, for some, 50), a 'golden handshake' of FFr 50,000 for those taking voluntary redundancy, nor the offer of retraining schemes on full pay. Opposition to the closure plans was particularly marked in Longwy for, as part of its plan, Chiers-Châtillon was to close capacity there in order to complete a new plant at Neuves-Maisons (south of Nancy: see FPCFMM, 1978). The iron-ore mining and iron and steel unions in Lorraine responded by calling a 24-hour general strike in the region for 12 January 1979, to oppose the announced closures. The strike was extensively supported within Lorraine: a major demonstration against the closure plans was held in Metz, attended by 60,000 people, and many smaller protest meetings were held in other towns. Transport and communications were also disrupted: trains between Paris and Luxembourg were stopped, as well as all traffic into and out of Hayange and Rombas. Parallel to these manifestations of direct action and civil disobedience, more conventional democratic political channels were being explored in an attempt to pressure the national government and the President, Giscard d'Estaing, to abandon the planned closure or, alternatively, to guarantee fresh employment for those made redundant. There were in fact important differences of emphasis between the Communist and Socialist unions in terms of their objectives: for the Communist union the campaign was primarily one against steel closures ('. . . The first of these objectives, and our first priority, is clear – we say no to any redundancies, no to any scaling down of the industry' – cited in Gauche-Cazalis, 1979, 3), whereas the Socialist union placed more emphasis on the provision of alternative employment, and this division was later to assume some significance (Noiriel, 1980, 49 and 95–96).

Demands such as those for the introduction of 'new' manufacturing and service employment amounted to an attempt to establish a fairly standard regional reconversion programme such as had been implemented, without conspicuous success, in response to employment decline in the French coal industry in the 1960s. The problem of such demands, when seen from the point of view of the French government, was that precisely the same sort of macro-economic pressures (balance-of-payments problems, inflation, etc.), which in part had led the government to the need to restructure the steel industry, prevented resources being made available to meet the demands on a scale commensurate with the proposed job losses.

Nevertheless, in the knowledge of past protests and fears of future ones, on 16 January 1979 the government announced a plan to create 11,600 new jobs in Lorraine and the Nord by 1982; again, however, though the redundancies were certain, new jobs for steelworkers were not. In addition, for reasons that are not wholly evident, only 925 jobs were allocated to Lorraine, though there have subsequently been suggestions that the French government almost persuaded Ford to locate a large car plant 25 kilometres west of Thionville and accordingly allocated little new employment there in the published plan (see Ardagh, 1982, 60; Danset et al., 1979). Predictably, the reaction in Lorraine was to regard the government's offer as a derisory one, and a further general strike was called on 16 February. Less predictably, the effect of the government's proposals in the Nord was not to defuse opposition to the proposed closures, but rather to enhance it, and the steel unions there organized a general strike to coincide with that in Lorraine. In a further generalization of the conflict, these regional general strikes were to be linked to a one-day national strike throughout the entire French steel industry. Before these planned strikes could take place, however, protests erupted again in Longwy, with direct action and civil disobedience on the streets by autonomous groups of workers outside the unions growing in scale: for example, factories, banks, and offices connected with the steel companies were occupied. This extension of the protests into civil disobedience and physical violence, challenging the state's authority in these ways, saw the local bourgeoisie, in the form of the APH, begin increasingly to distance itself from the anti-closure campaign, and at the same time it marked the start of a weakening in the position of the joint union organization, the Intersyndicale, as outbreaks of spontaneous protest, over which it had no control and in which it had no involvement, led to the steel unions themselves acting more on an individual basis (Noiriel, 1980, 40–42). In an effort to placate such demonstrations and possibly avoid a second series of general strikes in the Nord and Lorraine, the government attempted to allocate more resources to the reconversion programmes there, but had quickly to abandon the attempt just prior to the date of the strikes, thwarted by outbreaks of protests against mounting redundancies and unemployment in other French regions. Consequently, a second round of general strikes in Lorraine and the Nord was launched, accompanied by major demonstrations and a marked escalation in direct action, blocking transport routes, disrupting trade and travel and closing the Franco-Belgian border; moreover, the national steel strike received total support.

It appeared that the French government was losing control of the situation in the two steel regions and that the steelworkers were becoming increasingly confident of their own strength and capability to paralyse the regions' economies. In an attempt to retrieve the situation, after the second round of regional general strikes the government announced that it would stand firm on its plan for steel – not to do so would lead to an internationally uncompetitive industry and endanger the performance of the entire French economy. At the same time, protests intensified, particularly within Lorraine – most notably on the evening of 23 February and in the early morning of 24 February when police moved in to halt the occupation of a local TV station by protesters against the closures: those occupying it were objecting to the station's coverage of the campaign. This led to 2000 more protesters – together with a bulldozer – besieging the local police commissariat. Then, early in March 1979, the most serious outbreak of civil disorder that had yet occurred broke out in the Nord, at Denain. Borrowing the tactics initiated at Longwy, steelworkers blocked several roads around the town, intending to attract attention to the negotiations then taking place in Paris concerning further redundancies in Denain. On this occasion, though – probably in part because of previous events in Longwy and Sedan – the CRS (riot police) was ordered into Denain on 6 March, subsequently attacking and beating pickets blocking a road. To protest about these developments, a major demonstration against CRS involvement took place outside the police commissariat on 7 March, and, for the rest of the day and most of the night, a pitched battle was fought on the streets: seven policemen were wounded by rifle fire, 30 demonstrators by tear gas. This led to the withdrawal of the CRS on 8 March and, although sporadic streetfighting continued, most of the steelworkers occupied the plant in Denain. Later in March, there was a major demonstration in Paris against the closure plans, which culminated in violent clashes between demonstrators and police in the Place de l'Opéra; five days later government and unions began to negotiate on the closure programme.

The events of March 1979 represented the most serious breakdown in public order in France since May 1968, although, seen retrospectively, they formed the peak of the protest. Moreover, it seemed for a time that the French state was caught in the grip of forces which made it impossible for the state to resolve the problems facing it. For it could neither abandon the steel plan nor increase the resources available for reconversion programmes in steel areas to appease opposition to this plan; hence the only way in which it could contain opposition to the plan was by physical violence and repression, which in turn only served to heighten resistance to the planned closures. Moreover, and perhaps even more seriously for the French government, opposition to the steel closures was becoming generalized into widespread protest against the whole deflationary tenor of its economic policies, and in particular against the continued rise in unemployment to which this led. This situation seemed to pose a threat, at least to the authority of the French national government, if not the French state; in Habermas's terms, it threatened to trigger a 'legitimation crisis' (Habermas, 1976).

Yet only six months later, as protests against continuing steel cuts rumbled on, this major threat had disappeared. In part, this came about because the coherence of the protest movement itself began to break up. Further national strike action, having begun in April, ended in bitter defeat for the steel unions early in May, and the result of this was to further heighten the divisions between the various steel plants and the steel unions over whether to fight the closures or campaign for alternative jobs (Noiriel, 1980, 45–47), with the result that in May 1979 the Intersyndicale broke up. A further demonstration, broken up by the CRS on 17 May, marked an additional step in the winding down of the campaign. Thus increasing divisions between the steel unions opened the door for the French state to further defuse the threats posed by opposition to the steel closures, although at considerable cost: 'bit by bit, during the five months of exhausting negotiations, the protestors were bought off until only a few isolated pockets of dissent remained' (*Financial Times*, 1980b). Essentially, the French government stitched together a package of measures over a period of more than two years which eventually helped divide the steel unions and bought off mass protests against closures. There were two major elements in this. First, a long-term programme to create new employment, the main but by no means sole element of which was the FSAI. Second, and much more politically important in the short term, were the measures taken to cushion the effects of job losses on steel workers. These fell into three main categories. The first of these was the special grant of FFr 50,000 for any steelworker agreeing permanently to leave the industry, which was in addition to the usual redundancy payments which guaranteed workers sliding-scale payments starting at 75 per cent of their last pay during their first year out of work. In addition, for migrant workers, of whom there were many in Lorraine – some 16 per cent of the population being immigrants (Danset *et al.*, 1979) – agreeing to leave, a further FFr 10,000 was added. In all, about 6500 workers accepted voluntary redundancy on these terms (Ardagh, 1982, 61) – and it is not without significance that these grants were announced on 8 March 1979, in the middle of the Denain riots, by the Minister of Labour, Robert Boulin. The second set of measures was concerned with compulsory early retirement: all those aged 55 years and over in the steel industry were retired on 70 per cent of their previous salary, and a large percentage of those aged 55 or over (particularly those in physically demanding jobs) were retired on 79 per cent of their previous salary; in addition, a monthly minimum payment of FFr 2400 was set. Some 12,000 men fell into these two retirement categories (Ardagh, 1982, 61). The third set of measures dealt with retraining: the 4000 workers to whom they applied had the right to refuse two alternative job offers, but on the third refusal their case was examined by a special committee and they could be made redundant. If they took a new job that paid less than their job in the steel industry, their former employer had to make up 60–80 per cent of the difference and, if the difference was 15 per cent or more, the person had the right to a grant of an extra FFr 10,000. This clearly appeared a generous package to many French steelworkers, for by March 1980 over 50 per cent of the 21,000 workers that the closure programme wished to get rid of had left the steel industry, and the programme still had 15 months to run, until June 1981. In these terms, then, the tactics of the French government were successful, the financial incentives offered to individuals to leave the steel industry undermining union attempts to fight the closures and/ or to guarantee alternative employment: in the final analysis, for many steel-workers, a concern with their own welfare overrode a collective responsibility to fight for work in the steel localities.

In another sense, though, this very success was leading to other problems, for the massive costs of the restructuring programme were seriously exacerbating the macro-economic problems which the closure programme was supposedly helping to solve. These costs arose in three main ways: FFr 3 billion for the FSAI; FFr 10 billion over the period 1980–1985 for restructuring the financial position of the steel companies; and some FFr 7 billion on the various short-term ameliorative social programmes. In all, this FFr 20 billion associated with the closure programme amounted to about 50 per cent of the total French annual budget deficit for 1980.

Nor was the reorganization obviously and immediately successful in terms of steel production. In the case of the Usinor group, for example, in September 1979 it reported losses of FFr 2.5 billion in the 16 months up to April 1979, during which the government's reorganization of the industry was progressing. Two months later, it forecast losses of FFr 1 billion for the period May–December 1979, a forecast that was subsequently confirmed in 1980 when the actual losses for this period, FFr 993 million, were announced. In December 1979, 5000 workers were laid off at Usinor's Denain plant after a dispute between management and the Communist-led union about proposed job losses there; unlike earlier protests this did not become generalized, however. It essentially remained a local dispute about how many jobs should be lost, not about *whether* any should be. In the first half of 1980, Usinor made a small (unexpected) profit, but this was rapidly reversed by the European price-cutting war which saw Usinor slump back into making losses which totalled FFr 1.2 billion for 1980 overall: that is, before the government-backed reorganization and shedding of jobs was completed. If anything, the position of Sacilor was even worse, with losses of FFr 2 billion for 1980. Clearly, despite cutting 30,000 jobs in the 18 months from mid-1979, the two major steel groups remained deep in crisis as the collapse in sales and prices within the EC undercut the attempts of the French government to restructure them and restore profitability. In an attempt to restore profitability, Usinor announced substantial planned output and employment cuts in the last quarter of 1980, unaccompanied by, and in sharp contrast to, the violent opposition to the earlier reorganization of 1979; Sacilor did likewise.

Both companies continued to make heavy losses in 1981 despite their own attempts, and those of the EC, to restore profitability by regulating prices and production levels after the declaration of a situation of 'manifest crisis' in the industry in November 1980. As a result, the Chairman of Usinor, Claude Etchegarry, though a 'fervent advocate of private enterprise' (*Financial Times*, 1981), was forced to admit that nationalization was inevitable, because it was the only way of meeting the companies' financial needs. The election of Francois Mitterand as President and of a Socialist government in 1981 ensured that Etchegarry's forecast was a correct one, and, indeed, the widespread disenchantment with the previous Barre government was in no small part the result of the effects of the earlier opposition to steel closures becoming translated into more general opposition to government policies. Under these circumstances, the steelworkers might have reasonably expected a better deal under the new Socialist administration: in fact they did not get one, and the discrepancy between expectations and outcomes was important in reviving violent opposition to plans for further employment losses in steel. In February 1982 it was announced that, although the government would turn its back on market forces as a way of rationalizing the steel industry, nevertheless, Usinor and Sacilor would have to consider ways of achieving this goal. Thus in an attempt to increase labour productivity, the by-now nationalized Sacilor and Usinor announced plans in July 1982 to cut a further 6000–7000 jobs from their total workforce, then about 100,000; these would be mainly concentrated in Lorraine, in the area south from and including Longwy. These plans provoked further outbreaks of direct action and sometimes violent demonstrations in the Nord and Lorraine, particularly the latter (the headquarters of Usinor's special steels division was burned down, for example), despite government efforts to defuse the explosive atmosphere by making more money available for reconstruction in the worst-hit areas. Indeed, such offers, when seen from the point of view of the steel communities, missed the point and had not been the reason for helping to elect a Socialist government: the situation as seen by them was neatly summarized by Juleps Jean, the first Communist Mayor of Longwy (scheduled to lose a further 2000 jobs under the mid-1982 proposals) when he wrote to Mitterand 'these plans are quite contrary to your thinking and your wishes' (quoted in the *Financial Times*, 1982). Even if this was an accurate analysis of Mitterand's views (see also Ardagh, 1982, 117–118 on this point), it is clear that the new Socialist government was no more capable of halting the decline of steelmaking in Lorraine than was its predecessor; nor could it deliver the alternative jobs which it promised. As a result, disillusionment with Mitterand and his government deepened; as Galey-Berdier, Communist Mayor of Morfontaine, a small dormitory town near Longwy, put it (cited in *The Sunday Times*, 1982):

> I voted Mitterand and I cannot hide the fact that I am totally disillusioned. Not only did the left promise to save what remains of the French steel industry. It promised to improve it. But it is applying the same policies as the right.

Although the steel communities had ultimately been willing to settle for increased resources to counter steel closures from a right-wing government, the demand made of a left-wing government switched to the preservation of the industry, but this overestimated the room for manoeuvre open to the incoming government and the capacity of a change of government to effect a dramatic change of policies within the context of a capitalist state. Far from nationalization solving the problems of the French steel industry, it constituted simply one moment in a continuing process of restructuring and rationalization which threatened continuing work and life for particular steelmaking localities as out-migration and unemployment rose sharply: the parallels with the UK experience are self-evident.

9.3   Region, class, and the politics of steel closures: northeast England

The formation of the nationalized British Steel Corporation (BSC) in 1967 ushered in a period of considerable capacity closures and job losses in parts of northeast England; at the same time, however, south Teesside was selected as one of the five major coastal production complexes that were central to BSC corporate planning and strategy for capacity expansion. Thus the closures were effectively unopposed (though this was not the only reason for this: see section 9.4), especially after 1973; for the 1973 *Ten Year Development Strategy* (BSC, 1973) accorded a particularly high priority to the Redcar/Lackenby complex on south Teesside and appeared to guarantee secure employment in a massive, modern, fully integrated plant (for a fuller account, see Hudson, 1985). Only with the announcement, as part of the BSC's 'business strategy', of the proposal to close the BSC's Consett works, made on 11 December 1979, did any substantial form of organized protest against steel closures emerge in the northeast (for a fuller analysis of this, see Hudson and Sadler, 1983b).

In one sense, this announcement came as no surprise, however. Even in the context of a rapidly expanding steel industry, which the 1973 *Strategy*

Table 9.3   The British Steel Corporation's financial results, 1974–83

Accounting year	Profit/loss (£ million)	Accounting year	Profit/loss (£ million)
1974/75	73	1979/80	−545
1975/76	−255	1980/81	−668
1976/77	−95	1981/82	−358
1977/78	−433	1982/83	−386
1978/79	−309		

Source: BSC, 1974–1982

assumed, Consett was regarded as no more than a 'marginal' works. Thus, when the forecast of rapidly expanding demand for steel failed to materialize, the BSC plunged into crisis, with falling output and mounting losses (Table 9.3). Furthermore, the Conservative government elected in 1979 correctly saw the steel industry as one riven with inter-union differences and so an attractive target against which to direct its doctrines of monetary control, privatization, and the reduction of trades union power: indeed the BSC was to be used as an example to private capital of the sorts of restructuring needed to produce an internationally competitive manufacturing sector. Thus the BSC was forced to switch from a strategy of expansion to desperate tactics of retrenchment in a bid to stem the haemorrhaging losses. Given the BSC's overall problem of overcapacity and the fact that Consett was regarded as having made a loss of £15.2 million between October 1978 and September 1979 (though it did make a marginal profit of £124,000 in September and October 1979), it was a prime target for closure: moreover, both employment and capacity had been steadily eroded at the plant prior to the eventual closure announcement. In 1974, 6700 people were employed at the works, compared to 3700 in December 1979, with the most significant loss of capacity involving the closure of the Hownsgill plate mill in October 1979. After this, David Watkins, Labour MP for Consett, led a delegation to meet Charles Villiers, then the BSC's Chairman, to discuss the future of the Consett works – a move indicative of the mounting local concern, prior to the closure proposal being announced, over the plant's future.

Initial reactions to the closure announcement were mixed. Watkins spoke of 'nothing less than a return to the depression . . . Consett could become the Jarrow of the 1980's' (*Financial Times*, 1979). Moreover, consciously or otherwise echoing reactions to steel closures in France, Ernest Armstrong, the Labour MP for Durham Northwest, warned the Prime Minister in the House of Commons that the closure might provoke the normally law-abiding people of the North to militant action (*Newcastle Journal*, 14 December 1979).[3] With greater perspicacity, the reaction of the local community was reported as 'one of anger and bitterness, with more than a measure of resignation' (*Financial Times*, 1979). Echoing this, Watkins seemed immediately to accept the inevitability of closure, and the solution to the problems this would pose as being the attraction of new industry, the traditional Labourist response (see

Hudson, 1983b): 'the closure means the murder of the town [again alluding to Jarrow in the 1930s: see Wilkenson, 1939 – our insertion] and the whole community must get together to attract new employment. That means churches as well as trade unionists. Everyone must be involved.'

Rather than everyone being involved, a mainly union 'Save Consett' Committee was formed to fight the closure. But the local issue of opposition to the closure was rapidly overtaken by the start of the national steel strike on 2 January 1980, a strike over pay rather than in opposition to job losses and plant closures (and which was to have a significant impact on the campaign to prevent Consett closing: see section 9.4). Nevertheless, on 8 February, local officials of the Iron and Steel Trades Confederation (ISTC) organized a peaceful protest march in Consett, the first in what was to be a series, attended by about 300 people; this was followed by a major rally in Consett on 14 March, attended by over 3000 people and addressed by the Chairman of the ISTC, Bill Sirs.

On 2 April 1980, production started up again at Consett after the ending of the national steel strike. Eight days later, the BSC promised to provide the detailed reasons for the closure decision within four weeks. Two months later, on 11 June, local BSC management at Consett agreed to recognize a 14-man Union Committee to deal with the closure issue. On the next day, 12 June, the BSC presented its arguments for closing the works in a document entitled *The case for Consett closure* (BSC, 1980): these apparently hinged around the proposition that Consett was unprofitable. Responding at this level, and thereby accepting the legitimacy of profit as the criterion on which the closure decision should be taken, the Union Committee presented the key findings of their economic audit to a press conference; at this it was alleged that figures leaked from within the BSC forecast a profit of £7.5 million for Consett in 1980, and it was claimed that the plant was one of the lowest-cost and most productive steelworks in the United Kingdom, with an annual productivity (240 tonnes/man) considerably in excess of that of the BSC as a whole (140 tonnes/man) and even that of the West German steel industry (238 tonnes/man). The commitment to this approach was such that John Lee, Secretary of the Union Committee, argued that 'If BSC refuse to listen to a rational, sound economic case, then we will demand a meeting with the Government' (*Newcastle Journal*, 20 June 1980) – seemingly in the belief that the BSC would listen. The day after this press conference, 20 June, was the occasion of a second major rally in Consett, with about 2500 people packed into the local football ground to protest against the BSC's plans; they voted unanimously to travel to London to protest there. Two days later, one of the non-union members of the 'Save Consett' Committee (John Carney), echoing Watkin's earlier allusions to the 1930s, reportedly claimed (*The Sunday Times*, 22 June 1980) that Consett would become a political symbol – 'the Jarrow of the 1980s' – finally touching off a spark against mounting unemployment, as it had done in France (see also Carney, 1980).

On 24 June 1980, members of the national Union Steel Committee met the local campaign organizers in Newcastle, the day after the first phase of a

£12 million government aid package for Consett was announced. This would create, at most, 250 jobs, and was dismissed by steel-union leaders as 'so inadequate it is not worth comment' (*Newcastle Journal*, 24 June 1980). After this, on 9 July, about 800 Consett workers travelled to London in a specially hired train for a peaceful protest march to the House of Commons to lobby MPs and to hand in to the Prime Minister's official residence at 10 Downing Street a petition with 20,000 signatures opposing the closure. Again, the thrust of the argument was that Consett steelworkers were reasonable moderate men and that the steelworks was profitable. Carney described the proposed closures as a 'grave *commercial* error'. We are an efficient and cooperative workforce. We say we have answered the criterion [of profitability] and responded with a responsible attitude' (*Newcastle Journal*, 10 July 1980 – our emphasis). Again, though, allusions to France and the threat of more direct violent protest followed, for, as Lee added, the government was pushing the workforce into a position 'where we shall have to act in ways other than responsible ones' (*Newcastle Journal*, 10 July 1980).

On 23 July 1980, the 'Save Consett' Committee presented its response to the BSC's (1980) document, setting out its case against closure to the Corporation. This response, *No Case for Closure* (JTUCS, 1980), was presented at the first joint meeting involving the BSC, the local 'Save Consett' Committee, and the national Union Steel Committee, held at Teesside. Two days later, on 25 July, what was to be the final major protest meeting, attended by about 2000 people, was held in Consett. On 12 August, the BSC presented its evaluation of the argument put forward in *No Case for Closure* at a second joint meeting. The Corporation's view was that no fresh evidence had been produced and therefore the closure would go ahead as planned on 30 September; the Union Steel Committee were unable to refute the BSC's objections to its case. At the first meeting between the Union Steel Committee and the BSC's Chairman, Ian McGregor, the former was told that the date for closure at Consett would have to be advanced to 6 September.

All seemed lost, until the final bizarre episode, simultaneously tragedy and farce, of the campaign to save Consett's steelworks began. On 2 September 1980, it was reported that a consortium, headed by John O'Keefe (managing director of Chard Hennessy, a Gateshead engineering firm), aided by John Carney (formerly of the 'Save Consett' Committee and who, it was announced, was to become Derwentside District's Industrial Development Officer in October) and Christopher Logan (of Logica, a computer consultancy company), wanted to take over the Consett steelworks (*Newcastle Journal*, 12 September 1980). The consortium was reportedly named the Northern Industrial Group (borrowing the name, accidentally or deliberately, of an earlier grouping of capitalist interests in the northeast: see Carney and Hudson, 1978) and exploratory talks were held between Carney, Logan and Department of Industry officials in London (*Financial Times*, 1980c), although considerable mystery surrounded the identity of the consortium members: 'the search continued yesterday for the members of the elusive consortium' (*Newcastle Journal*, 2 September 1980).

Seemingly no approach was made to the BSC, however, and the Corporation announced that, unless a firm approach was made by 12 September 1980, the furnaces at Consett would be allowed to die down (*Newcastle Journal*, 11 September 1980). Watkins criticized the consortium for failing to identify its members, adding, 'Frankly, I will believe in the consortium when I see the colour of their money' (*Newcastle Journal*, 11 September 1980). With no response forthcoming from the consortium, which remained as elusive as ever, the last batch of steel was produced at Consett on 12 September and it was now just a few days before the blast furnaces cooled to the point where, to all intents and purposes, they could never be reused.

Four days later, the first details of the consortium and its plans became public (see *Newcastle Journal*, 16 September 1980). It allegedly consisted of 10 (unnamed) UK businessmen, with £70 million available; nevertheless, they planned to buy Consett steelworks for £3 million (the BSC was asking for £100 million), restoring it as a private steelworks with a reduced workforce of 2700, Western Europe's highest productivity rate of 320 tonnes/man/year, and a forecast first-year profit of £20 million. Thus the price of the privatization solution would have been substantial job losses, sharp changes in working practices and a severe weakening in the position of organized labour at the actual point of production for those still employed as the conditions necessary to allow profitable production and attract private capital. Further details were released two days later at a press conference in London (*Newcastle Journal*, 18 September 1980), where it was revealed that the consortium now had 11 (unnamed) members with a combined annual turnover of £700 million; the offer price for Consett had declined to £1.5 million, however. Further significant information became available the next day (*Financial Times*, 1980a); in particular, that the consortium members had not committed themselves to putting cash into the business and the significance of the £1.5 million latest offer was that this sum was what Logan was able 'to lay his hands on', and that O'Keefe was described as someone 'who runs a scrap business in Gateshead'. The BSC then 'handed a tough list of conditions of sale to the consortium of northeast businessmen hoping to revitalise Consett', not least because the 'BSC are believed to be privately cool towards the takeover bid which, if successful, could make the corporation look inept and lose them business' (*Northern Echo*, 20 September 1980). The developments taken together considerably further diminished the already low level of credibility in the consortium as the saviour of Consett steelworks.

Whatever residual credibility the consortium had disappeared the next day when it was revealed that it simply did not exist (*The Sunday Times*, 21 September 1980). On the following day, the BSC turned off the gas supply to the coke ovens and let the furnaces begin to cool, and a day later it was reported that (*Newcastle Journal*, 25 September 1980):

Any remaining hopes of a private takeover of Consett steelworks were shattered last night. The businessmen said to be interested in buying the works met for four hours and later issued a tersely worded statement severing their interests in the plant. A spokesman for the group, ISTC official Keith Bill, blamed British Steel for the failure of the takeover bid. Their decision to blow out the plant blast furnaces and coke ovens had left 'insufficient time' to follow through its plans. . . . The names of the companies which made up Northern Industrial Group (Holdings) would never be revealed, he added.

In fact, the names were to be revealed a few days later (in *The Sunday Times*, 28 September 1980), and it appeared that only two of the companies named, Cronite Alloys and British Benzol Carbonising (with a combined turnover of £20 million), had shown any serious interest. O'Keefe's own company, Chard Hennessy, was revealed to have made a pre-tax profit of less than £3000 in 1977 and 1978, and its subsidiary, Potts and Sons, losses of over £50,000. More revealing still, perhaps, given his prospective role of Chairman of the Northern Industrial Group (Holdings) consortium, was O'Keefe's revelation that 'I'm not really good at running companies. That sounds stupid, but I tend to get involved in the start of things and leave them to run themselves. If they don't go very well, I close them.'

In fact, Consett was already *de facto* closed. The most serious and urgent issue on the agenda for Derwentside District was the provision of alternative employment. Not only had Consett closed, but one of the District's other two major employers, Ransome Hoffman and Pollard, had closed its ball bearings factory at Annfield Plain in November 1980, with a loss of 1300 jobs; and the other, Ever Ready, had reduced employment at its battery plant in Stanley by several hundreds. Prior to the Consett closure, there had been virtually no serious consideration of how alternative employment could be provided. Indeed, it would be fair to say that, from the outset, at least a significant element in the local steel unions had been more concerned with the levels of redundancy payments to cushion the immediate and short-term closure impacts than with the issue of alternative employment. As Aidan Pollard, a member of the 'Save Consett' Committee, pointed out: 'We must try our best to gain the greatest amount of compensation for our members to equip them for survival and mount a socio-political campaign to bring real investment to the Consett area' (*Newcastle Chronicle*, 15 December 1979). In fact, as the closure drew more imminent, attention became increasingly focused on the issue of redundancy payments, and in September 1980 relatively generous severance terms were agreed with the BSC: an initial payment of 25 weeks' wages at normal rates, plus 10 weeks' holiday pay, plus up to 12 weeks' pay in lieu of notice dependent on length of service (one week per year of service). In addition, a further 25 weeks' pay at normal rates was to be given one year after the closure. Thus, for example, an employee with 10 years' service received £8000. This preoccupation with the level of redundancy payments in one sense no more than reflected more widely held attitudes within the steel unions, however (see Sadler, 1982b). Thus the question of alternative employment was effectively put aside by the unions until after the closure had taken place. Moreover, the various branches of the state were equally indifferent to this issue prior to the closure, as, in a sense, they could afford to be, as it had not formed a central plank in the unions' anti-closure campaign.

After the closure, the local authority appointed its first Industrial Development Officer (John Carney), and a whole range of assistance from the European Communities, the national government and local authorities became available to give 'what is, according to Mr Carney, one of the best incentive schemes in the country' (as reported in *The Times*, 1981). In particular, as well as generous capital grants, rent- and rate-free periods, etc., companies locating in Derwentside District were eligible for the employment premium scheme financed by the European Social Fund, through which an employer could obtain a grant of up to 30 per cent of the total wage bill for the first six months of its operations. Despite all this and the substantial labour reserves recreated within the District by the closure decisions, only 500 new jobs were created in the 12 months to the start of November 1981, less than 10 per cent of those lost in the District in the 12 months to November 1980. By July 1982, it was claimed that 1200 jobs had been created in the District since the steelworks closure, but clearly many of these were for women, not ex-steel workers (see *Newcastle Journal*, 13 July 1982): given the level of state grants and subsidies, together with the substantial labour reserves existing within the District, it would have been truly remarkable had no capitals taken advantage of a situation that, from their point of view, offered considerable opportunities for profitable production, at least in the short term.

9.4 Concluding comments

Clearly, the character and strength of the opposition to steel closure plans differed markedly between northeast England, on the one hand, and Lorraine and the Nord, on the other; in turn, these different forms of opposition influenced the actions of the United Kingdom and French states in devising tactics to defuse opposition and proceed with closure plans. It should be remembered that these are only two of several cases: there have been significant protests against steel closures in Wallonia and the Ruhr. In this section, we seek to identify some of the important similarities and differences between the two regions and to explore their theoretical and political implications.

One obvious contrast is the different ground on which closure plans were contested. In the case of Consett, the local union's entire case rested on an economic argument that it would be a 'grave commercial error' to close a profitable steel plant. Rather than challenge the legitimacy of a narrowly conceived profitability criterion, the unions chose to accept this and to fight the BSC on terrain that the BSC initially selected: in fact, the BSC could define whether or not Consett was profitable by its own management practices (via internal pricing arrangements, selective loading of particular plants, etc.).

When the BSC moved its ground from that of individual plant profitability to wishing to close Consett because of an overall excess of steel-making capacity (see Hudson, 1985; Sadler, 1982b), the local unions apparently failed to perceive this shift and continued to fight on the basis that Consett was a profitable plant. Thus the whole Consett campaign was fought in such a way as to depoliticize the issue: the hints at direct action and protest as used in

France, coupled with the peaceful law-abiding character of the protest campaign and the failure of such action to materialize, merely emphasized this point. In contrast, in Lorraine and the Nord, the closures were contested, on broader social and political grounds, perhaps best symbolized in slogans such as 'Vivre, étudier et travailler à Longwy'[4] (see Carney et al., 1980) and opposition to the closures did frequently spill over into direct action on the streets and the use of physical force and violence.

To some extent, these differences in the forms of protest in the two regions can be related to the differing national political contexts and also to the very different situations in the steel unions and wider trades union movements in France and the United Kingdom. For in France the Left was engaged in an increasingly vigorous campaign to get rid of Barre's right-wing government, and the protests against steel closures became a pivotal focus from which protest against that government's deflationary economic policies, and the mounting unemployment that these brought, increasingly spread. When this campaign succeeded, a sense of growing disillusionment and cynicism with the policies of Mitterand's Socialist government towards the steel industry quickly set in once it was realized that these policies differed little from those of Barre's, and this led to a further upsurge of violent protest. In contrast, the campaign to save Consett was conducted against the background of the recently elected Conservative government under Prime Minister Margaret Thatcher, which was bent on the pursuit of the goals of monetarism, rolling back the boundaries of the state and extending the scope of the private sector and the smashing of union power, and seeing, correctly from the government's point of view, the steel industry as a prime target.

As to why it was seen in this light requires a consideration of the strength and coherence of the steel unions and their ability or otherwise to generalize the issues of plant closure beyond the steelworkers immediately affected. In France, for a crucial period in the campaigns against closure, there was a considerable degree of unity within the small number of steel unions, which allowed regional and national steel strikes to be coordinated, and, in addition, a much broader social base to the protest campaign than simply steelworkers and steel unions, which enabled regional strikes to be extended into general strikes, cutting across and involving a whole series of groups in Lorraine and the Nord indirectly affected by plant closures. This contrasted markedly with the situation in the case of Consett. For here, there was dissension and disagreement over the tactics of the closure campaign between the plethora of unions involved at plant level, divisions which were exacerbated by the national steel strike in 1980. The existence of inter-union disputes at Consett cannot be understood without some reference to the structure of trades unions within the steel industry nationally, however. For the steel industry is characterized by a multiplicity of unions, organized labour within it being fragmented into no fewer than 18 of them. In addition to the two process unions, the Iron and Steel Trades Confederation (ISTC) and the National Union of Blastfurnacemen (NUB), there are 14 craft unions, loosely linked via the National Craftsmen's Coordinating Committee, and the Transport and General and the General and Municipal Workers' Unions both have members in the steel industry (for a fuller analysis, see Sadler, 1982b; Upham, 1980). This structure, which has evolved over a long period, has had important effects from the point of view of organization in the face of closure proposals. It has led to a rivalry between the process unions (ISTC and NUB) and the craft unions over issues such as wages, and the internal structures of the process unions, especially the ISTC, led to a degree of intra-union divisiveness as a result of their rigid hierarchical character, itself a reflection of the organization of the labour process between the various phases of steelmaking.

Furthermore, Consett was isolated as a works within the BSC, a situation that took its sharpest expression with the attempt to privatize Consett and take it out of the BSC, and so into a situation in which it would be in direct competition with plants in the BSC, but even prior to this there was little effective coordination or linkage between the unions there, at other plants, and nationally. Indeed, it could be argued that, more generally prior to this, the unions allowed themselves to be drawn into a situation where *de facto* individual plants within the BSC were being played off against one another in a frenzied competition for jobs, drawn into an inter-plant competition with one another in terms of labour productivity and the reorganization of work practices at the point of production, for example – perhaps not altogether surprisingly, given the unions' histories and the regionalized organization of the largest union (the ISTC) and the long-established atmosphere of chauvinism and rivalry between steelworkers in different regions (for instance, over the location of new fixed capital investment). The net result of the combined effects of inter-union, intra-union and territorial divisions within that fraction of the working class involved in steel production was 'to produce a pronounced form of intra-class sectional interest to an extent not apparent in other major industries' (Morgan, 1983, 191).

Moreover, the mainly union 'Save Consett' Committee was either unable or unwilling to generalize support for Consett regionally throughout the northeast: the protests were confined to Consett or Consett people, and the fact that only 20,000 signatures could be obtained for a petition protesting against the closure plans in a region with a population approaching three million is indicative of the localized nature of the protest. Thus Consett was isolated, both in terms of region, lacking effective support from other classes and indeed from other groups within the working class in the northeast, and in terms of class, being in competition for employment with steelworkers in other localities.

These differences in form, generality and strength of protest were important in influencing the tactics of the respective national governments in attempting to deal with them. In the case of Consett, the government conceded very little prior to the closure, even in terms of promises of resources to try and provide alternative employment; in a very real sense it had no need to, as the 'Save Consett' Committee was more concerned with the level of redundancy payments than with new jobs. Thus when the privatization plan collapsed in September 1980, it was easy for the government, via the BSC, to smooth over the actual closure with what seemed relatively generous redundancy and severance terms, but which in fact amounted to a small price to pay for the permanent destruction of 3700 jobs. Even so, the acceptance of redundancy payments at Consett was symptomatic of a broader acceptance among steelworkers in the United Kingdom of this individualistic solution to their and the industry's problems, which must at least in part be related to the absence of credible alternatives for the industry being put forward from within the steel unions themselves or the Labour Party (see Manwairing, 1981), as well as the fragmentation of organized labour within the steel industry itself. In contrast, in Lorraine and the Nord, the ferocity of the protests and the government's fears of their becoming a more generally destabilizing element in the political situation led to offers, not only of generous redundancy payments but also of substantial resources for reconversion programmes to provide alternative jobs in steel closure areas. Despite seriously threatening the deflationary strategy, of which the steel cuts were themselves a part, and exacerbating the fiscal crisis of the French state, these programmes were not enough to save Barre's government; equally, though, Mitterand's new Socialist administration was seemingly incapable of pursuing alternative policies for the steel industry, and the steel closures continued, without the promised jobs appearing, accompanied by sporadic, but less generalized, protests.

What then of the wider significance of these protest movements? It has been claimed (Carney, 1980, 54–55) that:

> the new nationalist and regionalist movements associated with regional crises may not be able to acquire sufficient support to gain power, but . . . are likely to prove increasingly capable of further curbing the effectiveness of state intervention in accumulation. If so this is certain to threaten the coherence of late capitalism in at least some of the most advanced countries in Europe.

Clearly, this rather mechanistic interpretation of the inevitability of the disruptive effects of regional crises oversimplifies and overstates the case – it is certainly true in France that regional protest movements in Lorraine and the Nord were important in bringing about a crisis for Barre's government, governmental change as between Parliamentary parties, and deepening of the fiscal problems of the French state, but none of this necessarily meant a curb on the effectiveness of the state in promoting accumulation. Indeed, one could argue that the devalorization of capital, which the *de facto*, then *de jure*, nationalization of the French steel industry entailed, freed capital to seek higher profits elsewhere, and that the very attraction of substantial labour reserves and generous state financial aid in steel closure areas would itself serve to promote accumulation by at least some capitals. The same point could be made of Consett after the closure there, but it is extremely difficult to see how the preceding protests attempting to prevent the closure could in any sense be interpreted as curbing the effectiveness of the state's ability to intervene so as to promote accumulation: if anything, the wider political significance of Consett was to demonstrate the ability of the Conservative government to act in the interests of capital, irrespective of the costs of these actions on particular localities and, more generally, on working people. Furthermore, the resilience of the French and United Kingdom states in containing the protests associated with steel closures has effectively prevented criticism of the supranational EC and its role in promoting the restructuring of the steel industry in those localities most severely affected by closures (indeed, arguably the most persistent protests about EC restructuring plans have come from private capital in steel production in Germany, which views them as artificially cutting its share of the market and profits). The acceptance of the EC's involvement as a legitimate one by the two national states and the identification by steelworkers and others in the closure areas that, *de facto*, the main locus of state power remains located at national level, has been crucial in moulding the course of the struggle in terms of which localities would suffer most as a result of devalorization within the two national territories rather than leading to any fundamental questioning of the EC's role, its definition of capacity and production targets for each national industry, and the sense in which there is overcapacity both at the national and at the EC level. Thus the issues of the EC's authority to act in these areas, and the potentially divisive split of state power between supranational and national levels, have not been effectively raised by those in areas which suffered as a result of closure decisions or, indeed, more generally within France or the United Kingdom.

Notes

1 In addition to the sources cited in the References, the chronologies of events in northeast England and north and east France are also based upon newspaper sources, in particular in the French case, *Le Monde* and *Le Figaro*; in both cases the *Financial Times*, *The Sunday Times* and *The Guardian*, and in the case of Northern England, the *Newcastle Journal*, the *Northern Echo* and the *Consett Guardian*. Where the chronologies directly quote from these reports, the specific reference is given, otherwise not.

2 The billion used in this chapter is the US billion which is equivalent to one thousand million.

3 Specific references to *The Sunday Times*, the *Newcastle Journal*, the *Newcastle Chronicle* and the *Northern Echo* are located in Newcastle City Library, Princess Square, Newcastle-upon-Tyne NE99 1MC, England in the form of a bound volume entitled *Consett Newspaper Cuttings* (reference 942.82C755C).

4 'Live, learn and work in Longwy'.

Chapter 10

Institutional change, cultural transformation and economic regeneration: myths and realities from Europe's old industrial areas[1]*

10.1 Introduction

The last three decades or so, but especially the last 10–15 years, have been a period of continuing decline and deindustrialization in most of those areas that were the original birthplaces of industrial capitalism in western Europe. For many years prior to this, these were locations of not only national but also global significance as centres of capitalist production, associated with a range of industries such as coal mining, chemicals, and metals production and transformation. These old industrial areas were characterized by the employment of waged labour, in big plants owned by big mining and manufacturing companies; associated with this, there was typically a strong gender division of labour between male waged work outside the home and female unwaged domestic labour within it.[2]

The interests of capital were pursued largely unfettered by consideration of those of labour in the initial phases of establishing capitalist social relations in these areas. With the passage of time, the dominant capitalist firms developed institutional forms and linkages through which their interests could be organized and relationships between them regulated. In this way, as well as through political and social relationships developed within the sphere of civil society, the emergent class of industrial capitalists formed itself in these areas (see Grabher, 1990; Hudson 1989a). Over the subsequent years, too, slowly, haltingly and often as a result of bitter capital/labour conflict, an embryonic working class began to establish itself in its emergent places and institutional arrangements were constructed to represent the interests of workers and their families counterposed to those of capital (see, for example, Thompson, 1968). In part, this involved the creation of trades unions to represent the interests of labour as wage-labour *vis-à-vis* capital in the workplace, characteristically associated with ongoing struggles over the conditions, terms and wages associated with work. In part, and generally linked to this, it involved the formation of

* First published 1994 in A. Amin and N.J. Thrift (eds) *Globalization, Institutions and Regional Development in Europe*, Oxford University Press, Oxford

political parties to represent more general working-class interests, both as producers and also as consumers, often articulated through claims for state provision of housing, educational, health and social services. Indeed, typically over the years there was a growing national state involvement in regulating the conditions of reproduction of labour power in such areas and industries and, on occasion, those of production itself. In this way, national states established regulatory frameworks that mediated relationships between local areas and an international economy and set the limits within which local institutional forms have been created (see, for example, Lash and Urry, 1987). A history of dependency upon wage-labour and service provision by major private-sector industrial enterprises often resulted in an easy slippage to dependence upon the state as a source of jobs and services. Indeed, the balance of private and public sector involvement in the provision of jobs, housing and services has varied precisely because such a switch was often an explicit objective of working-class political parties and trades unions in these old industrial areas.

The net result of this history was the creation and reproduction of a culture of dependency in these old industrial areas, in three senses. Firstly, upon waged labour provided through major capitalist enterprises and/or the state; secondly, upon the provision of housing and services through these same channels; and thirdly, upon ways of understanding the world, and possibilities for changing it, that were profoundly shaped by these particular forms of social relations of production and consumption. This culture of dependency was thus sustained through the particular and thick 'institutional tissue' of such areas and through widely accepted beliefs as to what was possible by way of change and through practices as to what was regarded as appropriate behaviour within such areas.[3] The sense of social order conferred by trades unions within the workplace diffused more widely through the institutions of local civil society in such areas, but within the context of a hegemonic culture of waged labour. As a result, the often profoundly paternalistic social relationships within the workplace became extended into mechanisms for social ordering and control in the community beyond the workplace, transmitted and communicated through the institutions of local civil society and/or the state. In this way, a social fabric of community evolved, with its constitutive institutional tissue, through which people came to know and understand their area, the world around them and their place within it as wage workers or dependents of wage workers.

As long as effective demand for commodities produced in and through such areas remained high – or at least exhibited only cyclical fluctuations rather than secular decline – the economic, social and political basis of such areas could be reproduced, albeit within the defining parameters of capitalist relations of production. From the late 1950s, however, it became increasingly apparent that secular decline rather than cyclical fluctuation was becoming the order of the day. An emergent new international division of labour and global shifts in geographies of production were accompanied by deindustrialization, capacity cuts, job losses and decline in old industrial areas that had once been at the forefront of the process of capitalist industrialization (see, for example, Dicken, 1992a; Hudson and Sadler, 1989; Sadler, 1992). From the mid-1970s the pace of secular decline accelerated and came to characterize a widening range of industries (see, for example, Mandel, 1978). This raised important theoretical questions in terms of how to understand such changes and also practical ones in terms of what might be done to reconstruct an economically viable future for such areas. *Prima facie*, for example, it seems unlikely that such old industrial areas, with their particular heritage, would be feasible locations for growth strategies based around expanding activities such as professionally based knowledge-intensive business services. The particular *types* of thick institutional structure that had been evolved within them were simply not appropriate, repelling rather than attracting such activities.

What sorts of activities could therefore feasibly provide a new economic basis for such areas? What sort of cultural and institutional changes might be necessary within such areas to enable them to reposition themselves more advantageously within a global political economy? To what extent are such necessary changes possible? To transpose a point made by Offe (1975a) into a slightly different context, to what extent are necessary cultural and institutional changes impossible, or impossible cultural and institutional changes necessary, if a viable alternative economic basis is to be constructed for these old industrial areas?

There are, however, two further and crucial points. First, even if new institutional forms are necessary for economic regeneration, it by no means follows that they will be sufficient for it (see also Martin, 1988). Secondly, if the old forms of localized institutional structures were to be deconstructed and new ones successfully created in their place in pursuit of this goal of economic regeneration, which class interests and social groups' interests would they favour? What, in this sense, would be the social context and content of a recreated, locally specific institutional tissue? The remainder of this chapter considers a number of alternative solutions proposed for such areas, a range of options that must be understood in relation to the limits set by the national regulatory frameworks as they themselves have been radically altered over the last 15 years or so in most of the advanced capitalist states of Europe. In addition, it also examines the extent to which these presuppose particular localized forms of cultural change and involve innovations in institutional arrangements. Some comments will also be made about the feasibility of these varying options, identifying the implications of this for the future of these areas.

10.2   Productionist solutions, I: small and medium-sized manufacturing firms and the enterprise culture

The social class legacy of a dependence upon wage-labour in large plants of major companies has often been seen as a major cultural barrier to the economic transformation of the old industrial areas through the fostering of an enterprise culture, centred around new small firms and self-employment (see, for example, Sengenberger and Loveman, 1987). From the late 1970s, however, as many national governments retreated from direct confrontation with problems of spatial inequality, one strand of policy saw the emphasis in urban and regional regeneration shift dramatically to approaches that centred on endogenous growth of new small firms in problem areas.[4] For such small-firm policies to be successful pre-supposed, *inter alia*, a decisive transformation within the dominant working-class cultures of old industrial areas to embrace the values of entrepreneurship, self-employment and the enterprise culture. A key intervening variable in producing this cultural change in attitudes and practices was seen to be the creation of new mechanisms and institutions to nurture the formation and growth of new small companies; *a fortiori*, to persuade ex-wage workers to become self-employed or, better still, themselves become employers of wage-labour. To this end, local governments had to acquire new skills in place promotion and in formulating local economic development measures while a plethora of new institutions, often bridging the private and public sector divide (such as local enterprise agencies), were established in old industrial areas to help create the conditions in which new small firms could be born and then grow.

An emblematic example of these processes of change is provided by the town of Consett and the surrounding Derwentside District in northeast England (see Hudson and Sadler, 1991). For almost a century and a half, the District's economy and society had revolved around the iron and steelworks and coal mines of the Consett Iron Company. The company built the settlements of the District to house its workers and its deeply paternalistic influence penetrated local life far beyond the workplace. Over time, with the growing state involvement in housing and service provision, and the nationalization of the coal and steel industries, this became transformed into a deep dependence on a statist form of Labour party politics that was later rudely challenged by the closure policies of the nationalized coal and steel industries. Thus the way was opened for an attempt radically to transform the area's economy and the dominant cultural legacy of its industrial history. The much-publicized closure in 1980 of the British Steel Corporation's Consett works symbolized the deindustrialization of the District, although in fact this had been preceded by over three decades of industrial decline there. Following the steelworks closure, Consett became the focus of a vigorous campaign to create a newly diversified local economy, to replace the jobs lost in large plants with new ones born of a flourishing enterprise culture in new, local small firms. This reindustrialization policy required radically different local institutional arrangements, involving cooperation between local (Labour-controlled) government, central (Conservative-controlled) government, new agencies such as British Steel (Industry) and British Coal (Enterprises), subsidiaries of the nationalized industries, and the private sector. These institutional innovations were represented most clearly in the creation of the Derwentside Industrial Development Agency (DIDA), a local enterprise agency launched in a blaze of publicity and intended to facilitate the emergence of this new economy. There is no doubt that, in the 1980s, Derwentside witnessed the vigorously publicized promotion of a 'reindustrialization strategy', centred around the existence of 'probably the best project package in the UK'. Thus, whatever the new institutional arrangements in Derwentside, the key point was that they sought to

Plate 10.1   New institutions to help create new industries? The Derwentside Industrial Development Agency.

promote an enterprise culture through the creation of a (temporarily) support-ive and highly state-subsidized environment for potential entrepreneurs. DIDA took a high profile role in publicizing this fact and in stressing its involvement in the preparation of Business Plans for potential new businesses.

At one level, it would seem that institutional innovation proved highly successful in bringing about the desired local economic transformation. According to DIDA, £50 million of public expenditure was undertaken in Derwentside District between 1980 and 1988. This in turn triggered over £70 million of private sector investment, involving more than 200 businesses. Moreover, by 1989 it was claimed that this had resulted in the creation of 4500 new jobs, mostly in a diverse range of indigenous, locally owned and controlled, competitive and profitable new companies. Such claims, however, have to be understood not so much as a description of what has happened but as part of a process of place promotion and, in some instances, institutional self-justification. For those involved in the new institutions have a vested interest in promoting their efficacy and the success of the policies that they promote and implement.

In contrast to DIDA's claims, other data, from central government's own *Censuses of Employment*, suggest a much more modest net growth of 1270 manufacturing jobs in Derwentside between 1981 and 1989, compared to a decline of over 8000 between 1978 and 1981. Other data from the District Council's own 1988 *Progress Review* suggest considerable volatility of employ-ment, with a marked turnover of births and deaths of new firms (see Hud-son and Sadler, 1991). Moreover, it is highly debatable whether many of the new firms, especially the relatively larger and/or more successful and well-publicized ones, can be appropriately described as 'local'. Most of these were established not by 'local heroes' but by 'entrepreneurial immigrants' (Caulkins, 1992), attracted by grants, loans and the mass of unemployed people desperate for waged work. Indeed, it is richly ironic that by far the most successful 'local hero', Bob Young, is an ex-coal miner whose company, R and A Young, has its core activities in opencast coal mining. The old coal economy apparently remains the most fertile breeding ground for such a new enterprise culture as may be emerging in Derwentside!

There is, then, little evidence that the new institutional arrangements put in place in Derwentside in the 1980s have had anything more than a cosmetic and superficial effect in bringing about a cultural transformation there that embraces the values of entrepreneurship and enterprise. There is little evidence of local people setting up their own companies; those few that have done so have typically been motivated by a defensive fear of unemployment rather than an offensive embracing of the enterprise culture. Far from being unique in this, Derwentside is simply one of many examples of European old industrial areas in which the legacy of a culture of waged labour has proved highly resistant to change, though often because unemployed former workers recognize only too well that the economic climate in the areas in which they live is a far from promising one for the would-be entrepreneur (see, for example, Rees and Thomas, 1989). Moreover, most of such new small firms as are established are in low-tech services rather than high-tech manufacturing (see Storey, 1990; Hudson et al., 1992), And for those who do set up their own small companies, 'enterprise means . . . a twilight world of hard work, low pay, casual labour and insecurity as . . . people plod along trying to secure a decent living through enterprise' (MacDonald, 1991). Once again, this suggests that the prospects for an emergent enterprise culture are less a matter of individual psychology or institutional innovation than a consequence of economic depression and precarious labour market conditions in old industrial areas.

It is certainly true that in some old industrial areas there is evidence of burgeoning small firms formation, though there is much less evidence of a recreation of the regional economy through Marshallian industrial districts of linked small firms. In Germany, for example, almost 12,000 new companies were registered in North-Rhine Westphalia (NRW), which includes the Ruhr area, in the first six months of 1991 alone, some 25 per cent of the national total (Parkes, 1991). But such success stories only emphasize the importance of the national and local economic climate and the national regulatory framework and public expenditure policies on education and training. Before 1965, there was not a single university in the Ruhr area. There are now six, so that the region currently has one of the densest university landscapes in Europe. There are also six polytechnics, 11 technology centres, four centres of the Max Planck Institute and two Fraunhofer centres (Hassink, 1992b, 55–56). Moreover, major private sector companies, coming together as the Initiativskreis Ruhrgebiet in 1989, had invested 5.0 billion marks, committed another 4.5 billion and earmarked a further 12 billion marks of expenditure on industry and infrastructure in the Ruhr by 1996. In this way, established major private sector interests in the area are stepping in to fill a gap left by the switching of federal government expenditures to the territory of the former GDR, thereby sustaining an economic climate in the old industrial area of the Ruhr that encourages the formation of new small firms.

It is also true that new institutions intended to foster the emergence of an enterprise culture in old industrial areas have had some effects as intended, albeit effects that are limited by their cultural history and the contemporary political-economic environment in which they exist. Such institutional innovation is, however, of much less significance than the investment and spending policies of big private sector companies and central government in such areas in creating and sustaining an economic environment supportive of small-firm formation and growth.[5] While the economic impacts of institutional innovation may be limited, the greater significance of such changes may well be – and be intended to be – ideological and political, seeking to create the illusion that an enterprise culture can be conjured up in even the most unpropitious of circumstances. For if it can be seen to be created in old industrial areas, then surely it can be created anywhere.

10.3   Productionist solutions, II: big firms and the branch plant economy

The enterprise culture project essentially aimed to replace large firms with small ones as a source of waged labour and/or waged labour with self-employment, or even emergent capitalist status, for some people as part of a radical class restructuring of old industrial areas. This implies a simultaneous reworking of the boundaries and composition of the working class and the creation of a new fraction of the old middle class of petty capitalists. An alternative solution, with different implications in terms of cultural and institutional change, is to seek to create a new economy around the branch plants of multinational companies but in different industries from the old ones and employing waged labour within a reformed working class on very different terms and conditions from those 'traditionally' found in the old industries. Thus industries that are 'new' to these 'old' areas, which have become characterized by high unemployment, are attracted by their supplies of 'green' labour from which they can carefully select their workforces as the route to trouble-free production, high levels of labour productivity and profits. This involves not so much an abolition of the working class but rather a radical reformation of it in terms of its age and gender composition and attitudes towards work and trades unionism.

However, to attract such multinational branch plant investment in a global place market requires new regional and local institutions to promote the virtues of areas as spaces for profitable production. In part, this has been linked to a territorial decentralization of powers within some national states so that the creation of new promotional institutions is itself part of a restructuring of the national state; in other cases, such institutions have been grafted on to existing state structures, often in an *ad hoc* way. But in both instances the intention is to create new institutions that can act as 'one-stop shops', selling the attractions of areas to multinational companies in terms of the availability of grants and loans, the quality of the built, social and natural environments and, often crucially, the adaptability, flexibility and passivity of the labour force. These new promotional institutions typically take a pseudo-corporatist form, with carefully selected trades union representatives who promote the virtues of the area's workers, usually in the hope of securing the right of representing them in new single-union deals.

A typical example of these sorts of institutional changes can be found in northeast England, where the prime role of the Northern Development Company, established in the 1980s as the latest in a line of such agencies, has been to promote the attractions of the 'Great North' to multinational companies, especially those from Japan (see Hudson, 1991b). The NDC and its predecessors have played an important role in persuading Japanese companies such as Fujitsu and Nissan to invest in branch plants in the northeast in fierce competition with other areas within the United Kingdom (Dicken, 1990; Spellman, 1991), elsewhere in Europe and, indeed, in the world. The process of inter-areal competition is a pervasive one and this in itself creates pressures for innovations in the institutional structures through which old industrial areas are sold – or sell themselves – to multinational capital.

Now, in many respects, this contemporary pursuit of multinational investment is similar to policy responses to the onset of industrial decline in old industrial areas from the late 1950s. Then, too, the attraction of branch plants of multinationals, above all those based in the USA, was seen as central in creating an alternative economic basis for these areas. In addition, attracting such new inward investment was and is seen as something that required institutional changes to secure the conditions that would entice it to old industrial areas. In the United Kingdom in the 1960s, for example, this involved the creation of new corporatist Regional Planning Councils and a restructuring of local authorities and a redefinition of their role *vis-à-vis* local economic development. Equally, it was seen as necessitating profound changes in working-class culture, both in terms of trades union attitudes towards industrial relations and in terms of lifestyles. But the pursuit of such policies in the 1960s and 1970s created, for some, fears of external control, vulnerability and dependency in a branch plant economy (Firn, 1975) as formerly coherent regional economies became 'global outposts' within corporate production chains (Austrin and Beynon, 1979). Moreover, it was precisely the failure of such policies to deliver sufficient new jobs to old industrial areas that helped open up the space in which very different policies of self-reliance, endogenous growth and the encouragement of small firms could evolve. These policies in turn failed to create even a fraction of the necessary new jobs, hence reopening space for the renewed emphasis on attracting multinational investment and on constructing new regional and local institutions to enable old industrial areas to compete more effectively for it. Hence, such areas embarked on further circuits around the mulberry bush.

In other respects, however, the pursuit of multinational investment in the 1980s takes place in a very different context from that of the 1960s. For, whilst it is unlikely that the regional economy will be reconstructed around spatially agglomerated complexes of companies linked into just-in-time production systems (see Mair, 1991), there is evidence that multinational companies are now pursuing more 'regionalized' versions of their global strategies, with an organization of R&D, production, distribution and sales on, say, a European rather than a worldwide basis (van Tulder and Ruigrok, 1993). Clearly, if branch plants cease to be the 'global outposts' of transnational production chains but become more embedded (see Dicken, 1992a), in a variety of ways, into local and regional economies, the potential for stimulating local economic development may be greatly enhanced. This raises questions as to what sort of state policies may be necessary to attract such investments, with a much greater concern for quality in terms of the environment, labour force skills and so on. However, while such embedded branch plant investment may provide a greater economic coherence at one spatial scale, it does not necessarily remove the uncertainty surrounding the future of particular old industrial areas. This uncertainty in turn increases the pressure for institutional innovation and to invent fresh competitive advantages in marketing one area against others. Moreover, against a background of high unemployment, multinational companies can exercise great selectivity in whom they choose to employ, creating new divisions within a redefined working class. Such companies effectively exclude the vast majority of people in such areas as part of their potential workforces. Nevertheless, it may well be the case that attracting major branch plants, especially those which concentrate on the more technically sophisticated, high value-added segments of the production chain, offers the best opportunities for some employment creation in many old industrial areas. If this seems a pessimistic conclusion, it may also be a realistic one.

10.4   Consumptionist solutions, I: from working–class production spaces to tourism based on the heritage of working–class production

Given the generalized switch towards a tertiarized economy, it is no surprise that in some deindustrialized old industrial areas, attempts have been made to identify potential service sector activities that might be developed there. Nor, given the environmental characteristics of such areas, often blighted by the scars of industrial production followed by decay, is it any great surprise that in some of these areas attempts have been made to transform the legacy of earlier industrial production into the basis for contemporary employment and economic activity (Hudson and Townsend, 1992). Thus the heritage of deindustrialization is not to be completely cleared away as part of the process of seeking to attract fresh industrial investment but is to be selectively preserved and transformed. The past is to be restored in a partial and sanitized form as a tourist attraction for people who wish nostalgically to engage with a

Plate 10.2   Re-presenting old industrial areas: South Tyneside becomes Catherine Cookson Country.

typically romanticized version of an earlier industrial era. Indeed, whilst in some cases it is the preservation or reconstruction of actual industrial landscapes, of old ways of work and living, that is the focal point of attraction (as in Bradford, or in museums such as Beamish in northeast England or Big Pit in South Wales), in others the selling point is a fictional world that itself portrayed life in the industrial past through rose-tinted lenses (as, for example, south Tyneside becoming Catherine Cookson country). In short, this is the old industrial areas' particular version of the heritage industry (Hewison, 1987).

To represent old industrial areas in this new guise again has necessitated important changes in local and regional institutional arrangement. New regional Tourist Boards have been created and local government economic development departments have had to develop new skills and expertise so as to invent and then market the attractions of their areas. Such heritage sites have often been linked to the formation and growth of local small-scale service activities, such as pubs, clubs and hotels. In these ways the tourist industry is again associated with important gender and age changes in the composition of the local working class, with a switch from full-time to part-time and casualized precarious work, much of which is non-unionized (Hudson *et al.*, 1992). Such radical changes both presuppose and help bring about significant changes in local culture, in the understanding of what waged work means. For working for a wage by serving other people carries with it very different connotations from working for a wage by cutting coal or producing steel or ships. Areas that were formerly centres of industrial production become economically dependent upon incomes earned elsewhere, both locally and more widely. However profound the changes in local culture, in attitudes to work, and however innovative the marketing strategies of promotional institutions, tourism remains a profoundly risky basis on which to rebuild a local economy, subject to the whims of tourists as much as to the effects of economic crises in other areas.

There is no doubt that the turn to tourism in many old industrial areas represents a sort of politics of despair, born of a lack of alternatives in terms of manufacturing industry or other types of service sector activity. It is a recognition of economic marginalization in areas that were once focal points in an accumulation process that now increasingly bypasses them, with deep implications for class structures and social relationships within them. There may be circumstances in which promotion of such activities might appropriately form one element in local economic development strategies, but to rely upon them as their main, or sole, basis is a very risky course of action. For it places such areas in competition with the vast array of locations, scattered around the globe, that seek to sell themselves to tourists in various ways. There is no guarantee that a tourism based on nostalgia for a departed industrial past will prove more attractive to consumers than one based on sun, sea and sand in southern Europe or the mystic delights of distant continents as the tourist industry becomes increasingly globalized.

10.5 Consumptionist solutions, II: from working–class production spaces to middle–class residential and consumption spaces

The enterprise culture, the branch plant economy and the turn to tourism, whilst in many ways drastically differing, do share one common characteristic: they are all strategies that seek to employ people in an area as waged labour, as workers involved in the production of goods or provision of services as commodities. Certainly these various options imply drastic local labour market restructuring, by age, gender and so on, with deep cultural changes in norms as to the meaning of work, working conditions and practices, and lifestyles and living conditions. But at least they were predicated upon capital's interests in these old industrial areas as locations for the production of profits, in at least some of their populations as wage workers. In this sense, they all presuppose the reproduction of a working class culture of waged labour, albeit of different kinds.

In other old industrial areas, there have been radically different responses to deindustrialization, decline and decay. Above all, these hinge around a profound and selective demographic and social recomposition as, through a mixture of gentrification and comprehensive physical redevelopment, these areas are reworked as sites of middle-class residence and consumption. This process is most sharply symbolized by waterside locations within the old conurbations, but it is by no means confined to them. Nevertheless, the visual images of former areas of derelict docklands, warehouses, shipyards and associated working-class housing areas converted by a combination of refurbished former buildings or brand-new residential ones on sites cleared of their former occupants, occupied by the new middle classes and surrounded by equally new retailing and leisure complexes targeted at meeting their demands and those of the more affluent fractions of the working class, are very powerful ones. There are sometimes passing gestures to 'high-tech' production, but these are essentially cosmetic, for the rationale of these new complexes, shimmering with reflective glass and adorned with the trappings of postmodern architecture, is consumption. The dominant themes are those of visibly conspicuous consumption, social recomposition and political change, physically symbolized in the contrast between the old and new built forms.

These are very dramatic transformations and have typically required very drastic institutional changes to facilitate them. Above all in the United Kingdom, these have been identified with the establishment in the 1980s of the Urban Development Corporations, non-elected bodies appointed and generously funded by central government. The UDCs are a type of authoritarian corporatist institution, driven by the interests of particular sections of the private sector with token representation of the interests of organized labour and local residents. Local authorities, typically Labour-controlled, in areas where UDCs operate have tended to accept the situation, albeit unwillingly, and work with them, not least because they represent a substantial source of extra central government expenditure in an era when this has generally been shrinking. They have a brief to transform selected old industrial areas within the conurbations via a speculative, private property-led redevelopment strategy. The task for the UDCs is to cut through the regulatory controls formerly exercised by elected local authorities and to provide an environment in which private capital will invest in spectacular redevelopment projects of the desired type, deflecting local criticism by claims that such projects will result in the creation of thousands of jobs in the midst of areas blighted by high unemployment and widespread poverty.

The activities of the Teesside Development Corporation (TDC) are by no means untypical of the UDCs outside London.[6] The TDC sees its mission, in the words of its Chief Executive, as 'market led and about creating jobs and confidence' and its portfolio of projects reveals starkly what this means in practice. They include a £10m offshore technology park on the site of a former shipyard; a £40m marina, residential and leisure complex in the old

Hartlepool coal docks; and an £80m project to convert the derelict Stockton racecourse into a leisure and retail centre. The precise extent to which, and the forms in which, these proposals will come to fruition will depend upon private sector investment decisions. Equally, while there is considerable uncertainty about the precise employment figures and types of jobs, suggestions that all this will create 15,000–20,000 jobs help defuse local opposition and persuade local authorities in the area to try to work with, rather than against, the TDC. But there is no doubt that it is the TDC and not local authorities that sets the agenda for development, for what is to be defined as development, on Teesside in the 1990s: private sector-led speculative property redevelopment resulting in increasingly sharp social and spatial differentiation within a post-industrial, post-production Teesside, juxtaposing middle-class affluence with working-class poverty and unemployment. In so far as such investment generates new jobs, they are typically part-time or casualized, unskilled, poorly paid, and non-unionized.

The sort of future being created on Teesside is far from unique, as the other UDCs pursue similar sorts of policies. In other areas, different institutional mechanisms have been devised to pursue the same objectives, but all derive essentially from a US-based urban redevelopment model imported into Europe (see, for example, Judd and Parkinson, 1990). The recent (1991) explosion of riots in some English inner cities suggests that the effects of such visibly divisive policies may well not be consistent with the establishment of a stable localized mode of regulation in such areas.

10.6   The welfare state solution: from industrial workers to clients of the welfare state

For some deindustrialized old industrial areas, there is no foreseeable future role as locations of production, no feasible possibility of reconversion into places of middle-class residence and consumption. For them, the only foreseeable future is one of dependence upon state transfer payments as the vast majority of their populations become clients of the welfare state that typically has been drastically cut back in scope.

In one sense, this is not new. There have long been people dependent upon such transfer payments. What is new, and especially stark in former one-industry settlements, is the generalized and simultaneous occurrence of mass unemployment in many such places. This in turn has required new forms of institutional arrangements, new forms of state organization specifically focused upon social containment and control through varied training, retraining and temporary employment schemes. In the United Kingdom, this has been most clearly manifest in the activities of the Department of Employment/ Manpower Services Commission and schemes such as the Youth Training Scheme, the Community Programme, and Employment Training. These new national schemes and state organizations have come to play a pivotal role in social regulation in many old industrial areas, centred on the allocation of places on training and temporary employment schemes and the management of a substantial surplus population. In Derwentside District, in the late 1980s, for example, over 2800 people were engaged in such schemes. To put such figures into context there were around 14,000 people formally employed in the District at this time and around 3500 registered as unemployed. Of the latter, 40 per cent had been continuously unemployed for 12 or more months, 25 per cent for 24 or more months, with the recorded incidence of such long-term unemployment depressed by the impact of temporary job schemes. As a consequence of the low levels of welfare payments, there has often been a parallel growth in the black economy (in Cleveland County perhaps 1 in 10 of those registered as unemployed are involved in the black economy, for example) or even in criminal activities in these areas (as became clear in the Meadow Well riots in 1991 on Tyneside) as people seek to supplement inadequate incomes from transfer payments in a situation in which they are effectively excluded from the formal labour market. In old industrial areas in other parts of western Europe, the formal limited citizenship rights of temporary – or, still more so, illegal – migrant workers both limit their welfare rights and exacerbate the pressures pushing them into black work or criminal activity.

As Pugliese (1985) some years ago wrote of much of the population of the Mezzogiorno becoming clients of the welfare state, so too can one now write of a very substantial fraction of the populations of many old industrial areas of the United Kingdom in these terms. Increasingly, however, they are clients of an emaciated welfare state. Such people are often consigned to the worst of the remaining stock of public sector social housing, subjected to high levels of surveillance by police and social security officials, especially so in those locations in which they are spatially adjacent to the transformed consumption spaces of the new middle classes. Once again, such a pattern of sharp socio-spatial segregation is one that is increasingly common within western Europe's old industrial areas.

10.7 Conclusions

The five scenarios or alternative futures for old industrial areas are put forward here as analytically distinct options. In reality, elements of more than one of these will be found in particular times and places, interlinked in practice as the uneven development of capital is reproduced in new forms. Moreover, while they have been presented here as analytically distinctive options, they must be understood as linked together as elements of, and experiments within, a wider political vision of neoliberal policies, predicated upon selectively cutting back the state and promoting competition and the market as the main mechanism of resource allocation within society.[7] As central governments aggravated, if not provoked, local economic crises in old industrial areas via their macro-economic, monetary and fiscal policies, they simultaneously withdrew from serious engagement with their resulting localized and regionalized consequences. Rather they used these as a further opportunity to promote their own political projects through new policies which sought to encourage competitive localized solutions to such problems; and, moreover, they sought to secure generalized acceptance of a view that there was no alternative way of tackling such problems. The result was a series of experiments at local level, varied in form and content but unified by their origins in broader national, and at times supranational, political projects as Thatcherism and Reaganomics, for a time, sought a global hegemony.

Nevertheless, the fact that there is a range of possibilities within a particular matrix of opportunities, albeit one that is far from optimal for old industrial areas, raises important practical questions concerning their future. These areas had their origins as workshops of the world, as gateways to the world. They were significant centres of capitalist production and trade, with characteristic social class structures, institutional and cultural traditions and practices associated with particular mixes of industries. And now? The cohesion that these areas formerly had is collapsing or has collapsed. They are increasingly characterized by internal fragmentation and dislocation as what had previously been relatively coherent – within the limits set by capitalist social relations – local and regional economies shatter and local and regional societies are subject to intense pressures under which they threaten to rupture, perhaps irrevocably.

How then can such areas position themselves, or be positioned, more favourably in the context of a rapidly shifting global economy, further deepening of integration within and widening of the European Community, and the failing capacities of national states to regulate social and economic life and maintain socio-spatial inequalities within 'acceptable' limits? What are the most appropriate and realistic objectives for such areas – and what are the most appropriate institutional arrangements at global, European Community, national and sub-national levels to try to ensure that they are attained in an uncertain and fluctuating global political economy? As the preceding pages of this chapter indicate, despite often intense local efforts and innovativeness, aimed at creating fresh new localized and thick institutional tissues, many of the 'solutions' attempted so far have been at best partially successful. Is it, then, the case that the institutionally necessary really is impossible and the institutionally impossible necessary? It would, on the evidence presented so far, be easy to conclude, echoing voices heard in the depression of the 1920s and 1930s, that in fact there is no solution for the problems of such areas, bar complete abandonment. It should certainly be emphasized that there are no 'local' solutions to 'local' problems and no amount of localized institutional and cultural change *on its own* will guarantee local economic prosperity. Local institutional innovation and local pro-activity can never in this sense be *sufficient* to guarantee economic regeneration and social progress. The real questions are much more to do with the relationships between local changes, corporate strategies and national/supranational state strategies and with *whose* solutions are to be implemented in such old industrial areas.

In this context, it is important to remember that the localized thick institutional structures that evolved in the past have often become a mechanism to stifle dissent or hinder opposition to what was regarded as the conventional wisdom of orthodox solutions. During the 1960s in northeast England, for example, militant miners who wanted to take industrial action against colliery closures were threatened with disciplinary action by their own trades union because it would embarrass a Labour government that was closely identified with the organized labour movement (Douglass and Krieger, 1983). More generally, attempts to contest plant closures and job losses have often been hampered by inter-union, or even intra-union, disputes (see, for example, Hudson and Sadler, 1983a, 1986a). Moreover, as the 1984–85 miners' strike in the United Kingdom dramatically demonstrated, regional chauvinism and territorial division within trades unions can decisively influence the trajectory of economic and social change within different areas built up around the same industry (Beynon, 1985). In these circumstances, the organizational structure of labour itself becomes a barrier to effectively contesting, and posing a credible alternative to, the futures proposed for particular local areas by capital and/or national states. During the 1980s, attempts to establish grass-roots local community-based solutions to the economic and social problems facing Derwentside in the wake of the closure of Consett steelworks were effectively marginalized as the local Labour-dominated District and County Councils laid heavy emphasis upon a conventional reindustrialization programme. Resources were channelled into this through the state while they were simultaneously denied to local grassroots initiatives. This was by no means a process unique to Derwentside. In all these cases, albeit in different ways, the legacy and residue of a former thick localized institutional tissue suppressed the exploration of alternatives to, and resistance to, the conventional solutions. Furthermore, it is not just the legacy of the past that stifles the exploration of alternative futures for such areas but also the way in which attempts are made to create new institutional structures which will provide part of a regulatory framework for new trajectories of local economic development. For example, the incorporation of key individuals in the organized labour movement into territorially based coalitions that seek either to seduce the branch plants of transnational capital and/or to promote the growth of small firms and a localized enterprise culture serves to define an agenda which excludes more radical, alternative policies in and for such areas. Under these circumstances, it would seem that localized institutional thinness may have held greater emancipatory and radical transformatory potential.

There is a further point. It would seem that particular combinations of local economic growth models and localized institutional structures, which play an important regulatory role, evolve in such a way as to provide a stable localized basis for production and social reproduction. While this stability may be sustained for longer or shorter periods of time, in the end serious disjunctions emerge between the growth model and the regulatory framework, and a localized crisis erupts. This is resolved – or more accurately, attempts are made to resolve it – by searching for a new stable combination. What remains an open question is whether localized institutional structures can be developed which would allow more of a smooth and incremental adaption of local economic and social life to the broader exigencies of national state policies and global political-economic change. The evidence to date suggests that this is unlikely, at least in the vast majority of old industrial or deindustrialized areas.

What then can we conclude? Perhaps the real task is to accept that as long as there is capitalism there will be uneven development, and so to seek to find the most appropriate strategies within these limits to allow people to live, learn and maybe even work in the old industrial areas in which they feel most comfortable. But, in seeking to define such strategies, there is a deep ambivalence surrounding the role of local institutions. Certainly, there are no simple relationships between the existence of local institutional thickness, guarantees of local economic regeneration and growth and the socially progressive and transformatory content of the politics pursued through it. Rather, the relationships between localized institutional structures and localized economic and social change are ones that both are reciprocal and vary between places and over time. Local institutions may be important in some circumstances in fostering change, in others in resisting it. Whether encouraging or opposing change is politically progressive or regressive is, however, another matter. There are those who place great faith in a constitutional reform and decentralization of power to regions as the way of furthering working-class interests in old industrial areas (see Byrne, 1990). This is to adopt a very partial view of the issues. In other circumstances, local institutions may be, at best, of marginal significance. The key theoretical questions are, therefore, to do with understanding this complex interplay between those circumstances in which local institutional structures do or do not matter, the ways in which they matter, and the political implications of their influence. The key practical questions are then to do with specifying the most beneficial – or least damaging – local strategies in the light of this knowledge.

But what it does seem reasonable to say is that there can be no localized solution that impacts equally on all social classes and groups in an area. In this sense, the myths of socially undifferentiated pro-active localities seeking to pursue their own interests are dangerously misleading ones. What it is necessary to do is specify *whose* interests are to be prioritized in seeking to formulate and implement local solutions and create appropriate localized institutional structures to facilitate this process. If the intention is to promote a selective social and political recomposition through transforming old industrial areas into new middle-class consumption spaces, or into islands of an enterprise culture, then it must be recognized that such solutions are unavoidably locally divisive as many more people will be excluded from them than will be included in these particular projects.

Clearly, the welfare state solution is more inclusive but it is premised on accepting marginalization, accepting generalized expulsion from the working class to the ranks of the surplus population. While there are clear dangers in fetishizing waged labour, there are at least equally serious ones in passively accepting the poverty, poor living conditions and ill-health that follow from prolonged unemployment. In this sense it ought to be challenged, consistently and strongly. There are those who suggest that organizing marginalized social groups so as to make increasing demands for improved living conditions through increased state transfer payments offers a route to a more generalized progressive transformatory politics (Byrne, 1990). While there is a very strong case for increasing the quality of life of the members of such groups, it is unlikely that this would be translated into a wider radical reformism.

Solutions which are predicated upon a recomposition of the working class are certainly more inclusive, which is not to deny they are also divisive. In terms of both economics and politics, a future based on commodification of the remains of an industrial past for present-day tourists is less attractive than one based around new forms of production in the branch plants of new industries. Nevertheless, at least in the old industrial areas, in an era of increasing globalization this sort of local economic reconstruction via inward investment would require deep and radical policy changes at national and supranational level as well as local institutional innovation. It would, however, have to be acknowledged that this sort of reindustrialization policy would offer little chance of a return to 'full employment' in the local economy. The ability carefully to select workers against a background of high unemployment is typically a decisive element in such investment decisions. However, policies to improve the quality of the local productive environment, not least in terms of a more skilled and qualified workforce, would increase the chances of attracting factories that would require a wider range of occupational skills and, maybe, greater local multiplier effects. Given that such reindustrialization policies will not lead to 'full employment', however, there will be a powerful need for strong welfare policies to provide for those who will remain unemployed, and this will require decisive changes in national and perhaps supranational policies. If this seems like a pessimistic conclusion, postulating a scenario that, even if it came about, would offer only a limited and far from ideal solution, then this is no more than a reflection of the limited choices open to most people in the old industrial areas faced with the realities of globalization and the structural constraints of the contemporary political economy of capitalism.

Notes

1 This chapter draws upon a number of years of research into the causes and consequences of industrial restructuring in old industrial areas within western Europe. In particular, it draws upon research in northern Britain but reference to other areas is made as appropriate.

2 In some areas and industries, such as the textile districts of northern England, women were of greater significance as industrial waged workers and this had important ramifications for gender divisions of labour, institutional forms of representation of working-class interests and social relationships within civil society (see, for example, McDowell and Massey, 1984). In general, however, the stereotypical pattern of waged male labour outside the home and unwaged female labour within it was dominant in these old industrial areas.

3 The concept of an 'institutional tissue' refers to the existence of institutions in the spheres of both the state (for example, local government) and civil society (for example, trades unions and political parties). Indeed, one of the characteristic features of old industrial areas, typically under the sway of social democratic politics, is that the boundaries between state and civil society were both blurred and

permeable. Not least, key individuals typically were active in institutions on both sides of the state/civil society divide. It is this partial overlapping of institutions, in terms of membership, interests and function, that characterizes the institutional tissue of such areas as 'thick'. This dense mutual interpenetration of key individuals, interests and institutions is crucial in shaping views of what is desirable and possible by way of stasis and change in such areas and so in reproducing those same institutional structures, especially in conditions of economic well-being.

4 In some cases, such as the United Kingdom Enterprise Zone experiments, new central government small-area policies were specifically created with the intention of promoting localized solutions, buttressed by central government expenditures, while presented as evidence of an emerging enterprise culture.

5 Certainly the evidence from areas of successful small-firm growth elsewhere in Europe would support such a conclusion (for example, see Hudson and Williams, 1995).

6 The London Docklands UDC operates on a different scale from the remaining nine. In 1990–91, for example, its budget from government was £333m, while that of the remaining nine was £550m. Moreover, its proximity to the City of London offers a very different environment from those of the other UDCs.

7 In practice, the distinction between state and market is at best one of degree; markets are social constructions, which depend heavily on state activity and regulation, as the events of the 1980s in the United Kingdom and elsewhere sharply revealed.

Chapter 11

Making music work? Alternative regeneration strategies in a deindustrialized locality: the case of Derwentside*

11.1 Introduction

During the 1980s there was considerable academic and political debate about the extent to which and the circumstances within which localities could determine their own developmental trajectories in a rapidly changing and increasingly volatile globalized economy. There were those who were optimistic about the possibilities of local self-determination (see, for example, Cooke, 1990). Equally, there were those who were much more sceptical as to the extent to which this both would and could be possible, especially for those localities that were already marginal to the main currents of the accumulation process (see, for example, Hudson and Plum, 1986; Beynon and Hudson, 1993). Beyond this, there are a series of questions relating to cultural and political – as opposed to some narrow view of economic – change that are related to, but ought not to be reduced to, those of change in the economy and its determinants.

In this chapter I want to consider the extent to which it was possible to construct such a locally controlled economic and social development path in one particular deindustrialized locality in the northeast of England – the town of Consett and the surrounding District of Derwentside. In particular, I want to examine, critically, the extent to which it was possible to pursue a genuinely radical alternative to the conventional reindustrialization strategy, encompassing cultural and political as well as economic change, centred around the triple strands of truly endogenous regeneration informed by the interests of a majority of local people, cooperative development as an organizational form through which to pursue this objective, and music as the substantive basis of this alternative approach.

The origins of this alternative can be traced back to the period immediately following the failure of the campaign to keep open the Consett works of the British Steel Corporation (BSC) in 1980. Belatedly, the threat to the place,

* First published 1995 in *Transactions of the Institute of British Geographers*, **20**(4): 460–473, Royal Geographical Society, London

which the closure of the steelworks represented, had generated community opposition to the BSC's proposals. There was a strong desire to channel this concern constructively and to explore possible options for the future rather than let it simply fizzle out in bitter recrimination, apathy and resignation in the aftermath of defeat. What might represent a viable cultural, economic and social regeneration strategy that challenged the conventional wisdom of rein-dustrialization? A small group of activists, who had fought to keep the works open, had one view as to how the articulation of a very different alternative vision of the locality's future could be encouraged and facilitated. Building upon a long tradition of popular music as part of a working-class culture there, which extended back virtually to Consett's establishment well over a century ago, this search for an alternative initially began as a voluntary project in a derelict Miners' Welfare hall. By the end of the decade, however, it had grown in influence and become formalized in the Making Music Work programme, centred around a new cooperative, Northern Recording, funded by the state and the subject of considerable local and national media attention.

The structure of the chapter reflects this sequence of events. First of all, the historical geography of the origins and subsequent evolution of industrial and urban development in northwest Durham as a direct consequence of the establishment of capitalist social relationships there will be briefly sketched out. Then consideration will be given to the transition from a locality dom-inated by a paternalistic capitalism to one in which both production and reproduction were increasingly either regulated or directly controlled by the state – for this was and is an intensely state-managed locality in a region that itself can be characterized as state-managed for much of its history (see Hud-son, 1989a). From the late 1950s the material basis of the old coal- and steel-based economy of the locality began to be steadily eroded as a consequence of choices over national political strategy and changes in the international economy, reaching crisis point in 1980 with the closure of Consett steelworks. This served as the catalyst that triggered frenzied attempts to construct an alternative economic basis for the area via a conventional reindustrialization programme on what was claimed to be a 'bottom-up' rather than a 'top-down' basis (see, for example, Stohr and Fraser Taylor, 1981; Stohr, 1990). It also, however, stimulated an attempt to construct a radically different vision of the locality's future, based around a genuinely 'bottom-up' strategy that reflected the aims and aspirations of local people, and sought to empower them, with popular music at the heart of a strategy that was at least as much political and cultural as it was economic in its aims.

There was a clear recognition that activities other than music both could and should form the basis of such an alternative approach, both within north-west Durham and in other areas. There was no necessity that music form the focal point but rather a recognition that it could – and did – in northwest Durham in the early 1980s. As John Kearney (1990) has emphasized, a key characteristic of the working-class culture of that area over the last century or so is the development of a tradition of music being used by communities not only to provide recreation and entertainment but also to express dissent and challenge the orthodoxies of prevailing modes of thought. One implication of this in the early 1980s is that the alternative vision of the future for deindustrialized localities that was then taking shape in, around and through Consett was emphatically *not* cast in the mould of fierce competition between places for mobile investment but rather reflected and respected their specific cultural histories and identities.[1]

11.2 From nineteenth century work camp to state-managed locality

Iron production began in northwest Durham in 1841, based on locally available raw materials and immigrant labour, much of which came from Ireland.[2] The ethnic mix of the migrants and the way in which they settled in what was then not so much the one town of Consett but two separate work camps was very influential in shaping the initial pattern of capital–labour relations and internal social and spatial differentiation within the emergent working class of Consett. Gradually, however, the divisions within the workforce became more muted as labour began to organize into trades unions. At the same time, capital–labour relations were moulded into profoundly paternalistic forms as the owners and managers of the Consett Iron Company managed to convince their workers that they had compatible, if not identical, interests. In addition, gender relations became deeply patriarchal as women's unpaid domestic labour became central to the reproduction of the Iron Company's male labour force. With some cyclical fluctuations, this laid the basis on which Consett remained a profitable location for private-sector iron and steel production until the 1960s while the Company's collieries in the surrounding villages continued to provide coking coal for steel production.

In time, the Company itself became more directly involved in the reproduction of its workforce through the provision of educational and medical facilities as well as company housing. It also became prominently involved in local politics, its representatives standing as the Progressive Party, espousing liberalism and dominating local politics for over three decades. In this way it was able to ensure that its interests were represented in the actions and decisions of local government once this was established in the latter part of the nineteenth century. As Tony Kearney (1990, 22) has graphically put it: 'In short, the Company didn't dominate Consett. The Company was Consett.'

It is significant in the present context that one of the few spheres in which the influence of the Company was relatively minimal was in the world of clubs, pubs and popular music which emerged and subsequently remained as an integral part of working-class culture in northwest Durham from an early stage. The music that was then made in the locality reflected two dominant influences. First, the Irish origins of many of the recent immigrants to the area were an important influence. They brought with them their own folk music while their Irish origins also became a topic for exploration in the new popular music of the area. Secondly, it was influenced by the character of work in the coal mines and iron works and of life in the surrounding communities; this encompassed issues ranging from death at work, in dangerous jobs and bitterly contested strikes (in which the Irish immigrants were often initially pilloried as strike-breaking blacklegs), through the gendered differences in roles in the community, household and workplace, to the joys of a hard night's drinking in the local public house or a day at the races on neighbouring Tyneside.[3]

Popular music and the world of clubs and pubs of which it was an integral part therefore formed a rare segment of working-class life in civil society that was relatively autonomous and insulated from the generally pervasive influence of the Company. The implication of the Company's general domination of life both inside and outside the works is that as the locality became transformed into a meaningful place for its residents, it did so in a partial and one-sided way. The meaning of the place was refracted through sets of institutional filters that reflected the interests of the Consett Iron Company; people there defined their own interests in this way. As a consequence, a deep and lasting 'culture of dependency' upon the Company and its view of the world diffused in a powerful and profound way into an emergent working-class culture and people's understanding of their place and what it meant to them. In contrast, making music provided a medium through which opposition and resistance to this culture of dependence and the power structures that underpinned it could find a dissenting voice through music. It provided a medium through which the concerns, fears and interests of the members of an emerging working class could find expression in their own terms and language. In this way, in the public setting of pubs and clubs which helped make the experience collective rather than simply individualized, making popular music helped keep alive an understanding that there were alternatives to the dominant local social order and the wider relations of capitalism of which it was an integral part.

Gradually, over a period of decades, however, the state replaced the Company in the structuring of social relationships within the locality. This process began in the inter-war years as the Labour Party rose to political ascendancy, taking control of the local councils and the services that they provided. This provided both institutional channels through which working-class needs for more and better quality public-sector housing provision and services of various sorts could be articulated and limited powers and resources to respond to these expressed needs. In effect, the state, locally, within the boundaries defined by national legislation (which were themselves considerably pushed back after 1945 as the new Labour government implemented its package of radical social reforms), socialized an increasing share of the costs of reproducing the Company's workforce but at the same time helped effect a more generalized improvement in living standards there. In redefining the social relationships that influenced housing and service provision in this way, people's relationship to their place and its meaning to them potentially changed. It is, however, important not to overemphasize the extent of this change in practice. For the ascendant Labour Party practised very moderate reformist politics, grounded in a conservative form of Labourism that was deeply coloured by the preceding liberal tradition, the paternalism of the Company and the incorporation of senior trades unionists into the Company's view of the world. For the world of senior local Labour Party notables and senior trades union officials who worked for the Company was a small, interpenetrating and overlapping one. If previously it was the Company that knew best, now it was the Party and local politicians, committed to a deferential form of politics in which the Company's interests were never far from the forefront of the agenda and in which local politicians would tell their constituents what was going on, what they ought to think about events, and what they ought to do. Nevertheless, while there were strong strands of continuity, the switch from Company to state and Party as the dominant institutions shaping the local mode of regulation had important effects in shifting the political focus in struggles about localized conditions of reproduction.

It was not until the post-war period that the state became directly involved in production in the locality, initially through the nationalization of the coal mines in 1947. Then in 1967 the 14 largest steel companies were nationalized, including the Consett Iron Company. One effect of these nationalizations was to move key strategic decisions affecting the future of the locality to corporate headquarters in London in a state-managed version of a branch plant economy. Again, this had significant impacts in relation to key decisions about local levels of employment, investment and output. As it became clear that nationalization was simply a prelude to savage rationalization in both coal and steel (see Hudson, 1986b), it also became evident that the economic basis of northwest Durham was becoming increasingly precarious and unstable. A locality that in many ways was emblematic of not only Labourism but, more generally, the post-war One Nation project, was increasingly teetering on the verge of crisis as that grand project itself reached its limits (see Hudson and Williams, 1995).

11.3   Localized crisis: the closure of Consett steelworks and the collapse of the old order

The announcement of the closure of the steelworks in 1979 was not, therefore, entirely unexpected yet in many ways was seen as posing a severe threat to the locality, removing the last major component of the old industrial economy. Whereas the earlier colliery closures had been accepted because of the promise of alternative employment in branch plants attracted via central government's regional policy, such investment was very unlikely to materialize in northwest Durham in the very different environment of the early 1980s, not least because the new Conservative government was intent on a severe reduction in regional policy expenditure. Indeed, several of the branch plants that had previously been attracted to the locality also closed around 1980. As a result, total industrial employment fell by 8000 between 1978 and 1981, some 63 per cent of the 1978 total, as the effects of the steelworks closure were added to those of other factories. Under these circumstances, not surprisingly, the closure of the steelworks was seen as threatening the very rationale of the place and so was contested but without any real chance of the anti-closure campaign being brought to a successful conclusion (for an analysis of why this was so, see Hudson and Sadler, 1983a, 1986a).

Above all, the closure raised profound questions about the locally dominant forms of political thought and practice and their adequacy in the new harsh climate of Thatcherism's Two Nations project. For decades, local government and the trades unions had been deeply influenced by a conservative Labourism – predominantly the domain of the middle-aged and male, operating via deference and patronage. Very close links had developed between the trades unions, the local authority and the local Labour Party; many Labour councillors were or had been union officials. The trades union hierarchy and the local Party élite formed a tightly cohesive group in terms of shared attitudes and values. There was, therefore, an easy slippage from Company paternalism to state dependency as the politically dominant Labourism put its faith in a statist interpretation of socialism within the framework of deferential politics.

Working-class interests were to be advanced and protected through state ownership of key industries, a centrally directed regional industrial policy to replace any (male) jobs lost in the nationalized industries, and the provision of public services and housing. In this way, refracted through the lenses of dependency upon the state rather than those of the paternalism of the Company, people came to understand their place and what it meant to them, living, learning and working within parameters defined and guaranteed by the relationship of the locality to the state. The closure of the steelworks revealed, in very dramatic fashion, that what had previously been taken for granted and seen as the 'solution' now appeared to be the 'problem'. Nationalization turned out not to be the means through which the interests of working-class people in their places could be defended and promoted but to be a mechanism of state-sponsored industrial restructuring that was just as savage as any private sector restructuring, if not more so (see also Beynon *et al.*, 1991). Recognition of this swept away many of the previous old certainties about life in Derwentside. Working in industries such as coal mining and steel making had provided a material basis for a working-class culture which provided security and an enduring sense of place through positive attributes such as collectivism, discipline and social solidarity. This culture also had its darker side, however: for example, in terms of a deeply gendered division of labour and in the construction of communities that were socially conservative, introspective and resistant to change in the local established order. Deindustrialization clearly brought major cultural dislocation and very considerable human costs as old certainties melted away. In doing so, however, the collapse of the old order also opened up space for alternative 'solutions', alternative definitions of the locality's future, to be put on the agenda. It was a matter of considerable practical importance as to whether these would accept or challenge the parameters of the political economy of Thatcherism and the competitive politics of the global marketplace. This, however, raised questions about the availability of local cultural and political resources that could be drawn upon and mobilized in constructing these alternatives. In particular, it raised the question of the extent to which, and forms in which, such resources existed that were not enmeshed in the traditions of dependency and Labourism that had come to a position of dominance in the locality.

11.4   Constructing an alternative development trajectory, I: the reindustrialization strategy

Following the closure of Consett's steelworks, Derwentside District became the focus of a vigorous campaign to redefine its role as a successful centre of industrial production. This bore the marks of the locality's history in its prioritization of private sector interests, in its subjugation of community interests to national ones, and in its insistence on the need for cooperation across the private–public sector boundary as well as the influence of Thatcherite rhetoric and its particular view of the enterprise culture and the efficacy of the market (see Hudson and Williams, 1995). The reindustrialization programme set out to create a very different new local economy. In strong contrast to that which had so precipitately collapsed, which had revolved around past dependence on employment in the big plants of nationalized industries and multinationals, it was to be a much more diversified local economy, replacing the jobs lost with new ones born of a flourishing enterprise culture in locally owned small manufacturing firms. New institutional arrangements involving cooperation between local and central government and the private sector, exemplified most clearly in the creation of the Derwentside Industrial Development Agency (DIDA), were established to facilitate the emergence of this radically different local economy.

Implementation of this programme involved increased public expenditure within Derwentside, providing a familiar mixture of infrastructure to provide enhanced general conditions of production, and the provision of loans and grants to private sector companies, to which were added direct wage subsidies and free information and advice, especially in formulating Business Plans. In short, the reindustrialization package offered '. . . probably the best project package in the UK' (DIDA, n.d.). Clearly, despite the rhetoric about the need to cut back the scope of the state and promote the market and private sector investment, if an enterprise culture *was* successfully to be established in Derwentside, it was to be heavily underpinned by public expenditure. In part, this was because there were clear ideological and political advantages to be gained by central government if it could plausibly claim that an enterprise culture had taken root and was flowering in Derwentside. For if it could be implanted there, in a locality that was a bastion of Labourism and dependency upon the state, then it could surely be established anywhere. For this reason, central government was prepared to commit additional public expenditure to Derwentside, a commitment that is also indicative of the wider contradictions and tensions within the Thatcherite view of enterprise (see Hudson and Williams, 1995).

In some respects, the programme could be seen to have been successfully implemented. Between 1980 and 1988 £50 million of public expenditure stimulated £70 million of private sector investment. It was claimed that over 200 businesses were involved in the programme, with a substantial representation of both 'high tech' and locally owned new companies (Carney, 1988). In fact, very few of the new companies could in any meaningful sense be described as 'high tech' while the most highly publicized have been set up by 'entrepreneurial immigrants' rather than 'local heroes' (to use the terminology of Caulkins, 1992), indicative of a reproduced dependence on inward investment and another set of decision makers with interests and roots outside the locality. Moreover, if the District has become a profitable location for such activities, it is also one characterized by great volatility as companies were established and then shortly afterwards collapsed and died – including some of the highest profile investments. This was so despite the availability of generous financial support and labour market conditions that permitted a very careful selection of workforces (see Hudson and Sadler, 1991). While claims were made by DIDA that several thousand new manufacturing jobs were created in the 1980s, central government labour market statistics provided through the National Online Manpower Information System show that there was in fact a net growth of some 1300 in manufacturing employment between 1981 and 1990. Thus the 1980s saw the replacement of fewer than 20 per cent of those jobs lost between 1978 and 1981 alone.

One consequence of this is that, despite a small fall of 400 in service sector employment over the same period, Derwentside has become relatively much more dependent on the service sector as a source of employment (see Hudson et al., 1992). This also reinforced its dependence upon the state as an employer since, despite central government restrictions on the scope of local government activities, the main source of full-time service sector jobs (over 4000) was in local government. A majority of the private service sector jobs were part-time and poorly paid, largely taken by married women. Thus there was also a significant change in the gender composition of the District's waged labour force, without any corresponding change in the gender division of labour within households, even when married women had become the main or sole wage earner. The old gender division between waged and unwaged labour may have changed somewhat but the legacy of the past remains a powerful one.

Moreover, the great imbalance between jobs lost and jobs gained, between labour supply and demand, meant that a greatly increased number of people in the District became surplus to the labour power requirements of both private sector companies and the state within the 'formal' sector of employment. This was expressed in several ways: net out-migration; rising registered unemployment, still at a level of around 3500 in 1990, with a very high proportion of these effectively permanently unemployed; and, most significantly, a growing number of people absorbed by central government temporary employment and training schemes, which temporarily removed them from the ranks of the registered unemployed – almost 3000 by the end of the 1980s. For many people, then, the 1980s simply meant a change in the dominant form of state regulation and control within the locality rather than newly found freedom and empowerment within an emergent and flourishing enterprise culture. This starkly revealed the intensely socially divisive impacts of pursuing a Thatcherite notion of 'enterprise culture' in a locality such as Derwentside. It also emphasized the extent to which the political economy of

Thatcherism involved selective re-regulation just as much as it did deregulation. Certainly there was no longer the former dominance of the nationalized industries as the main source of waged labour in the locality, or of the social structuring mechanisms that flowed from them and ordered local society. Their role, however, has been taken over by other branches of the state, concerned with social control and the management of unemployment, and a continuing dependence upon state transfer payments as the main form of monetary income for many of the inhabitants of the area.

11.5   Constructing an alternative development trajectory, II: Making Music Work and cooperative development

Clearly, many people within Derwentside found no place as workers within the new economy but became clients of the welfare state (to borrow a term from Pugliese, 1985). In this way, they remained tied into a web of dependency relationships that shaped and constrained what they were able to do. They remained imprisoned by rather than became empowered by the changes that were wrought there in the 1980s. The Making Music Work programme sought to address these issues directly and to *exemplify* a radical alternative to the conventional wisdom about regeneration in two crucial respects. Firstly, premised on a belief that the reindustrialization programme provided at best an inauthentic and cosmetic 'bottom-up' approach, it was to be a *genuinely* and *authentically* bottom-up approach, with its aims and content defined and pursued by local people in their own terms. In so far as it involved institutional forms, these were to have a role that was to be enabling and facilitating rather than directing and controlling. It embodied a radically different conception of the meaning of releasing the abilities, talents and enterprise of local people from that of the officially dominant reindustrialization programme. Secondly, it was to focus upon music both as an activity around which economic reconstruction could occur and as a vital living element of working-class history and contemporary working-class culture in the locality. Music was a medium through which continuity with the locality's cultural history could be maintained and through which political issues relevant to its future could be raised.

It is very important to stress that Making Music Work (MMW) was seen as an exemplary project, *demonstrating* the possibilities of an alternative approach rather than itself seeking to solve all the profound problems of the District. From the outset, for example, those involved with MMW were at pains to stress that there was no intention of seeking to solve the District's unemployment problems via MMW. More generally, they sought to damp down quantitative expectations that could not be satisfied, for this would only serve to deflect attention from the *qualitatively* different aspects of its approach that were seen to be of the greatest importance. It was also emphasized from the start that there was no implication that music could or should provide the focus of local regeneration strategies in other places (although in fact it has done so in Liverpool, Newcastle and Sheffield). The point that was emphasized was that local strategies should reflect local strengths and be sensitive to local circumstances.

There have been a number of distinctive phases in the evolution of the MMW programme, in part a consequence of the difficulties of attracting resources to support it. These are briefly outlined below. One consequence of this history is that while the general guiding principles of the programme have remained invariant, both the specific objectives and to a degree the means chosen to achieve these have altered in response to the requirements of funding agencies (with the clear implication that to a degree at least this must compromise attainment of the programme's objectives). Thus the history of the project will first be sketched out before attempting some evaluation of the extent to which it has been able to offer a radical challenge to the conventional orthodoxies concerning regeneration.

11.5.1   The genesis of the project, 1980–88

The closure of Consett steelworks in 1980 was the catalyst for the formation of Consett Music Projects (CMP), a voluntary community group organized by former steelworkers and unemployed young people, guided by a very different vision of the locality's future from that embodied in the conventional reindustrialization strategy. It was, in its origins, very much a working-class project, though by no means one with which all of the locality's working-class population identified closely. At the same time, it was a project that attracted more widespread support, socially and spatially. It became an emergent focal point of opposition to conventional orthodoxies about regeneration strategies in deindustrialized areas. CMP sought to draw upon the traditions of popular music in northwest Durham as a bridge between the past and the future. Central to this was a recognition that grassroots, bottom-up cultural and social development would be crucial to the successful regeneration of the area in the context of the profound economic and political changes that were occurring there and more generally within the UK in the 1980s. As one young woman participant in MMW put it:

> Basically all attempts by outsiders to regenerate the community have failed. They've promised us all sorts of things but nothing has been delivered. Now it's our turn.

In turn, for them to have 'their turn' implied making available the capacities and resources that local working-class people would need to pursue this approach. This would be necessary to enable them to provide their own analyses of their personal problems and possibilities, as well as those of their community, and develop appropriate institutional forms and arrangements to enable them effectively to pursue alternative developmental trajectories.

Such resources were not readily forthcoming, however. Working, therefore, of necessity with a minimal budget, CMP provided community access to rehearsal and basic recording facilities in a virtually derelict Miners' Welfare hall and developed forms of educational practice which encouraged members of the wider musical community to make use of these facilities in developing their own ideas. From the outset, CMP consciously and deliberately also involved itself in collaborative projects with local and national trades unions and other campaigning groups, addressing a wide range of social issues and avoiding a narrow territorial parochialism in its approach. Perhaps the most prominent of these wider activities emerged during the 1984–85 coal miners' strike, with the 'Heroes' album produced in conjunction with local miners' wives to raise money for miners' support groups. The album was promoted via a tour which began in east Durham Miners' Welfare halls and ended in a sell-out concert in London's Royal Albert Hall. In 1986 CMP was instrumental in the establishment of the Tommy Armstrong Memorial Trust, established with the aim of restoring Armstrong – 'the pitman's poet' – to his rightful place in the working-class culture and history of northeast England (see Forbes, 1987). In these ways, with its activities firmly grounded in popular music and using live and recorded music to focus on a wide spectrum of working-class issues and causes, CMP established a national reputation as a politically progressive and educationally innovative organization. While it was undoubtedly place-based and firmly grounded in the realities of life in northwest Durham, it was certainly not regressively place-bound (see Beynon and Hudson, 1993).

11.5.2   The formation of Northern Recording and the Educational Support Grant (ESG) phase, 1988–91

In 1988 two separate, though closely related, developments took place which were of major significance within Derwentside. Firstly, funding was obtained from the Department of Education and Science's ESG programme, with co-funding from Durham County Council, which allowed the Making Music Work (MMW) programme to be established. Although the budget was very modest – £227,000 over 30 months between 1988 and 1991, much of which was required for refurbishment of the building and purchase of recording equipment – it represented an enormous increase in resources as compared to the previous eight years of CMP. Indeed, over the life of the ESG phase, it was supplemented by another £52,000 in grants and loans from Northern Arts, Derwentside District Council and Durham County Council. Secondly, four founder members of CMP formed a cooperative company, Northern Recording, as the most appropriate way of contracting with the County Council to deliver the MMW programme whilst at the same time demonstrating that there were alternative forms of social relations of production to those that were central to the small firms of the conventional reindustrialization programme.

Building on the previous experiences of CMP, the emphasis within the MMW programme was, from the outset, placed very firmly on 'user-led' and 'user-centred' learning. The content of the various components of the programme was to be determined by users to meet their various perceived needs. Their success was to be assessed in terms of their practical achievement in solving real problems. It would perhaps be too simplistic to claim that MMW simply found out what people wanted to do, then found ways of helping them to do it within the constraints of the situation in which it found itself – although that was clearly the intention of those involved in delivering the programme. It is, however, undoubtedly true that the direction and content of the various components within MMW were heavily influenced by users in negotiation with members of the cooperative. One consequence of this, given the very wide range of people, male and female, of all age groups, who became involved with MMW, is that an equally wide range of types of music were made there. Some participants in the MMW programme, often older ones, had interests in performing the traditional folk music of the area for no other reason than the enjoyment that it gave them, or to ensure that there was a permanent record of local songs and the stories that they told. In contrast, young people experimented with writing and performing their own modern music, often reflecting on their own lives in the locality in the lyrics of their songs. Older local professional and semi-professional rock and pop musicians used MMW's facilities for rehearsal and recording. Both of these latter groups then took their music out of the rehearsal room and the recording studio to wider audiences, predominantly in local clubs, pubs and concert halls but also by producing tapes and albums for dissemination to more distant audiences. Those involved with MMW typically filled a crucial facilitating role in making the varied production arrangements necessary for live performances and recordings.

What MMW did *not* do – because those who sought to use it did not see music in these terms – was to foster the 'traditional' conception of serious classical music and the arts and this was to lead to friction with elements of the educational and arts worlds (see, for example, Harris, 1988). For MMW did not see its educational role as indoctrination within a different – though equally valid – musical culture but rather saw it as enabling people to make their own music on their own terms. The diversity of musical tastes which found an outlet for their expression through MMW was nevertheless truly impressive (and though empowerment for most in this way led to commercial success for some, for most it did not do so; this was not, however, seen as a problem – a point developed below).

Despite this diversity of musics and music makers, however, the primary focus in this phase, of necessity, was the Department of Education and Science's Category XIX ESG 'Learning by Achievement' programme. This was established to support projects which were targeted at young unemployed people at risk in deprived urban environments. The intention was to focus on the 14 to 21 years age group, with 90 per cent of those involved aged between 16 and 21 years and no longer in full-time education and 10 per cent aged 14 to 16 years who were expected to shift into the former category on leaving school. The objective of the ESG projects was to help establish new relationships between Local Education Authorities and the voluntary sector to help respond to the needs of these disadvantaged and deprived young people. In turn, the ESG activities were intended to assist the mutual development of innovative work with socially disadvantaged older adolescents who had been unresponsive to the efforts of the mainstream youth services, who had dropped out of government training schemes or had been unable to find a place on such schemes. In this way the Department of Education and Science (DES) intended to encourage innovation in community educational provision, and it attached some importance to the role of the creative arts in this. This emphasis on people rather than institutions and upon the creative arts as a medium was perfectly consistent with MMW's own emphases and philosophy. Accordingly, MMW sought to use ESG funding to further its objectives and to build on existing innovative projects within the locality based on popular music.

The ESG and other resources enabled the derelict Miners' Welfare hall to be transformed and upgraded, with the provision of state-of-the-art recording facilities. From the outset, the MMW programme exceeded the quantitative targets specified for it. The DES specified between 15 and 20 individual users each week in the revamped studio facilities. This was exceeded in every week in the two years that the programme was operational, peaking at 260 different users in a single week. As well as unemployed young people, MMW developed links with the six comprehensive schools within Derwentside, so that pupils from them worked in the studios as part of their school studies, experiencing not only the processes of making music but also of working within the context of the social relationships of a cooperative. In all, over 16,000 young people were directly involved in the various activities of MMW – which included concerts and live music venues as well as work in the studios – in these two years. In part, this involved MMW in continuing to collaborate and liaise with other organizations to develop a range of other initiatives and projects within Derwentside District (for fuller details, see Hudson, 1991a). As before, MMW became a focus of considerable national as well as local attention as an innovative music project.

There was, however, another implication of the transformation of the voluntary CMP into the state-funded MMW. For while this brought financial resources, it also led to a degree of formalization in the running and organization of the project that took it beyond the control of those directly involved in delivering it in a way that had not been the case in its earlier voluntary phase. State funding brought with it a requirement for monitoring of the MMW programme and the effectiveness of its use of public expenditure. In response to this, a Steering Committee was established to oversee the development of the MMW programme while an impressive list of some 40 influential supporters drawn from the worlds of education, music and politics was also assembled.[4] While a few individuals from the latter – such as Alan Hull and Tom Robinson – were regularly involved with MMW's activities on an ongoing basis, their primary role was to add authority, respectability and legitimacy to the work of MMW, especially in relation to the wider world outside northwest Durham. On the other hand, the Steering Group had a much more frequent involvement in the evolution of the programme. It was made up of County and District councillors (including Don Robson, a Derwentside councillor and Labour leader of Durham County Council, and David Hodgson, Labour deputy leader of Derwentside District Council and also a County councillor), officers of Durham County Council (including John Kearney, who had been involved with CMP from the outset) and Northern Arts, the members of Northern Recording and another sympathetic academic, Huw Beynon (now Professor of Sociology at Manchester but then at Durham) who had been involved in a number of struggles contesting closures elsewhere in the region as well as in Consett itself. The Steering Committee met at monthly intervals, in the Old Miners' Hall. As well as holding a formal brief to monitor the progress of MMW, it also became a forum for debate about the direction in which MMW should develop, and the priorities that should inform its activities. While MMW remained true to itself as an essentially working-class project, there were persistent tensions within the Steering Committee, which echoed wider structural constraints upon projects that sought to empower local people and challenge the dominant orthodoxies of a capitalist economy. Most fundamentally, these tensions reflected a difference between a minority who saw the touchstone of success as commercial viability (and for whom MMW generating another Dire Straits would have been the ultimate vindication of the programme) and those who saw MMW primarily as a route to political mobilization and radicalization. Whereas the former group conceptualized empowerment in terms of success in the market, the latter group saw it much more in terms of facilitating the articulation of a different view of Consett's future that challenged the logic and rhetoric of the market.[5]

11.5.3   The post-ESG phase, 1991–93[6]

It was clear in 1988 that ESG funding from the DES would cease in 1991, in the expectation that the local authority would then increase its funding to compensate for this. While Durham County Council agreed to maintain its past level of funding over the period from April 1991 to March 1992, increasing central government control of its expenditure precluded any increase in its funding of MMW. The effects of this translated into a 55 per cent reduction in income compared to the previous 12 months. Under these circumstances, an urgent search for additional sources of funding began. It was not without success. An additional £48,000 of support was obtained, mainly from the European Social Fund but with contributions also from Derwentside and Durham District Councils. The ESF funding was for the calendar year 1992, however, so members of the cooperative took a substantial cut in salary in order that the MMW programme could continue to be delivered within Derwentside in 1991. There were two further important implications of these changed funding arrangements: MMW's activities were no longer confined to Derwentside District but extended to other areas of County Durham; and they were no longer formally limited to those aged 21 years or less. The target groups remained broadly the same, however: the existing unemployed, those in danger of becoming unemployed (particularly in Easington District, threatened with the closure of its remaining collieries; see Hudson, 1992d) and self-employed musicians.

In April 1992 the County Council approved further funding of £51,400 for the 1992–93 programme. Of this, 80 per cent was eligible for matching funding of £41,200 from the European Community's ESF and this was made available through the Durham and Cleveland Integrated Development Programme. Although Derwentside District Council reduced its support because of charge capping, additional resources were obtained from Durham County and District Councils which more than compensated for this.

Despite the 1991 cuts in funding, and continuing uncertainty over future funding, the MMW programme continued to have positive local impacts. A further 6000 people became involved with, or experienced through concerts or other forms of live performance, the effects of the activities of the programme in 1991. The quality of the programme was further enhanced from the high levels already established. This was particularly apparent in relation to work with school leavers and in the live music programme. The 1991–92 programme was described by the DES as 'of impressive standard and quality' and was instrumental in Durham County Council receiving the Diploma of Merit for 'excellence in music teaching' from the National Council of Local Education Authorities. This was one of only five of such awards (for fuller details, see Durham County Council Community Education Department, 1991).

This restored level of funding allowed MMW to develop more intensive work programmes, with a greater vocational emphasis. The ESF funding provided bespoke training packages tailored to the varied requirements of individual musicians and groups and was intended to encourage self-employment within the regional music industry. The programme ranged from introductory courses to more advanced support for existing self-employed musicians, by making available access to core services which included recording technologies and consultancy. Considerable emphasis was placed upon technology transfer and the deployment of new technologies to improve performance. The remaining 20 per cent of non-ESF eligible work included preliminary contact with people who were currently ineligible for entry to the mainstream programme but who might well become so at a later stage, as well as 'unsupervised' access to rehearsal facilities, the promotion of live music events and programme development.

11.5.4 Evaluating the Making Music Work programme

As with the conventional reindustrialization programme, there are competing views as to the effects and effectiveness of the MMW programme in Derwentside.[7] Its supporters argue that it does indeed provide a radically different approach to tackling the problems of cultural, social and economic regeneration in a severely deprived locality ravaged by deindustrialization and the effects of a political strategy that marginalizes such locations. There is undeniable evidence that supports such an evaluation of the programme. In this interpretation, MMW does indeed *exemplify* a radical 'bottom-up' alternative, based on responding to locally identified and self-expressed needs. Music is the medium through which this approach is pursued in this place and is integral to the programme precisely because it draws upon and provides a point of articulation with a long-established local working-class culture of music in clubs and pubs. At the same time, it provides a medium through which young people, especially those who are in various ways alienated by and from mainstream educational programmes, youth service provision and government training schemes, can express both their frustrations and aspirations in their own terms. Its 'user-led' approach was (and is) recognized as highly successful. As one local school teacher (not a music teacher), commenting specifically on the schools' programme, put it:

> The people here [at MMW] create a very relaxed atmosphere for the children to work in. They are encouraged rather than brow beaten, so to speak, and I'm sure that brings out the best in them. I've been quite amazed at some of the results the children come up with and some of the ideas they put down . . . The end product is amazing at times, because I know the same children in school and when I come up here I can see a different attitude in them because of the environment and the atmosphere that is created in this place . . . the method that's used here certainly extracts the best from the children themselves. . . .

Comments such as these illustrate the ways in which MMW helped to release the creative talents of local people in ways that they themselves defined. As a result, MMW became a focal point for resistance to the dominant conventional wisdom of economic regeneration and one around which an alternative, encompassing a more locally sensitive and broadly based cultural and political definition of regeneration, could be articulated. This may be of particular significance in providing a medium through which disaffected young people, with no direct experience of or identification with the place's traditional industrial past but who nonetheless equally forcefully reject the enterprise culture which is supposed to succeed it, may nevertheless be given a meaningful stake in their place. As one of the young people involved with MMW commented:

> The atmosphere in the studio is brilliant. My parents back me up and listen to my music. It's making me think about my future.

Another, reflecting on a course that he'd taken part in at MMW, had this to say:

> It makes me realise that it [the music business] is not just about fame and fortune that goes into records, it's also about a lot of hard work.

There may be lessons to be learned from this in relation to preparing young people for the realities of life in the 1990s and for appropriate regeneration strategies in other deindustrialized localities.

There are undoubtedly quite a few examples of young people becoming professional musicians or finding other paid employment in the music industry (for example, as sound engineers) as a direct consequence of their involvement in the MMW programme. Equally, the immediate direct impacts of the programme in providing employment or facilitating people becoming self-employed within the music business have undeniably been rather limited, but these were never intended to be the criteria against which the success or otherwise of the programme was to be judged. Nor is there any evidence that the vast majority of those who have been involved with the programme were simply or primarily motivated by economic self-interest but rather by a curiosity to see what they can achieve in and for themselves in other ways. The criterion of success was to be the extent to which people were empowered, enabled to carry through their own projects and create and assert their own identities. There is no doubt that this criterion has been met, and many times over. Consider, for example, the following comments from a member of the audience at a concert given by local children in Consett and organized by MMW:

> I'm sure that it's had a great effect on their lives. I could tell at the end of the show just by looking at their faces that they had got an enormous amount out of taking part in the show, the preparation, the rehearsal . . . the discipline that was involved in that . . . it takes an awful lot of courage to go on stage in front of an audience to sing a song that you have written yourself . . . and the quality of the songs was excellent, really good.

One ten-year-old involved in making a recording summed up the reaction of many in terms of a sense of both excitement and achievement:

> It's very exciting. We have had lots of rehearsals . . . and our recording sounds really good. It's much better than I expected.

Futhermore, participation in the programme has had important demonstration effects as to the possibilities of cooperative development as Northern Recording emerged as the vehicle through which MMW was delivered. Such cooperative approaches are in principle of generalized applicability, by no means limited to activities associated with music. In any case, (self)-employment creation *per se* is seen as of much less significance than the cultural and political effects of demonstrating the possibilities for truly endogenous development. The thousands of people who have been directly involved in MMW's various activities and the millions more who have learned of these at second hand via widespread national media coverage have come to see that there are at least possibilities of alternative approaches which, albeit within strict limits, are more within the control of those affected by them.

There is, however, another and much less optimistic interpretation of the MMW programme. One element of this alternative perspective would point to the limits to cooperative development strategies within the wider context of a capitalist economy. The salary cuts that members of Northern Recording necessarily imposed upon themselves in 1991 are but one very clear illustration of this. In the specific context of developing a music-based economy, Consett and Derwentside remain profoundly marginalized in relation to the centres of power and circuits of capital that are decisive in structuring the geography of the political economy of the modern popular music industry. Such disadvantages are unlikely to be overcome, irrespective of the volumes of praise and good publicity that the activities of MMW receive. Indeed, this difficulty of regaining some sort of effective control of local economies is by no means specific to those based around popular music but in fact is endemic and a more deeply embedded structural constraint (see also Hudson and Plum, 1986).

Perhaps more fuhdamentally, it is by no means obvious that in fact the MMW programme has succeeded in breaking the ties of dependency on decisions taken elsewhere and restored control of their fate to local people, even within its own highly circumscribed sphere of operations. For it remains crucially dependent upon decisions by the European Community and by national and local governments within the UK for such extremely limited funding as it has managed to obtain. In large part, this was as a result of a deliberate choice to avoid Northern Recording becoming reliant on private sector funding via the market and thus becoming like any other emergent small firm in Derwentside. This distancing from the resource allocation mechanisms of the market was central to the attempt to demonstrate that there was an alternative to the dominant, allegedly bottom-up but in reality top-down, reindustrialization strategy and a championing of the virtues of entrepreneurship that was grounded in one politically partisan interpretation of the enterprise culture, whilst at the same time recognizing that some source of monetary income was necessary. As a consequence of this deliberate dependence on public sector funding, with the Labour-controlled County and District Councils crucially important both directly and indirectly as sources of funding, it can be argued that the MMW programme has just been incorporated into the regulatory mechanisms of the locally dominant and conservative right-wing Labourism and, more generally, of the UK national state and an emergent supranational one in Brussels. While it makes minimal demands on local authority finance, it performs an invaluable legitimating role since it can be argued that the local Labour councils are indeed anxious to support radical alternative approaches to local regeneration. Whilst expressing commitment to alternative approaches such as MMW, these are effectively marginalized in terms of the levels of funding devoted to them as the local authorities continue to channel much greater volumes of resources into the conventional reindustrialization programmes (which central government proclaims demonstrate the success of *its* enterprise culture policies).

11.6 Conclusions

The closure of Consett steelworks decisively marked the end of an era in northwest Durham, as the last major element in the old coal and steel economy disappeared after 140 years. It served as the catalyst for devising and then implementing a new reindustrialization strategy, built around the 1980s neoliberal *leitmotif* of an enterprise culture embracing locally controlled, 'high tech', small manufacturing firms. An unholy alliance of Conservative central government, Labour-controlled local authorities and self-interested agencies such as DIDA then emerged – for their own and varying reasons – to assert that this strategy had been successfully implemented and economic and social conditions in the District had been miraculously transformed. Others were much more critical of the limited and partial effects of this programme in terms of both numbers and types of new jobs. Many of the new companies involved inward investment rather than indigenous entrepreneurship whilst large numbers of people failed to find employment in the new companies and became instead clients of a shrinking welfare state. For them, the new economy of the 1980s meant not freedom and empowerment but simply substituting one form of dependence on the state for another.

At the same time, another approach that *its* supporters claimed was radically different, directly addressing the needs of local people as they defined them, had been emerging in the 1980s. Initially based around the voluntary Consett Music Projects, which emerged from the unsuccessful campaign to keep the steelworks open, it became formalized in the Making Music Work programme, centred around the Northern Recording cooperative in 1988. It aimed to use the locality's traditions of popular music as a medium through which people could become empowered, specifying their own needs and designing projects to realize their own objectives, with MMW acting as a facilitating mechanism within this. Many thousands of people were directly and indirectly involved in the diverse activities of the MMW programme. For example, between 1988 and 1993 there were over 20,000 visits by users to Northern Recording's studio and a further 14,000 people were attracted to live musical events associated with the MMW programme. Quantitatively impressive though these figures are, however, they cannot convey the quality of these varied musical experiences or their impacts on people's lives. For the qualitative cultural and political demonstration effects of showing that there were in fact alternative ways of defining developmental trajectories for the locality were of great significance.

On the other hand, it is possible to argue that despite the radical intentions of those involved in it, MMW in fact simply became the latest in a long line of institutions through which the dominant local Labourist politics exerted their influence over people in the District. This was most evident in relation to the dependence of the programme on funding either directly or indirectly provided by the County and District Councils and more generally on state sources. The amounts of money involved were, however, minimal, especially compared to the volume of public expenditure on the conventional reindustrialization programme. In this respect there can be no doubt that the MMW programme was effectively marginalized in the concerns and resource allocation policies of the local authorities and, indeed, the state more generally. Nevertheless, the fact that the local authorities were seen to be involved at all was important, helping legitimate their involvement in the processes of change in the locality. In this way, the MMW programme became entangled with the politics of Labourism which it set out to challenge and, whilst it was marginalized within its priorities, in the end it helped legitimate and reproduce those same politics.

Clearly, there is evidence that can be adduced in support of both of these competing interpretations of the MMW programme. While there are elements of truth in both, perhaps the key question is which is the more appropriate view: radical alternative and challenge to the conventional wisdoms or incorporation into and legitimation for the dominant politics of Labourism? The answer to this question has implications for local economic and social development strategies that extend far beyond the boundaries of Consett and northwest Durham.

Notes

1 At this point it is perhaps appropriate that I declare an interest and an involvement in what I am about to write about. This is because the articulation of this alternative in northwest Durham revolved around Consett Music Projects and its successor, the Making Music Work programme, and for several years I chaired the Steering Committee of MMW. Clearly this involvement as participant as well as observer has coloured my views about changes in Derwentside in recent years and, in particular, my views about any evaluation of the MMW programme. One of the anonymous referees of an earlier draft of this chapter commented that '. . . its whole style was one of distance, of summarizing about others, but in this case surely the author's involvement might be a potential resource to supplement critical distance with grounded, *in vivo* materials'. This is a very fair point, and one that I considered at length. It would certainly have been possible to write a much more partisan and one-sided account. In the end, however, I decided against going down this particular route in this particular chapter, and not just because of the implications of doing so in terms of the resultant increase in the length of the chapter. More to the point, it would involve a rather different chapter from the one that I set out to write, which quite deliberately sought that 'critical distance'. It did so explicitly in response to a recognition of the potential impacts of this closer involvement with much of the story told in the chapter for the way in which that story might be told. For what I am seeking to do here is to give an account which recognizes the limits as well as the strengths and continuing possibilities (to which I remain committed) of the sort of approach pursued through MMW in terms of its more general implications for local regeneration strategies.

2 For fuller descriptions of the early industrial development of northwest Durham, see Daysh and Symonds (1953) and Warren (1990).

3 Undoubtedly the best single representation of this new working-class popular music in northwest Durham was in the songs of Tommy Armstrong, 'the pitman's poet', who was born in Stanley, a mining settlement a few miles from Consett and himself worked in the coal industry as a checkweighman (see Forbes, 1987). As is pointed out below, the paths of Tommy Armstrong's music and Consett Music Projects were to cross in the 1980s.

4 It included the Bishop of Durham, Dame Kiri Te Kanawa, Sir Harry Secombe, Phil Collins, John Dankworth, Mark Knopfler, Jimmy Sommerville, Kirsty McColl, Professor Laurie Taylor, Sir Asa Briggs and MPs Hilary Armstrong, Mark Fisher and Giles Radice.

5 My own position, then as now, was very much that MMW should continue to pose an alternative to the prevailing orthodoxies of the reindustrialization programme in exploring the possibilities for local regeneration.

6 This is not to suggest that MMW came to an end in 1993. It did not, though it continued to exist on an uncertain financial basis and increasingly the focus of its activities shifted to east Durham in response to local political responses to coal mining closures there, especially in terms of the priorities of Durham County Council.

7 No doubt there will be a similar debate later in the 1990s in relation to the impacts of MMW's activities in east Durham as compared to those of the reindustrialization programme that has been established there in response to, firstly, the rundown and, secondly, the total cessation of coal mining activity in Easington District.

Part 5

Production, environment and politics

Introduction

In the preceding chapters of the book, the focus of attention has often been on certain sorts of mining and manufacturing industries (coal, chemicals, steel) in which the processes of production have had (and often continue to have) a significant impact upon the natural environment. These environmental impacts extend along the whole of the production chain, from mining raw materials to their processing and manufacture into a range of products and beyond that to their consumption in various ways. Often these impacts are heavily localized around points of production, so that the places in which these industries are located, places which had been built up around these industries and through which these industries had developed, bear the environmental scars of industrial production. Even in places in which production has often ceased, former mining villages and manufacturing towns continue to carry the environmental scars of earlier phases of production.

The first chapter in this part of the book is in fact made up of extracts from three previous publications that considered various aspects of the environmental consequences of coal mining and chemicals production within northeast England (Beynon et al., 1986b, 1989, 1994). For decades, people in places that produced coal, steel and chemicals generally accepted that environmental pollution was an unavoidable cost of production, externalized by companies and borne by local people and their environment. These environmental impacts and their consequences for the health of local residents were occasionally a cause of concern to commentators such as Priestley (1934; reprinted 1994) or Bell (1907; reprinted 1985) but local people continued to accept them as a necessary price to pay for employment. Environmental pollution and damage to human health were seen as a necessary price to pay for jobs and the wages that flowed from them through the local community. Indeed, local residents often had no choice but to accept them in mono-industrial settlements dominated by a single company. Not least, this acceptance reflected the ways in which the influence of the coal, chemicals and steel companies permeated beyond the workplace and into local politics and civil society. Nascent environmental concern and protest were absorbed within the political machine and neutered or marginalized. Often it became difficult to draw a line between company and place. As Tony Kearney (1990, 22) has commented in relation to the Consett Iron Company and the town of Consett, '. . . the Company didn't dominate Consett, the Company was Consett'.

Plate P5.1   The environmental impacts of opencast coal mining: Haravij, Greece.

As it became clear that employment in these industries was in secular decline from the 1970s and 1980s, however, people began to question what had previously been taken for granted. As it became clear that old certainties could no longer be relied upon and the former trade-off between employment and environmental pollution no longer held, local people increasingly began to challenge the environmentally polluting effects of these industries. Women were often very prominent in these protests, expressing concern about the effects of pollution on the health of their children. The three extracts that make up Chapter 12 document the ways in which these concerns about the environment became increasingly prominent in relation to three industries in northeast England. It discusses the ways in which they emerged on Teesside as employment in chemicals and steel fell sharply there. First, however, it explores how such concerns emerged on the Durham coalfield and on other coalfields within Britain as deep mines closed and opencast coal mining expanded, often threatening environmentally sensitive areas (for a fuller account, see Beynon *et al.*, 1999). The political implications and the possibilities for new forms of politics that arise from the growing challenge to the environmental destruction wrought by these industries are also discussed in a preliminary way.

The growing recognition that the environmental impacts of industrial production and consumption could become global and not simply local in their reach further stimulated a concern to move to environmentally more sustainable ways of producing and consuming. One implication of this was an acknowledgement that exporting pollutants and polluting industries (Leonard, 1988) simply displaced the problem, rather than solving it, and this became sharply evident in circumstances in which the effects of pollution were inadvertently re-exported back (for example, see George, 1992). The search for more environmentally sustainable ways of producing involved seeking to develop 'clean' production technologies, more energy- and materials-efficient ways of producing, and moves towards a greater use of recycling and the substitution of ubiquitous for scarce materials. Dematerializing the economy was seen as essential in the search for a more sustainable development trajectory and methods of production. These concerns are taken up in Chapter 13. Sustainability is seen as a slippery concept, however, the meaning of which is far from self-evident. The argument in this chapter is that if sustainable production is to be taken seriously, then sustainability as a concept must be extended beyond environmental sustainability alone to encompass economic, political, social and territorial dimensions of sustainability. Sustainability must be seen as multi-dimensional rather than a mono-dimensional concept. Moving to a more sustainable production system and onto a more sustainable developmental trajectory implies shifting onto new models of capitalist production and economic organization and the construction of new modes of regulation that would ensure that this was both possible and reproducible. The legacy of uneven development (as registered, for example, in the distinction between the First and Third Worlds) poses acute problems in a search for such a transition.

While more sophisticated concepts of sustainability may be a necessary condition for a transition on to a new developmental trajectory, such a shift depends upon changes in practice and not just thought. In Chapter 14 the emphasis is much more upon the implications of this search for a new mode of regulation. In particular it focuses upon the extent to which searching for a new mode of regulation and model of economic organization could be made compatible with tackling unemployment and social inequality in Europe. The re-emergence of permanent high levels of unemployment over much of Europe (and indeed many other parts of the world) creates a danger that tackling unemployment may take precedence over a concern with environmental pollution and dereliction within the discourses and practices of public policy. Even if successful in tackling unemployment problems (and there is no guarantee that this would be the case, given the dominance of neo-liberal conceptions as to appropriate policy), such approaches could further undermine the longer-term sustainability of employment, production and forms of economic organization via increasing levels of environmental pollution and degradation. It is this realization that has led to the emergence of attempts to find ways of simultaneously tackling problems of environmental sustainability, employment creation and social inequality and injustice. Seeking to make these various objectives mutually compatible would – *inter alia* – necessitate changes in state public expenditure priorities and policies, taxation systems and patterns of welfare provision. It would necessitate shifting the burden of taxation towards eco-taxes and restructuring systems of welfare provision and income guarantee to facilitate and enable new forms of work concerned with environmental valorization and the provision of socially needed services to emerge and flourish. As such, these developments would imply radical changes in societal norms and values. The necessary changes are, in this sense, much more cultural and political than technical. Rather than seeing an emergent 'third sector' as an intermediate labour market leading back to formal employment, it envisages such a third sector as a genuine alternative to the labour market, formal employment and the orthodox social relations of capitalism. There is a nagging question that remains, hanging over any attempt to specify such a normative alternative, however. That question may be posed as follows: to what extent would *any* attempt to create such an alternative form of economic organization and new mode of regulation be able to escape the contradictions of capitalist development that have undermined formerly successful modes of regulation and regimes of accumulation? Would the necessary again prove impossible, the impossible necessary in seeking to create such an alternative within the parameters of capitalist social relationships? The way that this question is answered in practice will be of pivotal significance to the course of human development in the twenty-first century.

Chapter 12

The environmental impacts of industrial production*

12.1 Coal mining, employment and the environment: towards a new politics of production in Britain?

Corporate decisions made within both the British Steel Corporation (BSC) and the National Coal Board (NCB) have had effects which (directly and indirectly) have cut deeply into the social fabric of the northeast. They have (through colliery closures) had a major impact upon the lives of people in the coal mining communities and upon the District and County Councils which serve them. As mining has extended through opencast into rural locations, the effect has been equally dramatic in the non-mining parts of the area. The question of the relationship between the rural (opencast) districts and the deep mining districts is most often understood as an antagonistic one. Here the competition for jobs is seen to be compounded by antipathy between urbanized/unionized working-class labour areas and more rural/non-union areas, with a higher proportion of middle-class people and Tory voters. While there is something in this polarized model (and we shall return to it in a moment) it is, in fact, a highly exaggerated assessment of these political and economic differences.

For example, while the rural areas have gained jobs from the decisions of the BSC and NCB, they have also borne considerable costs. These costs have been 'environmental'. Valleys have been flooded and fields overturned and badly restored. Transportation from opencast sites is invariably by heavy lorries which often run through narrow roads and lanes. Given this, there has been considerable local objection to the manner and scale of opencast development in the northeast. It is important to point out that it was environmental groups and not the miners' union or the Labour-controlled councils that first organized a response to the chaotic developments which 'market requirements' were wreaking upon the region's coal economy. Repeatedly throughout the 1970s, as coking coal stocks rose, the NCB Opencast Executive attempted to open huge opencast sites in the Derwent Valley. There, the Derwent Valley Protection Society, in association with the Council for the Protection of Rural England,

* First published 1986 as 'Nationalised industry policies and the destruction of communities: some evidence from North East England' with H Beynon and D Sadler in *Capital and Class*, **29**: 27–57; 1990 as 'Opencast coalmining and the politics of coal production' with H Beynon and A Cox in *Capital and Class*, **40**: 89–114; 1994 as *A place called Teeside: a locality in a global economy* with H Beynon and D Sadler, Edinburgh University Press, Edinburgh

raised deep and serious questions about the logic which underpinned the extension of this form of coal supply. Increasingly they raised the question of 'need'. Why was this coking coal needed? For which markets? How certain was the NCB of the long-term stability of these markets?

These objections were outlined (and sustained) at major planning inquiries in the Derwent valley (Horsegate, Medomsley, Whittonstall, Woodhead) and beyond (Redbarns, Daisy Hill), and inspectors from the Department of the Environment found logic in the argument and in the County Council's attempt to create a 'no-go' area in the West. These inspectors also accepted – albeit implicitly – that the nature of coking coal reserves was such that they should be handled and used with far greater concern and foresight than was apparent in the Marketing Department of the NCB. Particularly worrying was the growing tendency to sell high grade coals to the Central Electricity Generating Board (CEGB) in the form of Durham Opencast Untreated Small (DOUS). As one inspector put it:

> I am not persuaded that the power station use of DOUS is necessary or warranted. On the contrary, I conclude that the use of coal whose heat-energy rating is considerably in excess of what is called for by the CEGB specification is wasteful, especially when valuable indigenous energy resources ought to be sparingly used.

It was against this background that the National Union of Mineworkers (NUM) withdrew its support for opencast mining. This support was obtained under the 1974 *Plan for Coal* and continued beyond 1980. Today (1985) the National Union has argued that, in line with the recommendation of the Flowers Commission, opencast tonnage should be cut back from 15 million to nearer 5 million tonnes. In Durham and Northumberland the miners' unions are committed to the contraction of opencast coal mining in both counties and a negotiated reduction of output from the current level of 3 million tonnes. In this context, the Durham unions have, for the first time, attended planning inquiries and presented evidence which argued the case for deep mined coal and a more rational approach to coal production in the North. Their arguments have made an impact, and they point to the possibility of a political alliance between the Union, environmental groups and, occasionally, tenant farmers. Certainly the prospect of such an alliance has emerged as a major consideration in the North, given that further expansion of opencast mining is dependent upon political decisions within local government and beyond.

The prospect of an alliance was accentuated in the summer of 1984. Then, during the year-long coal miners' strike, miners became aware (many of them for the first time) of the scale of opencast production in the area. As one young miner from Murton put it:

> We knew a bit about opencast, but we never knew the scale of it. We were amazed when we came over here and saw what was going on.

In part their understanding of the developments was couched directly in terms of economic interests – 'this mining is taking our jobs'. This understanding was assisted in Durham by the scale of the private sites, out of the control of the NCB. This became increasingly obvious during the dispute, as the private operators increased their output and sold coal directly to the power stations. It was those coal movements which brought the striking miners to the opencast areas. Banks' operation at Inkerman in Tow Low in the west of County Durham was a major scene for picketing in the first months of the dispute. Eventually the company took out an injunction against the Durham NUM and the court ruled that picketing of all of Banks's sites should cease. All this increased the feeling of antagonism between the striking miners, their union and the private operators:

> They are just out to make as much profit as they can, and they'll do anything to make more profit. Anything. They're just private capitalists. Ruthless capitalists.

These words – uttered by the erstwhile moderate Durham miners' leader Tom Callan – carry all the more significance against the background of the nationalization of coal in 1947 and with it the clear understanding that the private coalowners were to be removed from the industry.

This clear political sense of historical change is added to by the reaction of many of the young miners to the physical nature of the opencast sites. Those who visited the large sites around Buckhead and Wam were often deeply affected by the scenes of rural devastation that accompanied opencast mining:

> I was brought up in the Bishop Auckland area and I remember the valleys and the countryside around there. To see it now. It's like being on the moon down there. What a bloody mess. It's complete devastation. It's terrible.

Many of them commented upon the way in which the quest for profit (often referred to as 'greed') adversely affected both the countryside and deep mined employment. A man from the threatened Herrington pit put it like this:

> It's cheap coal. That's what those people and this government is after. Cheap coal and big profits. We can produce this type of coal – good quality coking coal – at Herrington. But they'd rather come out here and let those cowboys dig up the countryside because it makes more money. I don't think it makes much sense that.

Against this background the more assertive response by the NUM towards opencast expansion is understandable. It was this that initially led the union to oppose the NCB's plan for opencast mining in the Plenmeller Basin. The union argued that the high volatile coals from the Plenmeller site could be used to substitute for deep mined coals. On the announcement of the NCB's intention to proceed, the NUM strongly opposed the granting of planning permission and met with the planning committee of Northumberland County Council to press home its point of view. More interesting (or at least less orthodox) was their decision to send representatives to a public meeting called by the Parish Council of Bardon Mill, a picturesque village near the proposed site. The people there had been informed that the bunker and rapid loading system at the rail head would be virtually noiseless – 'like cornflakes falling into a bowl'. To this the Easington Lodge secretary explained how such a system operated, and the level of noise associated with it in his colliery. One local manual worker commented: 'This is the first time we've felt that we've been getting anywhere. You lads coming over.' This view was endorsed by the leading Conservative lady in the village:

> It's so good to sit and listen. You see, these NUM chaps are the experts – they know about the NCB; they know how it operates.

This was in September 1984, at the height of the conflict in the coalfields. It was one pointer of the potential for a new radical political initiative that could grow out of the disastrous experiences of public ownership in the United Kingdom. The coal and steel industries gave expression to a demand for the abolition of private ownership of industrial capital and a more rational approach toward the planning of production. Both industries, in their different ways, were hijacked by the political appointment of Ian MacGregor (successively Chairman of the BSC and NCB). His appointment made acerbic a reality that had long since existed. In the British experience 'nationalization' has been a form of 'state capitalism', and in their world of operations these publicly owned trusts have danced to the tune of the market. In this process what has been lost is an important critical sense of the purpose of production and the nature of a socialist or communistic alternative to capitalist forms of organization and life. This – in the face of a rampant privatizing tendency in the Tory government – is a major requirement for progressive forces in Britain. And the starting point for such a programme and a rethink is an open and honest appraisal of the experiences to date of the state sector industries. In such an appraisal the nagging reality from the North of England is that it is just these industries (state-owned coal, steel, railways, shipyards) and not the multinationals which have heaped most havoc upon the local economy and environment. This is the fact which socialists need to come to grips with and go beyond.

12.2 Opencast coal mining and its environmental and human impacts: implications for public policy in Britain

Opencast mining is recognized as having an environmentally damaging impact, although the scale of this is disputed by British Coal's Opencast Executive. The ongoing environmental costs of opencast mining also include noise and dust, plus the presence of heavy plant and haulage wagons on roads. Certainly the operation of an opencast site leads to an immediate loss of amenity value (loss of trees and original vegetation, hedges, wildlife, farm buildings and other surface features).

John Atkinson, former Land Commissioner for the Ministry of Agriculture, Fisheries and Food (MAFF) in the Northeast Region with responsibility for all NCB opencast restoration thought that 'restoration' (after opencast coal extraction) was a much abused word (Atkinson, 1986). He thought that it was nothing more than a salvage operation with the objective of mitigating, as far as possible, the enormous loss to the landscape, the diminution of soil productivity and the immeasurable loss to plant and wildlife and the natural environment. Mr Atkinson also added that if restoration is not supervised properly the results will be a 'cosmetic confidence trick'. He concluded that such tricks have been played in the past and that restoration of any site can only be as good as the supervision it receives.

The subject of restoration was ably discussed in the Inspector's Report following the Ellerbeck West Opencast Inquiry, in Lancashire (Ellerbeck West Public Inquiry, 1989). He thought that:

> . . . common sense insists that the restored landscape would for many years appear 'man made', more or less devoid of the countless natural features and eccentricities which are part and parcel of its present charm and result from the passage of time rather than man's artifice. In time, another landscape would begin to mature, but I consider that it would be very many years before this assumed an interest and sense of naturalness akin to that of the site's landscape.

Recent British Coal financial statistics show that site restoration accounts for only 2.6 per cent of total opencast costs (British Coal, 1988).

During 1988 several doctors in Glynneath in the Vale of Neath, in South Wales, produced a report linking high levels of illness in their community with the dust from opencast coal mines (see Watson *et al.*, 1986; Jones, 1988). The doctors carried out a two-year investigation among the 7000 people served by the health practice after high prescription costs had been challenged by the Welsh Office. The report established that a high level of asthma attacks occurred when the wind was blowing from the opencast site and that ear infections, which in many cases affected different people, tended to follow after a short time lag. The doctors had earlier expressed their fears for the local residents' health at the public inquiry into proposed opencast coal mining at Brynhenllys. British Coal attempted to belittle their concern and accused them of being 'anti-working class' because they expressed concern for the health of their patients. After lengthy considerations of evidence by the Glynneath doctors to the public inquiry into opencast mining at Derllwyn (during 1988) the Welsh Office has issued an unprecedented order to reopen the inquiry (with a medical assessor) to consider further medical evidence. There is growing interest throughout the country in this issue with further detailed research being proposed.

Where derelict land exists, these opencast sites can be seen to have a positive reclaiming impact. Such land is, however, declining rapidly on the exposed coalfields. In Scotland, in the northeast and the northwest of England, few applications in recent years for large or small sites have involved derelict land. Opencast operations are slowly moving onto land of higher agricultural potential and into areas of greater environmental sensitivity. This is likely to lead to increasing opposition to the expansion of this kind of mining.

The MAFF has tended to occupy a neutral position in most areas of dispute. In this it recognizes the damaging impact of strip mining, but also the economic interests of many farmers for whom 'coal is the best crop'. Environmental pressure groups such as the Council for the Protection of Rural England (CPRE) draw upon traditional support in the countryside areas which is often politically conservative. Such groups are often linked 'local residents', whose social composition is far from even. New arrivals, often professional and managerial couples, are usually in the forefront of action committees. Occasionally such opposition is supported by local business interests or development corporations.

In the old coalfield districts considerable care has been taken by planning authorities with their 'image' in relation to prospective new employers. Coal, with its connotations of dirt and grime – of old industry and the past – has been downvalued within this and a strong emphasis has been given to cleanliness and an ordered environment. Within these districts, therefore, there have been occasional signs of the supporters of 'economic' activity siding with 'environmentalists' against opencast coal mining. The politics of this are often convoluted, however, as local industrial development organizations seek to promote an 'enterprise culture' that seeks strongly to dissociate itself from the old collectivist culture of the coalfields, which some opponents of opencast seek to defend. Such has been the case of the Derwentside Industrial Development Agency which has presented evidence to several public inquiries opposing opencast activity (most recently at the Rose Hills public inquiry, County Durham, during 1986).

In this part of the chapter we have been concerned to comment on the variety of ways in which changes are taking place in the British coal industry, focusing upon the increasing significance given to opencast production nationally. While these changes have been expressed in terms of a dramatic increase in efficiency and competitiveness of the industry, such an emphasis fails to examine the costs and long-term consequences of these changes. Those can be viewed in terms of the loss of job opportunities in the established coal-mining districts of the country. In parts of Scotland, South Wales, the northeast and southwest Yorkshire and Nottingham increases in productivity have to be counterposed with accounts of heightening levels of unemployment and deprivation.

However, it is clear (at the time of writing, 1989) that the British coal industry will never again employ large numbers of miners and realistic projections need to recognize this. Equally, it seems inevitable that the political squeeze exacted by the Conservative government on the industry will continue up until the next election. Should the Tories be returned again, the industry will be privatized, and by that time the deep-mined sector will have been contracted even further than it has today. Equally it seems clear that the opencast sector will thrive, and that within it quite powerful private coal interests will emerge. The general impact of those changes upon the coalfield districts (strongly documented as Labourist and solidaristic) needs to be considered.

In making a judgement on the future of deep mines it is helpful to return to the writings of Ian MacGregor. In his book *The Enemies Within*, he makes clear the way he sees changes developing in the industry. Here he talks of 'the change from a labour intensive to a capital intensive economy' as being inevitable, and a force which 'no politician or union boss' can make disappear. His role he recognized as being one of managing the change, 'of being a sort of midwife to it' (MacGregor, 1987). In this respect MacGregor was quite clear, seeing that mining will be an increasingly capital-intensive industry. Modern faces produce as much coal as was, just recently, produced by efficient collieries. Superpits now produce greater tonnage with half the labour force they employed in the early 1980s, and it is likely that this trend will continue apace. The modern coal industry is clearly more complex than the industry of earlier decades. While the character of deep mining has altered, opencast mining and the presence of large numbers of private licensed mines will clearly remain a part of the industry in the foreseeable future. Here significant numbers of workers are employed on a quasi-casual basis. Some are unionized, but by no means all of them are. Many of the opencast operators have strong anti-union views. The relationship between these workers (and their trade union) and the deep mines is a perplexing question, as is their relationship with the people in whose locality opencast mines are established.

In an earlier article we tentatively pointed to the potential significance of environmentalist protests against opencast expansion (Beynon *et al.*, 1986b) – part of which is reproduced as section 12.2. It would seem that, currently, people's perceptions of their lives and the places in which they live are altering in important ways and in the coal districts this is of some significance. In these localities, for generations, people lived among the dirt and grime of industrial production, in the firm belief that the industries that led to environmental pollution were central to the livelihood of their area, that pollution was part of life. 'Where there's muck there's brass.' Yet surprisingly, and with some speed, people in these areas seem to be taking a radically different view of things. In the wake of the closure of collieries and steel mills, ex-miners and steel workers can be heard to talk of the 'human environment' and the future of their families. Men who, as children, slid down colliery waste heaps may reminisce about their past with some fondness but this is not what they look to as a source of entertainment for their children and grandchildren.

In the northeast it has been interesting to note how the deepest sarcasm of the barristers employed by the opencast sector has been reserved for the person who has presented evidence for the Miners' Support Groups. Here an active group (comprised of many ex-miners) which developed during the miners' strike has engaged positively with the issue of opencast mining. It has presented evidence which combines a view of the future of the deep mines with an awareness of the environmental loss involved in the operation of the opencast sector. These issues about the future pattern of coal and energy supply in Britain, and the environmental effects associated with it, are all contingent upon political decisions which have opened up the coal sector to the pressures of the international coal market. Changes in politics or in this market would clearly have their effects in the coalfields. Certain things (particularly certain certainties) seem to have changed for good, however. It is still far from clear where these changes will end. The shape of a new politics of production continues to be shrouded in mist and uncertainties.

12.3 Challenges to modernization policies: from unemployment to environmental concern on Teesside

The modernization consensus (in Teesside) centred on a series of political trade-offs, mostly relating to jobs. During the 1950s and 1960s, concern had

Plate 12.1   'Jobs not birds': chemical industry developments on Seal Sands, Teesside.

grown over the impacts of pollution arising particularly from chemicals and steel production on Teesside. This was effectively suppressed by local politicians, many of whom had close connections with the major companies involved (see Gladstone, 1976). 'Jobs' were the trade-off for a poor physical environment. By the mid-1970s, the opposition from environmental groups to the further reclamation of Seal Sands (intended to create still more land for the chemicals industry) was considerable and this was not so easily contained. Initially, it was triggered by the Tees and Hartlepool Port Authority (THPA) announcing plans for land reclamation in an area which had been designated one of Special Scientific Interest, because of its unusual bird population. The THPA lacked the necessary statutory powers to develop the land for 'port-related industry' and it sponsored a Parliamentary Bill for this purpose. This Bill was overtaken by events as the required powers were provided by the 1975 *Teesside Structure Plan*. However, strong opposition continued, and this protest came to a head at the *Structure Plan* examination in public. The report of the panel of this examination substantially accepted the conservationists' protests, recommending that the vast majority of the land should not be allocated for future port development.

This decision produced a furious reaction. The main trades unions and employers' organizations united to lobby in favour of the original *Structure Plan* proposals. Their platform was based on the slogan: 'Jobs not birds'. This lobby was influential, persuading the Secretary of State to overturn the panel's decision and heavily re-emphasize the central role of such developments on Teesside in relation to attaining national economic priorities:

The Secretary of State considers that the continued development of the port and port-related industry on Teesside is vital both at a national and at a regional level. He accepts that the need for port development in this area may arise unexpectedly and that, in this situation, it is very important that there shall be freedom for development to proceed. He has concluded that, despite the importance of the nature conservation considerations, it is of great importance for employment in the region and the overall national economic interest that plans should make provision for the potential need for deep water facilities by making this land available for 'port-related industry'.

The conflict continued to simmer, however. For example, in 1979 the Secretary of State for the Environment proposed making some 440 acres of land at north Gare/Seaton Channel, adjacent to the Seal Sands area, available for 'port-related industries'. The justification for this was an asserted need for further deep water facilities to meet 'national port-related industrial requirements'. This land had once more been designated a Site of Special Scientific Interest, on ornithological grounds. As before, the main trades unions and employers' organizations closed ranks: 'Jobs not birds'. Teesside Chamber of Commerce argued that 'Hartlepool must strike a balance between pollution and jobs', while the Middlesbrough District Secretary of the Transport and General Workers Union added that Hartlepool must not 'put birds in front of human beings'. In contrast, both Hartlepool Borough Council and Cleveland County Council objected to these proposals. They were concerned with environmental change, and disturbed by the fact that no case had been made for increasing the acreage of land already zoned in the *Teesside Structure Plan* for port-related industry. The worm had turned. No longer were environmental groups alone in opposition to big capital, organized labour and the state. The failures of the growth coalition on jobs had brought out divisions within the state between central and local government.

The pressures from the THPA to reclaim more land at Seal Sands continued, however. The powers that it possessed to reclaim land were due to expire in 1984. Accordingly, it sought to extend them beyond this date, and Cleveland County Council, Stockton Borough Council and various conservation groups objected. After much correspondence and consultation, the councils agreed to withdraw their objections. The THPA had agreed to include in the Bill assurances that it would consult with the councils and the Nature Conservancy Council before applying for consent to reclaim land at Seal Sands. Furthermore, it agreed that the Secretary of State would consult the local authorities before giving his consent under the Bill. However, there was no guarantee that the THPA or central government would then take their views into account in deciding on a course of action. This was far removed from the corporatist 'fix' of the post-war years.

Officers and councillors within local authorities were becoming increasingly concerned about environmental pollution and safety hazards. They were becoming suspicious of land use policies which prioritized the needs of major chemical, oil and steel developments. As these industries became increasingly associated with large-scale job losses, the 'jobs at all costs' philosophy became increasingly difficult to sustain. Local authorities began to re-evaluate their land use planning and economic development policies. In this they began to assess not only the environmental costs of further large-scale oil and chemical projects but also their 'employment effects'. As early as 1972 Teesside County Borough Council (1972, 103) had raised questions about the compatibility of growth in these sectors with other local government policy objectives:

> ... the image of Teesside as a growth area with an improving environment and widening job opportunities could be adversely affected by the persistence of a labour market dominated by chemicals and steel and an environment under continuous threat from pollution resulting from the development of large capital intensive plants.

The tone was conjectural, however, and at that time these reservations were set aside. More serious doubts were raised from the mid-1970s onwards, as it became clear that the *Structure Plan* documents had overestimated both the direct and indirect employment – creating effects of oil- and petrochemical-related investment (Cleveland County Council, 1975, 61). One consequence of this was a proposal to allocate more land for 'labour-intensive' manufacturing activities and for 'small firms'. The council suggested that, in view of the dangers and problems inherent in promoting chemicals, oil and steel expansion, 'it might be argued that diversification of employment should be the main objective of Structure Plan industrial policy' (Cleveland County Council, 1975, 64).

These proposals to emphasize other forms of industrial development were strongly resisted by the powerful lobby of the THPA, the major private sector companies and the trades unions. As unemployment continued to rise, however, it became increasingly difficult for this old coalition to survive. In its reassessment of existing *Structure Plans*, Cleveland County Council (1979) recognized that many objectives were far from being fulfilled. The rise of unemployment since the late 1960s had been very similar to that of the rest of the northeast, despite so-called growth zone status. In terms contrasting sharply with previous optimism, it suggested that 'the future outlook for employment' was bleak (p. 21) and argued 'it seems likely that high unemployment is here to stay, at least for some considerable time to come' (p. 23). It went on to consider the implications of a policy that prioritized heavy industry. In addition to the requirement for vast areas of land and the adverse visual impact, it concluded that such development might well deter other investors, and create a vulnerable local economy. It argued that 'there is some doubt that the presence of capital-intensive industry is beneficial in the long-term to the Cleveland economy' (p. 6).

By the mid-1980s (in the context of now soaring unemployment rates) Cleveland County Council prepared a new *Structure Plan* (see Cleveland County Council, 1985). The prospect of low economic growth and high unemployment had a pervasive influence on all aspects of the process of policy reassessment. In addition to the slump in manufacturing, the service sector had seen only hesitant growth. The equivalent of 12 major office blocks stood empty. Priorities changed.

In 1986 the Council set up an Unemployment Working Party. To the extent that this acknowledged that producing local 'full employment' through modernization policies was no longer a credible option, it represented an important ideological break. It was no longer viable simply to claim to address the needs of the unemployed indirectly through policies to encourage private capital within Teesside; there was a need for policies which addressed the needs of the unemployed, as citizens, directly. They did this in only a very limited and partial way, but in recognizing the need to do something they had registered an important political transition. A member of this committee, County Councillor David Walsh, explained what brought it about:

> The early 1980s saw a change within Cleveland County Council. There was a growing recognition of the shifts in power within the country to the southeast, and of a government ideology which felt that Teesside had played its historic role. There was a recognition that we could no longer rely on government intervention, that we had to take things into our own hands.

In 1987 the Working Party produced an 'unemployment strategy' (Cleveland County Council, 1987). Its priorities were to address the needs of the individual more effectively, improve communication with the unemployed, and establish 'Community Action Areas'. It was, however, barely scratching the surface of the problems of mass unemployment, constrained partly by the legacy of past Labour politics and partly by the real financial constraints on local authorities in the 1980s.

In this new political climate major companies on Teesside, such as ICI, began to cultivate a much more environmentally friendly image and to increase expenditure on pollution control measures. Whilst atmospheric pollution was reduced, however, discharge of pollutants into the River Tees and the North Sea remained a major issue. A report compiled by Greenpeace in 1992 listed ICI Wilton and BSC Redcar among the 20 industrial plants contributing most to marine pollution within the UK. ICI Billingham, BASF, Tioxide and the Tees Storage Company joined them in the top 50. Mr David McClean, Minister for the Environment and Countryside, responded to this by saying (quoted in the *Northern Echo*, 16 September 1992): 'It is unrealistic to suggest that we can simply tell industry to stop discharging waste to the rivers and seas overnight; the only way to do that would be to shut it all down.' Shortly afterwards, in November 1992, the Department of the Environment approved a proposal by Cory Environmental to build an industrial waste incinerator at Seal Sands. There was considerable local opposition to this project and environmental concerns were voiced increasingly loudly. They became enmeshed with the impact of unemployment to raise serious questions over the path which Teesside had taken in the 1960s, and of the political consensus upon which it was based. The environmental consequences of previous developments were pointed out and they became increasingly difficult to defend. This process was described as follows by Mrs Maureen Taylor, County Councillor and environmental campaigner:

It was always my argument that the sort of development which is on Seal Sands now wouldn't create a load of jobs, and that is why the trades unions got at me because they said, here lay the future of Teesside in terms of jobs. What tipped the balance in favour of the kinds of industry we have now was the argument that 'jobs had to come'. It was a false argument but nevertheless it was one which won the day. So the pollutant industries came to Seal Sands in the 1970s, spilling their muck. By the late '70s even some of the livelier proponents against what I was arguing were heard to say 'enough is enough'. And I think that was linked not only to the pollution but also to the point that in they come, they take the regional policy grants, and at the first sign of something that doesn't suit them, off they go.

It was a further sign of changing times on Teesside in the 1980s that a local campaign was successful in preventing the dumping of nuclear waste in old ICI anhydride mines, but the fact that the company embarked on such a plan is revealing in itself. Public opposition resulted in 83,000 people in the Billingham area alone signing a letter of protest to the Prime Minister, opposing the dumping proposals. No longer were such 'pollutant industries . . . spilling their muck' wanted there. It was no coincidence that the most vociferous campaigners against the polluting effects of industrial development were women. Claims such as those about the priority of 'people over birds' or that 'jobs had to come first' were, in essence, claims about the priority attached to male employment, to the need for 'jobs for the boys' in industries such as chemicals and steel. However, the impact of these industries extended beyond jobs and beyond the factory gates. They had ordered the division of labour between men and women for decades. They had also polluted the atmosphere and the environment of those who lived around their works. They had helped to damage health, and mounting evidence confirming this was becoming available in the 1980s. In the 1980s the residents of the area along the south bank of the Tees experienced some of the poorest health in the North. In a population of 86,000 there were 370 deaths per year of people aged under 65, of which 145 would not have occurred had this area experienced the mortality rate of England and Wales as a whole (Townsend *et al.*, 1988, 50). Only one other area in the Northern Region had such an extremely high level of premature mortality. Townsend *et al.* observed that 'there cannot be more than a very few such areas in the country'. While the concentration of high death rates and illness was partly due to severe poverty and deprivation, it also reflected 'the possible consequences for health of air pollution emanating – especially in the not so distant past – from the massive chemical and steel complexes' (*ibid.*, 127).

The collapse of the consensus which prioritized 'jobs at all costs' therefore posed a considerable challenge to established forms of politics, both nationally and, especially, on Teesside. The dilemma was neatly and inadvertently encapsulated in 1979 by Mr James Tinn, MP for Redcar. When further steelworks closures were being considered, he 'reminded the House' during a parliamentary debate (*Hansard*, vol. 973, 7 November 1979, co. 490) that:

over the last decade no less than 10,000 jobs [in steel] were lost on Teesside. In only one closure was there a massive and well-organised protest, and that was not in my constituency but across the river. The Teesside workers recognised that modernization was inevitable and that a price had to be paid for it. They paid the price before they got the new works.

Accepting the inevitability of job losses associated with modernization investment at Redcar during the 1970s, his concern was the possibility of further job losses. Such a trade-off had, of course, become central to the politics of Labourism, both nationally and locally. Yet subsequent events, and further massive reduction in BSC's workforce during the early 1980s, proved it to be flawed. Coming to terms with the failure of previous policies was to be especially difficult.

Throughout the 1980s the challenge to the old consensus emerged in a variety of ways. At one level, it led to some sharp changes in patterns of party political support; the local electoral map underwent some significant changes, for example. Within local government it resulted in a thorough-going reappraisal of previous conventionally accepted wisdoms. With the benefit of hindsight, many confident assumptions of earlier years were subjected to critical scrutiny. In the words of the head of Cleveland County Council's Research and Intelligence Unit, Reg Fox:

> The problem we've got is that we've been the forerunner of changes in manufacturing, and I don't think society generally and the national political scene has caught up with the fact that this is going to be happening in the rest of manufacturing. Until the country comes to terms with that, we'll not have the resources to tackle our problems.

and in those of the Chief Planning Officer, John Gillis:

> The sad thing is, for at least the last ten years we've been telling people this. We must cope with these people who are unemployed. What are we going to do with them?

Answering this question was one of the most pressing issues informing politics on Teesside in the 1980s and into the 1990s.

As Teesside entered the 1980s, then, much of its old economy had been decimated. The political consensus and modernization policies constructed around assumptions about future prosperity based upon restructuring the chemicals and steel industries and upon diversifying Teesside's economy became subject to growing pressures as unemployment rose and the environmental consequences of such forms of activity became better and more widely understood. As unemployment rose to, and remained at, high levels, previous constraints which had suppressed discussion of the undesirable environmental consequences of chemicals and steel production weakened significantly. Not least, environmental dereliction was itself increasingly seen as a barrier to economic and employment diversification. Moreover, the growing salience of environmental issues provided space into which gender relations within Teesside could be reappraised, as 'jobs for the boys' became seen as 'pollution for the girls'. But as an old order, political and economic, was thrown into deep crisis, questions arose as to what would – or could – be put in its place.

Chapter 13

Towards sustainable industrial production: but in what sense sustainable?*

13.1 Introduction

There seemed a time, extending for about 20 years after the Second World War, in which, at least over much of the advanced capitalist world, more or less steady economic growth, profitable production and full employment were mutually compatible objectives. In this 'golden age' of modern capitalism, the era of a Fordist regime of intensive accumulation and mode of regulation, the productivity gains of mass production were validated by changing consumption norms and the growth of mass consumption. This equilibration of aggregate production and consumption was above all facilitated by state policies which tended towards mild income redistribution, encouraging growing private consumption, along with other state policies which involved growing public expenditure on the welfare state and led to rising levels of collective consumption. While mass production and consumption were unavoidably dependent upon and had perceptible impacts on the natural environment, this grounding of economy and society in nature was not generally seen as problematic. It did not seem that there were ecological limits to growth. For a while, then, it did indeed seem as if a new golden age of capitalism had dawned – at least in the First World if not the Third, as uneven development grew apace at global level as an integral part of the boom in the First World.

From the mid-1960s, however, all this began to change. Partly this was because of the maturing contradictions within the Fordist model at a micro-scale within the factories. Workers became increasingly resistant to further intensification of the labour process and a proliferation of mind-numbing deskilled and unskilled jobs within a deep Taylorist technical division of labour. As a result the growth of labour productivity slowed and profitability fell. In due course, this led to a search for viable new micro-scale models of commodity production in all sectors of the economy. At a macro-scale, it became increasingly clear that the mode of regulation at national state level was becoming increasingly crisis-prone. In part, this was because it became evident that state involvement did not abolish the crisis tendencies within the

---

* First published 1995 in M. Taylor (ed.) *Environmental Change: Industry, Power and Place*, Avebury, Winchester

capitalist mode of production. In contrast, it transformed them and internalized them within the state. In due course, having simmered for a while, these crisis tendencies boiled over and dramatically appeared crises of the state itself and its mode of crisis management (issues that are examined more fully below). No longer could national states seek to maintain full employment via Keynesian demand management policies, continue to expand collective consumption and public sector provision of services such as health and education, or increasingly, even set a floor level to living standards through the safety net of a welfare state.

Recognition of these limits to state capacities in turn led to a search for new macro-scale regulatory models that accepted and respected the state's limited powers to counter the forces of the market (forgetting, perhaps conveniently, that markets are social and political constructions, not natural structures shaped by forces beyond social control). One implication of this was that various 'post-Fordist' regulatory experiments became characterized by much wider income differentials and a much greater degree of social inequality than had been the case in the immediately preceding period. If economic recovery did come about, it seemed certain that renewed growth would be achieved at the price of more unequal societies.

In addition, however, it increasingly became clear that the old Fordist model was remarkably blind to the ecological impacts of mass consumption and production. There is no doubt that industrial production had become a major cause of environmental pollution, both as a result of producing goods and services and then of consuming them. Indeed, it is the dissipative pollution generated by consumption that has a quantitatively much more significant effect on the natural environment than does production *per se*. Recognition of the links between mass production, mass consumption and environmental pollution, in principle at least, opened a window of opportunity to seek to combine economic recovery with greater environmental sensitivity in the organization of production, circulation and consumption of commodities. At the risk of some oversimplification, both prior to and during the Fordist era, the natural world was simply treated as a source of raw materials to be exploited with little if any thought about the impacts that this had on natural processes. In like manner, the natural environment was treated as a giant waste disposal site with infinite absorptive powers to cope with the pollutant effects of mass production and consumption. Innumerable noxious substances were carelessly discharged into the environment as unwanted by-products of production, through the transportation of goods and people, and through the emission of various pollutants into the environment in the course of consuming commodities or through dumping them once their socially useful lives had come to an end. This was all done, however, without any deep knowledge of the environmental implications of producing and consuming in these ways. Once it became clear that nature in turn had an 'impact back' on human societies, there was a growing ecological consciousness and 'green politics' of various hues from 'deep' to 'pale' emerged. It is certainly the case that strands of environmental thinking and advocacy had existed for a considerable period of time prior to the 1960s but these ecological concerns became more audible and visible in the latter part of that decade over much of the advanced capitalist world. Growing environmental consciousness was often tied to the causes of particular middle-class interest groups, who enjoyed materially comfortable lifestyles and were seeking to preserve their own, often localized, environmental 'positional goods' (Newby, 1980). At the same time, however, some of them were becoming increasingly concerned with more general and global aspects of environmental conservation (see, for example, Nicholson, 1970). In addition there was a more general growing concern with the 'limits to growth' as neo-Malthusian thinking grew in influence and there was, to a degree, coalescence around a common agenda of the need to protect the global environment (see, for example, *The Ecologist*, 1972). Despite the differences in immediate environmental concerns and in the depth of the shades of green, therefore, the emergent environmental groups all in various ways challenged the logic of Fordist mass production and mass consumption under the rubric of environmental conservation. By the 1980s the focus of attention had switched much more to questions about environmentally 'sustainable' forms of production and consumption.

The meaning of 'sustainability' is by no means self-evident, however; sustainability can be a very slippery concept actually to capture and pin down. Indeed, the concept itself requires problematizing. Not least, this is because a key issue is: To *whose* sustainability are we referring? To begin to answer this question, however, necessitates first conceptualizing the organization of the production process in general and of that within the social relations of capitalism in particular. Thus capitalist production can be analysed as simultaneously a process of value creation, a labour process and a process of materials transformation. Examining the interactions of these varying aspects of the totality of the production process helps elucidate some of the problems of defining 'sustainability', both conceptually and practically. So in what follows I will introduce the concept of 'sustainability' and its various dimensions from the perspective of differing social interests which are involved in or otherwise experience the totality of this production process. Thus 'sustainability' will be considered, for example, in terms of the social sustainability of the level and distribution of employment and of income; and it will be considered from the perspective of the ecological sustainability of the level and composition of output.

13.2 What is sustainable from the point of view of capital? —

Our point of departure here is a recognition of the crisis of mass production and mass consumption that grew increasingly severe from the second half of the 1960s in the countries of the advanced capitalist world. There were certainly various national variations around the basic theme of the old 'Fordist' model of equilibrating mass production and mass consumption, with labour productivity gains going hand in hand with increased working-class consumption norms. The fundamental point, however, is that this key macro-economic

Plate 13.1   A marker for an industrial district in northeast Italy: the aspiration of many old industrial areas in northwest Europe.

relationship became increasingly undermined as the internal contradictions of that regime of accumulation and mode of regulation became more and more exposed. The maturing crisis was reflected above all from the point of view of capital in a profound crisis of profitability.

In response to this, companies sought viable alternatives to the old mass production model: both alternatives to mass production and alternative forms *of* high volume (if not quite the old mass) production. The alternatives to mass production focused on small firms, skilled workers and flexible production. Such approaches placed considerable emphasis on networks and (allegedly) cooperative and non-confrontational relations between capital and labour and between companies linked in a mutually beneficial horizontal division of labour in a supportive regulatory environment. The 'rediscovery' of craft production was, however, heavily influenced by the emergence of the 'Third Italy' and the rediscovery of Marshallian industrial districts there (Bagnasco, 1977). This led to a renewed emphasis upon the character of place as a key, maybe *the* key, to industrial competitiveness.

Subsequently, there has been much debate about the extent, character and reproducibility of such industrial districts. There are, undoubtedly, clear limits to such an approach to production, in terms of products and markets (and also the number of jobs they will generate: see below); *a fortiori*, such forms of production are suited to niche market consumer goods from which the mass of the population are, by definition, excluded as consumers. There is more than a suspicion that the creation of these new market niches was linked in part to the more regressive taxation policies and unequal income distributions that came to characterize most of the advanced capitalist world in the 1980s. Nevertheless, this apparent evidence of a successful local growth model led to a frenzied search for other successful new industrial spaces all over the world. Unfortunately, this was sometimes done in a rather uncritical fashion as places experiencing industrial growth based on diverse constitutive processes were heaped together into an increasingly meaningless category (for example, see Scott, 1988a). Equally, against a background of high unemployment, it led many localities to seek to become 'pro-active' and to construct local economic development policies that would encourage the emergence of new industrial spaces in their territory (see Beynon and Hudson, 1993). This place-bound politics of local economic regeneration rapidly degenerated into a deeply divisive competition between localities in a global place market.

One consequence of the limits to the alternatives *to* mass production was, of necessity, the exploration of other approaches *of* mass – or perhaps more accurately high volume – production: just-in-time (Sayer, 1986); lean production which involves – or so its advocates maintain – an approach that '. . . can be applied in every industry across the globe' (Womack *et al.*, 1990, 6) with '. . . a profound effect on human society'; and flexible automation and dynamic flexibility (Coriat, 1991; Veltz, 1991) which represent further attempts to combine the 'best' (from capital's point of view) of mass production and craft production. Thus these approaches seek to combine scale economies with economies of scope and greater consumer choice in response to greater market segmentation and differentiation. They therefore involve a combination of concepts from mass and craft production – since they incorporate 'production to order' from a greater, but still relatively restricted, range of mass produced commodities. Mass customization is a further attempt to combine the 'best' (from capital's point of view) of mass production and craft production, which takes the tendencies visible in just-in-time, lean production, dynamic flexibility and flexible automation a stage further. For it seeks to combine the advantages of scale economies with the goal of batch sizes of one, uniquely customized commodities (Pine, 1993).

In fact although presented as alternatives – not least by their messianic advocates in the major Business Schools – there is in fact considerable overlap in what each of these approaches to production entails. Not least is this the case because they typically refer to the same sets of exemplar companies. For example, the 'Japanese roots' of lean production in Toyota's production strategies from the 1950s are reflected in many of the areas of similarity (if not identity) between it and just-in-time, though lean production certainly embraces *more* than does just-in-time.

Such 'new' approaches involve some significant changes in the organization of production, *inter alia*:

- a restructuring of capital–labour and capital–capital relations;
- a restructuring of work and of the working class by means of new recruitment and retention strategies;
- an intensification of the labour process within workplaces;
- a reshaping of the social division of labour based more on selective relations of trust and cooperation between companies as a part of their competitive strategies (although it is clear that major asymmetries of power remain in relations between companies);
- the introduction of new systems of distribution as the implementation of just-in-time concepts leads to the replacement of just-in-case stocks with inventories in motion, with important ecological consequences as a result of increased movement and transport activity;
- a recasting of spatial divisions of labour away from the stereotypical geographies of Fordism (in a variety of ways, with differing local economic development implications: see Hudson, 1995c) as companies seek to restore or enhance profitability in often volatile macro-economic conditions.

In brief, the emphasis is upon seeking to preserve the benefits of economies of scale while gaining those of economies of scope. This involves using existing and creating new forms of spatial differentiation as an integral part of this restructuring of production. It also encompasses introducing a greater element of product differentiation and consumer choice as the route to enhanced competitiveness and profits. One point though is worth stressing: in so far as these *are* alternative forms of *mass* production, variations around the basic mass production theme – and I would argue strongly that they really are – then there are strict limits, in terms of the material and social requirements of commodity production, that will limit the range of industries and products in which they *could* be applied. Moreover, despite all the experimentation, there is considerable debate as to whether these have revealed new ways of profitably organizing production, especially on a longer-term basis. And even if it can be argued that they have at the micro-level, there seems little doubt that as yet no stable macro-level combination of regime of accumulation and mode of regulation has emerged as *the* successor to the mass production and mass consumption combination of Fordism. In this sense 'post-Fordism' remains a clearly questionable proposition.

13.3 The social sustainability of the level and distribution of employment

The Fordist era was, at least for a while in much of the advanced capitalist world, characterized as one of 'full employment', albeit implicitly white adult male full employment. From the point of view of labour, the crisis of Fordism at macro-level has been experienced above all as one of high – and for many more or less permanent – unemployment. Equally, for those who succeed in finding waged work, there have been dramatic shifts in the sectoral pattern of employment, in the types of jobs on offer and in the terms and conditions on which they are offered in the labour market, and in the organization of the labour process within workplaces. These changes in turn have been linked with an increasingly uneven income distribution and an expansion of poverty. There have been marked industrial and occupational variations in the incidence of unemployment, and associated primarily with this but also with an increasing number of poorly paid jobs, a growing polarization between households with two wage earners and households with none (see Hudson and Williams, 1995). Growing poverty has been exacerbated by cuts in the welfare state as the restructuring of the labour market and of the state have reinforced one another.

The declining effective demand for labour is perhaps most clearly seen in relation to the new models of high volume industrial production. These employ far less living labour in the production process than did either Fordist mass production or small-scale batch craft production. This can be most clearly seen in relation to lean production and its claim to be 'lean' because – *inter alia* – compared to mass production it requires only half the human effort in the factory. Certainly, unless aggregate demand is exploding, 'halving the human effort' can only mean a drastic reduction in the number of jobs available; since a concern with economies of scale remains central in many industries, this almost certainly means fewer workplaces as well as fewer jobs. Unless the labour displaced in this way from such productive sectors is absorbed into other sectors of commodity production, the net result may well be a lower level of aggregate demand and greater uncertainty as to the prospects for corporate profitability. This in turn could well deter investment in newer and more environmentally sustainable production technologies. As a result, higher unemployment, lower output and threatened profitability could intertwine in a vicious downward spiral that locked production into environmentally unsustainable practices. It could even intensify these as companies sought 'spatial fixes' which would allow them temporarily to maintain profits at the price of yet more environmental pollution.

But if there are fewer jobs as a consequence of this shift to new high volume production methods, there are also strong claims that they are *better* jobs. There are claims from the supporters of the shift as to the beneficial re-emergence of multi-skilled workers, who are much more creatively involved in the process of production. Great stress is placed on teamwork, on reintegrating manual and mental labour, and on the 'empowerment' of production workers. Others dispute this. Critics argue that what is involved is not multi-skilling but multi-tasking, a very different attribute of a job, and the disempowerment of workers in new repressive regimes of control and surveillance of the labour process. As a result of the combination of these characteristics, in the new flexible high volume production methods, the line keeps running all the time. This does not mean that it *necessarily* runs at its maximum possible overall speed, nor necessarily at the maximum possible speed of each element of the labour process on the line. Indeed, the aim is not so much to maximize line speed as to minimize the number of workers needed for a given line speed (as dictated by the implementation of just-in-time principles of minimal or zero inventories). In *this* way the labour process is intensified, to the detriment of labour.

There are, then, increasing doubts as to whether the jobs on offer in the new production approaches are any better than those on the old mass production lines of earlier years. Indeed, there are definite suggestions that from the point of view of labour, these may involve greater intensification of work and greater stress than before. Not least, this is because team-working involves a micro-scale regulation of the labour process in which workers discipline themselves and one another within and through the rhetoric of 'teamwork'. Moreover, since there are fewer jobs, there is enhanced competition for them and consequently firms can be extremely selective as to who they recruit and equally precise about the terms and conditions on which they will offer employment. The clear implication of this *necessity* for selectivity is that only a fraction of the available labour force will actually be employed.

Such changes in the amounts and forms of available waged labour are by no means confined to such innovatory manufacturing companies. On the contrary, in broad terms the last two decades over much of the advanced capitalist world have seen dramatic and generalized labour market changes. The processes driving these changes are above all corporate strategies of restructuring in the pursuit of profit and state strategies of restructuring in the face of threatening fiscal crises. The resultant labour market changes have been expressed in diverse ways. These include a switch from manufacturing to service sector employment, shrinking public sector and expanding private sector employment, declining male and increasing female employment, declining full-time and growing part-time work, and a decrease in secure 'jobs for life' and a growth in marginalized and precarious jobs. Such latter types of jobs are often in the informal sector or in or on the fringes of the black economy in 'flexible' and more deeply segmented labour markets as class, gender, ethnicity and age combine in complex and subtle ways.

Without doubt, however, the most pernicious effect of labour market restructuring has been the seemingly inexorable tendency to rising permanent (long-term) unemployment and under-employment, with all the social tensions and political pressures that growing poverty and widespread marginalization bring. This is especially the case in the growing numbers of households with no wage earners, reliant on shrinking welfare state provision. The restructuring of production and labour markets has very visibly led to economic dislocation and the rupturing of stable reproducible social structures. The specificity of the interactions of general processes with the specific characteristics of particular places has generated different outcomes in different places (sometimes leading in turn to migratory pressures and questions to do with the regulation of migration, at various scales: Pugliese, 1991).

There is a real danger of a proportion of the population becoming permanently surplus to the requirements of capital; the old Marxist category of 'surplus population' may well be much more appropriate than that of 'unemployed' in describing this fraction of humanity. It would clearly seem that formerly valid concepts of 'full employment' – albeit often implicitly white male full employment – are no longer tenable. If, however, there is a continuing concern with issues of equity and social justice as a necessary condition for societies to remain civilized, democratic, reproducible and 'sustainable', then this suggests a pressing need to explore alternative ways of sharing out waged employment and of redefining citizens' rights to work.

Little more than a decade ago, the steelworkers of Longwy marched through the Place de l'Opéra in Paris demanding the right to 'live, learn and work in Longwy'. Now, so Lipietz (1992, 74) suggests, there is a need for a new social compromise in which the guarantee '. . . is not a "job" but "the right to live and work in one's own country"'. In other words, people must accept that there are no longer any 'jobs for life', *a fortiori* in one place for life, but that occupational and locational mobility must be accepted as the norm. The corollary of this is that the state must underpin such continuous change via sophisticated training programmes to preserve and enhance skills, especially transferable skills.

In itself, however, this will by no means necessarily solve the problem for it implies that aggregate demand for labour is sufficiently high to absorb the available workforce. For this to happen, it will be necessary to share out the available waged work more equitably, this in turn involving reductions in the working week and the growth of a 'socially useful third sector', providing work for around 10 per cent of the labour force (roughly speaking, the numbers unemployed at the start of the 1990s in the countries of the advanced capitalist world: Lipietz, 1992, 99). This reconceptualization could also extend to incorporate new models of relations between waged employment and non-waged work – alternative 'bundles' of waged and non-waged work packaged together into household survival strategies (Mingione, 1985). This in turn problematizes and forces a reconsideration of the links between production, reproduction and modes of regulation.

What then would be a socially sustainable distribution of work and employment? Put another way, what modes of regulation would be appropriate for different distributions of employment and income? These are key questions in the context of identifying socially sustainable models of production and forms of social organization, locally (see Tickell and Peck, 1992), nationally and supranationally.

13.4 The ecological sustainability of the level and composition of output

Emphasizing the importance of devising ecologically and environmentally sustainable forms of production has become increasingly popular in recent years, often in seemingly unlikely quarters. The World Bank (1994, 42), for example, recently pointed out that achieving environmentally sustainable development 'is a major challenge of the 1990s'. This may well be correct but part of the problem of evaluating competing claims as to what needs to be done is that there is by no means a consensus as to exactly what it means in practice. There are, for example, significant differences *within* the ecological movement between the various shades of 'green politics' as to what 'sustainable production' would entail. Implementation of some 'deep green' positions, for example, would require significant reductions in material living standards and radical changes in the dominant social relations of production (see Goodin, 1992; Jacobs, 1991). There is no doubt that such changes would be powerfully contested. In contrast, rather 'paler green' perspectives are conceived much more in technicist terms within the current relations of production, essentially trading off economic against environmental objectives (see, for example, Pearce *et al.*, 1989).

Perhaps the most quoted definition is that of the United Nations World Commission on the Environment and Development (1987, 43) which defined sustainable development as meeting 'the needs of the present, without compromising the ability of future generations to meet their own needs'. Thus defined, sustainability encompasses the relations between the environment and the economy, and a commitment to equity, intra-generationally, inter-generationally and spatially. It encompasses a vision of development that goes beyond quantitative growth in material outputs and incomes to include qualitative improvements in living and working conditions. In broad terms, it accepts rather than radically challenges the dominant logic of capitalist production.

There is no denying the impacts that human activities have had, and continue to have, on the environment, globally and locally. Of particular significance in the present context is the fact that industrial production, growth and transformation are the primary proximate causes of these impacts (see, for example, Commission of the European Communities, 1992). In turn, environmental degradation and pollution have major impacts back on economic activities in a variety of ways. Yet as Taylor (1994) has cogently argued, the links between the dynamics of the economy and the dynamics of the environment have been only poorly and very partially drawn.

In order better to understand these links between the organization of production and environmental change, we need to consider production as a process of material transformation. One way of doing so which allows some powerful insights into these relations between economy and environment is that of industrial ecology and industrial metabolism.[1] Industrial metabolism (Ayres *et al.*, 1988; Ayres, 1989) is an approach which, at its simplest, involves constructing a balance sheet of the physical and chemical inputs to and outputs (both desired and undesired) from production (although this in itself is a far from simple task, with extremely demanding data requirements). It centres on the notion of mass balance – that is, that the sum total of a particular chemical within a production process remains constant as it passes from production, to consumption, to disposal, with human activity both in production and in consumption providing the stabilizing controls. It therefore incorporates a conceptualization of the production process as one of flows of physical matter, with the identification of pivotal points at which key chemical and physical transformations occur and at which flows from the realm of social processes of production and consumption to that of the natural world take place. The crucial distinction between 'production pollution' and 'consumption pollution' is only implicit within the industrial metabolism literature but is nonetheless a crucial one to make, both analytically and in terms of exploring appropriate policy options (Taylor, 1995). There is considerable evidence to suggest that the latter is more pervasive and extensive than the former, not least because it has typically been less subject to restrictive regulation (see also below).

In brief, then, industrial metabolism describes the trajectory of flows of chemicals through an industrial economy and traces the discharge to and accumulation within the natural environment of the resultant pollutants. By tracing through the ecological impacts, in terms of inputs from and outputs to the natural environment, of particular methods and forms of organization of production, the implications of possible choices of production technologies can be clarified. By tracing through the ecological impacts of varying combinations of the production and consumption of different levels and compositions of output, the macro-scale implications of micro-scale choices can be clarified. This would at least provide a base point from which to review the ecological implications of these varying social choices about the how and what of production. Moreover, it could in principle be extended to consider the *where* of production, for example in terms of companies' attempts to find 'spatial fixes' for pollutant and environmentally noxious production.

A more precise identification of the ecologically damaging aspects of current methods of production and lifestyles and levels of consumption leads to a consideration of more ecologically sensitive approaches to production and consumption. One such approach is that which centres around notions of 'eco-restructuring' (see, for example, Weaver, 1993). In broad terms, this encapsulates the process of transforming modern capitalist industrial society from one characterized by high levels of use of virgin materials and fossil fuels, high levels of material consumption, and high emissions of wastes to one that is environmentally more benign. It could therefore be thought of as involving a transition to systems of 'clean production' (Allaert, 1994). The impetus for such changes once came from fears of exhausting non-renewable resources; of reaching and then breaching 'the limits to growth'. More recently it has been motivated by a recognition that there are limits to the capacity of the natural environment to absorb wastes. Consequently, there are limits to the robustness and reproducibility of natural processes as the variety and volume of pollutants rise.

One way forward is to examine the technical possibilities of and conditions for a greater degree of ecological closure in production and consumption (see, for example, Ayres *et al.*, 1988; Weaver, 1994). There are, however, dangers of seeking 'technological fixes' in circumstances characterized by uncertainty and partial knowledge. One has only to think of examples such as the unanticipated environmental impacts of chlorofluorocarbons (CFCs) on stratospheric ozone, the substitution of diesel for petrol combustion engines leading to increased emissions of fine particulates, or the recycling of paper leading to increased chlorine emissions. Such cases are sharp and painful reminders of the unintended production of unwanted and harmful environmental impacts as a result of well-intentioned attempts to ameliorate other environmentally harmful effects. What, though, would be the geography of a more ecologically sustainable production system? How would moves towards ecological sustainability be conditioned by current patterns of uneven development and asymmetries of power relations within and between enterprises and societies, at varying spatial scales from the global to the local? These are important but as yet unanswered questions (see, however, Taylor, 1995).

Clearly seeing production simply as a process of materials flows is at best a one-dimensional perspective. One crucial limitation is that it abstracts production from its socio-spatial context. It therefore ignores the fact that production is a social process which has a definite geography, and which has manifold societal implications. The industrial metabolism approach is based on a biological analogy (Ayres *et al.*, 1988) and, as such, at best incorporates a very limited conception of social process. In so far as it does incorporate a consideration of social process, it is in a very partial and emaciated form, with market prices seen as the only regulatory metabolic mechanism. The point that dissipative pollution as a result of consumption is a more generalized source of environmental pollution than that arising from production *per se* takes on added significance once one realizes that the industrial metabolism approach places sole reliance upon the market rather than the state as a regulatory mechanism. Furthermore, reflecting its connections with neoclassical economics, it incorporates a simplistic and naturalistic view of markets and the mechanisms of price formation. It thus fails to give due weight to markets as socially constructed and regulated institutional forms. Moreover, it is clear that within capitalism a consideration of markets while ignoring property relations, the social relations of production and social class structures gives at best a very partial insight into the organizational dynamics of production. As a result, it also provides at best a partial view of the links between those dynamics and the dynamics of human-induced environmental change.

At the same time, however, as a recognition of the socially constructed character of markets makes clear, it is important to avoid simplistic dualisms which posit the market and the state as dichotomous alternatives. Just as states are deeply implicated in the construction of markets, so too are they inextricably bound into the societies of which they are part. One implication of this is that state regulation does not necessarily offer a guaranteed and non-problematic resolution of the ecological problems generated by the industrial production system, Indeed, there is a considerable body of empirical evidence which supports such a conclusion and persuasive theoretical evidence which suggests that, with respect to a capitalist economy,[2] this is *unavoidably* so as state involvement cannot abolish economic crisis tendencies inherent to such an economy. These are simply internalized within the operations of the state itself. As the analyses of proponents of the French regulationist approaches make clear (see Dunford, 1990), the best that (national) states can hope to achieve is to 'discover' modes of regulation that are temporarily appropriate to particular economic growth models, or regimes of accumulation. As a result, these will for a time allow the conditions necessary for successful production to be more or less non-problematically reproduced. The economic contradictions displaced into the structures and operations of the state itself in due course appear in fresh forms, for example as rationality, legitimation or fiscal crises of the state (for example, see Habermas, 1976; Offe, 1975a, 1975b; O'Connor, 1973). There is, therefore, no *a priori* reason to believe that state regulation will be any more successful in solving the even more complicated problems of environmental damage and ecological destruction that arise as a consequence of the character of production under capitalism.

In summary, there are evident grave dangers in examining possible changes to more ecologically sustainable forms of production without full consideration of either the social conditions that this presupposes or its implications for economic and social sustainability. It is vital to consider possible moves towards ecological sustainability in terms of *their* economic and social sustainability. It would be profoundly dangerous to ignore such issues. For example, would the level, composition and distribution of what is produced under an eco-restructuring programme be seen as socially acceptable and/or politically legitimate within the parameters of a democratic (as opposed to enforceable within those of a dictatorial) state form? If not, what would be the ecological implications of what would be socially and politically acceptable?

13.5 Conclusions

So where does this leave us? It is clear that there are no easy choices that will allow the interests of capital, of labour and of environmental survival to be simultaneously satisfied; indeed, it is probably more realistic to recognize that there are fundamental incompatibilities between them but that at the same time they are interrelated rather than independent. It is worth recalling Harvey's point (1993, 22) that 'all ecological projects (and arguments) are simultaneously political-economic projects (and arguments) and vice versa. Ecological arguments are never socially neutral any more than socio-political arguments are ecologically neutral.' Indeed, he argues that the current preoccupation with concepts such as those of sustainable development represents a contemporary means by which capital is seeking the continuation of a particular dominant set of social relationships. This may well be an overly instrumental view, which underplays the significance of human agency in challenging the agenda of capital and its particular forms of industrial production. Equally, however, it is difficult to deny that the rhetoric of 'green products' and 'environmentally more friendly production processes' as a fresh source of profit for capital is an important input into the sustainability debate, seeking to reconcile capital's need for profits and labour's need for employment with a greater sensitivity to the natural environment. It is, however, difficult to see how such an attempted compromise can avoid the systemic and structural contradictions that beset all forms of capitalist production. This suggests that lasting solutions may well necessitate a close scrutiny of and radical changes in the dominant social relations of production rather than either tinkering with the problems at their margins by seeking various 'technological fixes' or seeking salvation via state regulation (supra-nationally or nationally) in an attempt to reconcile the irreconcilable. This may raise some very hard choices in balancing preservation of the undoubted benefits of generalized industrial production to the lifestyles of many people without violating the ecological conditions that make continuing human life possible.

Notes

1 Although the concepts of industrial ecology and industrial metabolism are often used synonymously, it is useful to draw a distinction between them. Metabolism refers to the examination of the inputs of energy and materials into a specific facility, industry or sector, and the waste products – heat and materials – that are released from it. Ecology in this context refers to the total process from the raw material extraction, transportation, manufacture, use and disposal of products and the interaction of these with the natural processes of the biosphere.

2 There is an equally impressive (if that is the most appropriate term) volume of evidence that ecological devastation was at least as bad and frequently even worse under the political economy of state socialism. I am not aware of the sorts of theoretical arguments that have been advanced in relation to the capitalist state being developed in relation to the socialist state, although it is quite likely that there were people who understood the ecological consequences but also the likely personal consequences (for example, by an enforced exploration of the Gulag Archipelago) of publicizing this knowledge. There seems little doubt, however, that more general propositions about agents acting in circumstances in which they have only partial knowledge of the consequences of their actions (for example, see Giddens, 1984) would equally apply in the circumstances of state socialism as they would in those of capitalism.

Chapter 14

In search of employment creation via environmental valorization: exploring a possible eco-Keynesian future for Europe*

14.1 Introduction

For two decades following the Second World War, over much of the advanced capitalist world, rapid economic growth, profitable production, rising material living standards and full employment appeared to be simultaneously attainable objectives. The main features of this Fordist regime of accumulation and mode of regulation are well known. Mildly progressive income redistribution encouraged growing private consumption. Growing public expenditure on the welfare state led to rising levels of collective consumption. State involvement along Keynesian lines was seen as central to ensuring macroeconomic growth and guaranteeing social justice within a 'full employment' economy. 'Full employment' was predominantly defined in terms of full-time jobs for life for male workers. The male 'family wage' plus the 'social wage' delivered via the welfare state, perhaps supplemented by the wages of married women working part-time, was seen as assuring continuing increases in material living standards for nuclear families, regarded as the normal form of household unit. This mass production and consumption economy was unavoidably dependent upon the natural environment in various ways, but this grounding of economy and society in nature was not seen as problematic. There were apparently no ecological limits to growth. There were, however, national variations around the basic themes as Fordism diffused unevenly over space and through time within Europe (see, for example, Albert, 1993; Hudson and Williams, 1995; Lash and Urry, 1987; Lipietz, 1987). This variability reflected – *inter alia* – the extent to which centre-left or centre-right politics were dominant and the legacies of uneven development. This uneven diffusion had important implications for the trajectory and character of development in different places.

From the mid-1960s, however, this Fordist model of growth (irrespective of national variations around its basic themes) became increasingly fragile.

* First published 1997 with P Weaver in *Environment and Planning A*, **29**: 1647–1661, Pion Ltd, London

This was partly because of the maturing contradictions of mass production at the micro-scale within workplaces. One consequence of this was accelerating deindustrialization as companies closed capacity or switched production abroad. The associated decline of full-time male industrial jobs for life and the further growth of paid work for women (largely part-time jobs in the services sector) problematized the notion of the male 'family wage' as central to household living standards. The changing gender composition of the labour force, and the growing variety of forms of labour contract, increasingly problematized the conception of 'full employment'. In many ways, the apparently relict forms of the household economy and informal sector of Mediterranean Europe (see, for example, Redclift and Mingione, 1985) seemed to offer a vision of the future.

At a macro-scale, the mode of regulation was increasingly and visibly crisis-prone. National states could no longer maintain full employment, continue to expand public sector provision of services such as health and education or, for many people, maintain living standards above poverty levels through the safety net of a welfare state. The Schumpeterian workfare state began to replace the Keynesian welfare state (Jessop, 1993). Neo-liberalism increasingly dominated the policy agenda as late modern capitalist states searched for a viable post-Fordist alternative. State involvement was increasingly seen as the proximate cause of, rather than part of the solution to, the problem of poor national economic performance. Renewed national economic growth, stimulated by a resurgence of enterprise released via deregulation and increasingly flexible labour markets, was seen as the solution to problems of unemployment. A concern with social justice was pushed down the policy agenda in increasingly and multiply divided societies. There was a greater awareness of environmental constraints and impacts, but the dominant motives informing neoliberal national economic policies were those of enhanced GNP growth rates. The transition to neo-liberal approaches was not uncontested but, nonetheless, by the mid-1990s, neo-liberalism had made its mark throughout Europe, not least in the east. At the same time, however, it became widely recognized that Keynesian regulation of the mass economy as the route to 'full employment' had severe and unsustainable ecological impacts. Simply returning to past practices and policies (assuming this was possible) would, therefore, at best provide a short-term solution to problems of unemployment (Hudson, 1995d).

As a result, European societies and states face several apparently irreconcilable dilemmas. They encompass issues of social cohesion, environmental quality, and international competitiveness and economic sustainability. Social cohesion is particularly undermined by a polarization of wealth associated with persistent high levels of unemployment and underemployment. Environmental sustainability is challenged by materials-intensive patterns of production and consumption. International competitiveness and economic sustainability is weakened by comparatively high production costs, technological backwardness, mature markets and oligopolistic supply structures. Globalization and market liberalization expose industry to competition from emerging market economies and/or developing countries with lower production costs, growing markets and fewer regulatory constraints, and from technologically more sophisticated advanced capitalist economies (notably Japan and the USA) and Newly Industrializing Country economies.

Each of these dilemmas must, however, be resolved in order to secure high levels of welfare and 'quality of life', and to discover a new model of development and mode of regulation within which this will be sustainably possible (Gibbs, 1996). Unemployment and related problems of social exclusion undoubtedly pose the most serious and immediate challenge, however, because of their human consequences and political significance. This has pushed concern about environmental sustainability down the political agenda. This demotion could clearly lead to further environmental destruction. On the other hand, this situation also presents an opportunity. Whether it represents a threat or an opportunity depends upon whether these issues are or are not seen as systemically related. Regarding them as unrelated leads to a prioritization of some problems and trade-offs between policy goals. Keynesian economic management *de facto* traded off pursuit of full employment against environmental damage and pollution, albeit unintentionally. Contemporary neo-liberal policy is less concerned with restoring fuller employment than it is with boosting profitability and output growth rates, but it remains largely insensitive to the environmental consequences of implementing such a policy approach. Seeing these problems as systemically related, in contrast, opens the possibility of searching for a model of development in which issues of employment creation and social cohesion, economic performance and environmental valorization can be addressed simultaneously. Such a new approach can be characterized as eco-Keynesian. Eco-Keynesianism seeks to create a regulatory regime which will facilitate fuller employment and a more egalitarian distribution of work (both unwaged as well as waged work in the formal and informal sectors) alongside enhanced environmental quality and environmentally more sustainable patterns of production and consumption. It seeks to restore the commitment of social democratic centre-left politics to greater social justice, inclusion and equality while combining this with a recognition that economy and society must respect their necessary relationships with the natural environment.

Development is, however, heavily path-dependent (Nelson and Winter, 1982). It remains an open question as to whether a revolutionary shift from one path to another can be achieved through incremental change and evolutionary reformist modifications to the existing developmental trajectory or whether it requires a rapid quantum leap from one trajectory to a qualitatively different one. The former is the more feasible option, however. It is difficult, for example, to conceive of the structuring principles of capitalism being replaced by those of some alternative principles of political economy, such as those of the centrally planned economy, in the foreseeable future. Moreover, if this were to come about, the historical precedents are far from propitious. On the other hand, while the dominant developmental model will remain capitalist, there are distinctive varieties of capitalism, and different trajectories of change, within a range of regulatory and political frameworks.

Major changes in practice presuppose as a necessary (but not sufficient) condition that the conceptual bases of such changes be specified. A major purpose of this chapter is therefore to explore the main lineaments of a possible developmental model that would combine employment creation with environmental valorization in Europe. It seeks to *envision* how the economy might be restructured (acknowledging that this is a multi-dimensional process: Simonis, 1994) onto the path of a more sustainable regime of accumulation, how a new distribution of work and employment might emerge, and to *conceptualize* models of transition processes that specify the regulatory and governance conditions under which desired changes might be realized. While environmental sustainability is a contested concept, moving towards sustainability will certainly involve much less materials-intensive patterns of production and consumption (Wuppertal Institute for Climate, Environment and Energy: Factor 10 Club, 1994). This will have profound implications for lifestyles and ways of living. Addressing normative and ultimately moral questions about the sort of society in which we wish to live and about the mechnisms through which desired societal goals might be achieved is therefore unavoidable.

The remainder of the chapter is organized as follows. First, the nature of the unemployment problem is examined and current approaches to job creation are evaluated. Then an alternative approach based upon a transition to a different development trajectory, to a more sustainable regime of accumulation and enabling eco-Keynesian modes of regulation, that could deliver higher levels of social cohesion and environmental protection, is outlined. The appropriate territorial basis of regulation within Europe is then discussed. Rather than seek to specify instruments and mechanisms to produce desired changes (which are discussed at length in Weaver, 1995a), the focus here is upon identifying and conceptualizing linkages between these desired goals both in restructuring processes and in outcomes from them.

14.2   The current impasse: clues about possible futures from the paradoxes of high unemployment and the limits to contemporary policy approaches

Persistent high unemployment in Europe reflects a deep paradox. On the one hand, it can be seen as an outcome of one of late modern society's greatest achievements: the capacity to produce so efficiently that the socially necessary labour time formally required (that is, in remunerated employment involving formalized and legally recognized relationships between an employer and an employee) absorbs only around 20 per cent of the potentially available working time. On the other hand, unemployment simultaneously reflects one of late modern society's greatest failures: the absence of corresponding adjustments in systems for redefining and allocating work and income across society – hence Britton's (1994) view that there is no shortage of work, only one of jobs. The welfare state is largely financed by taxes on employment. These discourage job creation. Equally, for many the benefits system provides disincentives and barriers to working for a wage. While the welfare state and

benefits systems have been eroded in Europe in recent years, these disincentives remain. At the same time, however, European societies have become more
polarized. Significant groups of people are marginalized and excluded as a
direct consequence of being denied the safety net of the welfare state and
benefits systems (Hadjimichalis and Sadler, 1995).

Clearly there are several facets to the problem, centred around notions of
economic democracy and citizens' entitlements, expectations and responsibil-
ities. The first concerns allocation of work and income. Paid employment is
'needed' – and unemployment is currently considered a problem – because
the employment system and the wage relation are the major mechanisms for
allocating entitlement to a share of GNP, directly through income and indirectly through related state transfer payments. There is, however, no necessary
correspondence between either socially useful (welfare-enhancing) work and
the availability of paid employment (jobs) or between the contribution an
individual makes to society through his/her work and his/her income and
status. Such work is often performed in the informal economy, or outside the
wage relationship (for example, within households). The second facet is that
welfare-enhancing 'work' remains undone because polarization of wealth
and income has resulted in a lack of effective demand or because it is bureau-
cratically barred to the unemployed by threat of benefit withdrawal. A third
facet is that exclusion from employment causes individuals rapidly to become
'unemployable'. They lose touch with the skills and contacts needed for gaining or creating a job (Britton, 1994). Moreover, unemployment can lead to
political and social, as well as economic, marginalization or exclusion. The
costs of this disenfranchisement are increasing and varied: growing criminality,
violence, vandalism, drug abuse, fear and ill-health (Schmidt-Bleek, 1994).
This may become systemically manifest as legitimation or motivation crises
(Habermas, 1976).

The impacts of unemployment are also experienced by those in employment through the intensification of the labour process because of economic
pressures for greater productivity and performance underpinned by fear of job
loss (Hudson, 1997a, 1997b). Changes (such as the growth of part-time
work) have increased labour market flexibility, but for many people labour
market rigidities still present stark choices: an all-or-nothing commitment to
full-time long-term paid employment, full-time long-term unemployment, a
life in the 'grey/black' areas of the labour market, or even, for some, criminality
as a survival strategy. Moreover, much of the 'flexibilization' of the labour
market and labour process that has occurred has emphasized the need for
strong progressive and egalitarian regulatory regimes if the restructuring of
work is to be compatible with enhanced social inclusion (Hudson, 1989b).

There is little chance of effectively tackling these problems within the
parameters of public policy as currently defined within Europe. Indeed, to a
considerable degree this policy framework is an important proximate cause of
the problems. With the demise of state socialism, capitalism became the dominant political and economic philosophy within Europe. More particularly,
there has been a widespread diffusion of neo-liberal policy approaches and a

particular normative conception of capitalist development. The latter presumes that prosperity is achieved through GNP growth based on productivity gains, investment in new technologies, and free trade. This emphasizes the production, trade and consumption of material commodities, driven by competition. The drive for profitability generates incessant pressure for productivity gains. Relative factor prices have strongly focused attention on increasing labour productivity, linked to increased fixed capital investment, mechanization and automation. Profits depend on the quantity of units of output sold, rather than on these constituting the initial element in a more comprehensive package of end-user services extending over the life of the product. The resultant 'throughput' economy is not oriented towards conservation and the efficient delivery of end-user services. Patterns of production and consumption are materials-intensive and 'leaky' (Weaver, 1995b). Although there has been considerable apparent business service sector growth, much of this reflects a changing social division of labour and increasing out-sourcing and sub-contracting (Sayer and Walker, 1992). Consumption of labour-intensive personal services is discouraged relative to consumption of material goods. Moreover, such employment growth in these activities as has occurred has often been of low-skill, poorly paid jobs in minimally regulated labour markets.

This developmental model is open to criticism on – *inter alia* – two grounds. First, increasing GDP per person does not necessarily enhance welfare. Adjusting GDP to give a more broadly based measure of economic progress, for example, Jackson and Marks (1994) derive an 'Index of Sustainable Welfare' which shows a negative correlation with GDP growth since 1974 (see also Nordhaus and Tobin, 1972). Wage-working and consumerism are no longer welfare-enhancing (Robertson, 1995) and do not correspond with the aspirations of many people in the affluent First World (Lenk, 1994; Kistler and Strech, 1992). Secondly, there are also severe limitations to job creation policies cast in the current dominant policy mould. Growth in output will not necessarily translate into employment growth (Fontela, 1994; United Nations Development Program, 1993). Jobless or 'job-shedding' growth is observable to a varying extent in all OECD countries. This is especially so as many of those jobs that have been created have been casual or part-time jobs. Economic growth (if sufficiently rapid) can temporarily create new jobs more rapidly than others are destroyed by labour-saving innovations, but the constant drive for efficiency (or 'technological rents': Mandel, 1975) under a competitive, profit-driven market system constitutes a compelling imperative to increase labour productivity, both absolutely and relative to the growth of output. Consequently, there are only limited opportunities to create jobs in the formal economy directly in activities that produce traded goods and services. Even without gains in labour productivity, no realistic growth rate could create sufficient employment in the formal economy to restore anything resembling 'full employment' and secure social cohesion. Growth in conventional formal sectors and activities can be, at best, a temporary panacea to unemployment problems.

There are also limits to the extent to which unemployment can be reduced by cutting direct labour costs. Because of technological and market changes, many contemporary products and services simply could not be delivered to the same standards of quality and complexity or within the same time-frame using human labour rather than machines (see Ayres, 1991). Consequently, there is a ratchet effect in much (though admittedly not all) of the labour market rather than reversibility in choice of production technique. While higher labour costs reduce manufacturing and traded-service sector employment, lower labour costs will not necessarily increase such employment.

Seeking to reduce unemployment within the prevailing neo-liberal policy framework would, therefore, be at best partially successful and further erode social cohesion. Extrapolations based upon current trajectories suggest the emergence or strengthening of dual societies (Fontela, 1994), involving either a coexistence of workers and unemployed (the continental west European model), increases in precarious and low-paid employment (the USA model), or a combination of the two (the UK 'Two Nations' model and the model increasingly observable over much of eastern Europe). Indeed, as factors such as age, ethnicity and gender interact with class, occupation and skills, to influence the distribution of employment and unemployment, a conceptualization in terms of multiple segmentation rather than simple dualisms may be more appropriate (Hudson and Williams, 1995). In addition, these scenarios imply continuing high levels of environmental impact and would further stymie an already inadequate pace of progress toward more sustainable development. This suggests a need to rethink the bases of public policy in the search for a new, more sustainable model of development.

14.3   Searching for an eco–Keynesian alternative: in pursuit of environmental valorization and a new distribution of work and employment

A key issue in the search for a new sustainable regime of accumulation is the creation of links between economic restructuring and the simultaneous resolution of problems of unemployment, environmental sustainability and competitiveness. Enhancing environmental sustainability requires greatly increasing materials productivity (Wuppertal Institute for Climate, Environment and Energy: Factor 10 Club, 1994). Such a shift has two components. One is to increase the materials productivity of goods and services and deliver useful end-services more efficiently. The other is to change the mix of the goods, services and technologies towards those with lower environmental impacts. Both components imply shifts in the sectoral structure of the economy, of output, and in patterns of work and employment. In this way environmental valorization and socially useful employment creation could be made compatible.

The growth of 'green' technologies and environmental protection has considerable conventional job creation potential (Jenkins and McClaren, 1994). Switching the emphasis from increasing labour productivities to increasing materials and energy productivities will necessitate innovative product and process redesign. This will generate markets for new, environmentally improved technologies and for people to design, manage and monitor them. It has, for example, been estimated that the effects of global warming alone will create a market of around $1800 billion over the next 40 years (International Institute for Energy Conservation, reported in Boulton, 1996). Jobs will also be created in emergent new sectors and in existing ones that will grow in importance, especially those involved with life-time materials management; for instance, jobs associated with organizing product sharing or leasing (increasing intensity of use), in product maintenance and repair (increasing product longevity) and in materials recycling and reuse (providing for materials cascading). Although jobs may be created in these ways, the links between economic, social and environmental sustainability must extend beyond job creation and loss in the formal sector (for there will surely be job losses, especially in the transport and travel, automotive, energy and chemicals sectors) to encompass work that currently is performed beyond the formal economy and to redefine the boundaries of work beyond employment.

Reconceptualizing employment and work, and changing the ways in which they are distributed between people, requires rethinking both demand-side and supply-side approaches to work and remuneration. Waged work in the formal sector and welfare-enhancing unwaged work outside that sector need to be linked through transfer payments and income redistribution to ensure effective demand for the latter. Efficiency and effectiveness in both sectors in the performance of welfare-enhancing work depend in part upon loosening the rigidities that currently prohibit or discourage socially acceptable flexible working arrangements and that deter employers from providing employment, workers from working and individuals from choosing how best to allocate their time between work (paid and unpaid) and leisure. Ensuring that welfare-enhancing work is efficiently and effectively accomplished and a balance struck between work and leisure that maximizes individual and social benefits requires a major shift in consumption profiles and in approaches to meeting end-use demands. It implies an enhanced emphasis upon the performance of individually organized work, particularly in non-traded personal services. Welfare-enhancing and labour-intensive activities currently precluded from the formal economy should be recognized and encouraged but within a regulatory framework that guarantees labour market conditions and standards and enhanced social inclusion. Lipietz (1992), for example, refers to the need for the emergence of a 'socially useful third sector' as one mechanism for ensuring that such work gets done while providing work and incomes for those currently unemployed.

The transition to a new, more sustainable regime of accumulation will require the creation of an enabling regulatory framework by a strong state. It will necessitate policy innovations and revised patterns of public expenditure. The critical element therefore is radically to revise the regulatory framework of state policies. The most important of these state policy changes would undoubtedly be a gradual but comprehensive and systematic root-and-branch revision of the tax and benefits system to one based ultimately on taxation of environmental resources used. The revenues raised would form the basis for a citizen's income and public expenditure. The idea of a citizen's income is not new (see, for example, Friedman and Friedman, 1980; Purdy, 1994). Linking this with a system for revenue raising based upon eco-taxes is more recent but is gaining increasing support in both the academic (for example, von Weizsäcker, 1994) and business (Business Council for Sustainable Development, 1994) communities. The gains from such a shift in the basis of taxation could be widespread and considerable. For example, Repetto (1992, cited in Robertson, 1995) reports significant possible economic gains (for example, higher environmental quality, reduced infrastructure needs and increased employment) from shifting the tax burden from incomes, employment and profits to environmental charges on waste collection and pollution in the USA.

Other policy changes would also be needed to facilitate the transition towards greater sustainability. Public policy initiative could help identify and support R&D on technologies to increase the materials efficiencies of end-service delivery, especially in growth sectors with potential applications in process and/or product redesign, such as materials science, electronics and biotechnologies. Infrastructure investment to underpin the new information- and knowledge-based economy would also be critical. Policy innovations to legitimize and strengthen the 'grey' and informal economy to ensure performance of welfare-enhancing work and maximize the multiplier effect of state transfer payments through the development of voucher or credit systems (local 'shadow' currencies) for trading community services could play a key role. So too could information exchanges and 'job' centres for the organization of community, voluntary and personal work.

A further set of policy changes might reduce the cost (financial and environmental) of comfortable living; in particular, measures to reduce unnecessary travel, improve energy productivities in the housing sector, encourage heat-recovery from waste materials, reuse or recycle waste materials, increase the longevity of products and structures through repair and maintenance, and increase food self-sufficiency. Such changes in ways of living and working imply significant shifts in individual and community attitudes. A further important area of policy innovation, therefore, would be investment in educational and information facilities for self-development (with both leisure and commercial applications) and for societal ends (to encourage shifts towards socially responsible behaviours, to develop social solidarity and to encourage shifts in consumption profiles).

Such policy shifts clearly imply both a qualitative and a quantitative expansion in the scope of state regulatory activity and expenditure. European governments currently face public expenditure constraints and are restricted in their capacity to raise income by taxing profitable enterprises because of the mobility of capital. This capacity to switch location has increased markedly as a result of liberalization and the pace of technological change in many industries (which contributes to higher depreciation rates and more frequent reviews of location), although it is important not to overgeneralize in this context. While there has been considerable emphasis on capital's growing 'hypermobility', there are counter-arguments that key activities and sectors are deeply embedded culturally and institutionally in specific places and, hence, are not locationally mobile. Much depends upon the activity in question. Shifting taxes from employment, income and profit and on to services received and resources consumed by firms and consumers is nonetheless one route to income redistribution in an increasingly borderless world (Ohmae, 1995). Greater levels of recycling, encouraged by tax reforms, would also help 'hold down' capital and fix it within the markets served, and reinforce existing tendencies towards 'glocalization' (see Amin and Thrift, 1994).

Whatever the nominal purchasing power of future incomes, their real purchasing power *vis-à-vis* consumption of products and services involving dissipative loss of materials and energy will decrease. This has implications for defining welfare-enhancing work. Meeting many *basic* needs, in particular for physical mobility and space conditioning (mostly heating), involves dissipative losses of materials and energy. Current delivery systems are inefficient or needlessly profligate (Löfstedt, 1995). These inefficiencies are embedded within present infrastructures and organizational patterns. They need to be changed. Buildings can be made to be almost or completely energy self-sufficient (Building Services Research and Information Association, 1994). Transport can be reduced through tele-applications (such as teleworking) and greater mixing of land uses (ideas presaged by Jacobs, 1961). Physical transport efficiency can be improved by modal switches linked to new public and private transport systems. The required changes all involve welfare-enhancing work in both private and (especially) public domains, and have strong equity implications since they help decouple the link between comfortable living and high monetary income.

Purchasing power in relation to conventional 'status' goods will decrease since these, too, are energy- and materials-intensive. In particular, recreational and optional travel and automobility will become relatively more expensive. Reduced physical mobility will put a premium on enhancing local environmental quality and providing leisure and recreational services locally. It will also foster greater local self-sufficiency as imported goods (embodying high energy taxes associated with transport) will become more expensive relative to local produce. Services provided at little or no cost to the environment, especially those provided mostly by human work, will become relatively more affordable. Health care and education will increase in importance and contribute to an enhanced quality of life. Education and learning within a 'learning culture' (Lutz, 1994) will become key elements of welfare-enhancing work.

This emphasis on a learning culture resonates with claims as to the enhanced significance of knowledge within a globalized late modern political economy and society (for example, see Giddens, 1990; Lundvall and Johnson, 1994; Strange, 1988). There is considerable potential for information and communications technologies (ICT) to contribute to local and regional economic revival, support alternatives to conventional employment and provide for environmentally benign development. Unlocking this potential will, however, require enhancing the life-cycle environmental performance of ICT (Ayres *et al.*, 1994). Moreover, rapid technological advance leads to equipment soon becoming technologically obsolete, although it is important not to confuse technological obsolescence with an ability to allow particular tasks to be undertaken. Moreover, ICT can enhance economic competitiveness, especially in respect of the important and growing traded business services sector. The critical point is that ICT means that commissioner and contractor need never physically meet in a global market.

The capacity to use ICT to replace travel, especially commuting, and to facilitate decentralized living and working arrangements, spreading environmental stresses more evenly in both space and time, could also be beneficial. While spatial proximity will remain crucial for many types of work, teleworking is likely to be encouraged in other occupations as firms seek to rationalize on office space, which accounts for an ever-higher proportion of overall costs and represents an increasing liability to many companies (Henkel, 1994). Time–space compression (Harvey, 1989), seemingly paradoxically, can enhance the importance of the specificities of *some* (but not *all*) places in a global economy – with a clear resultant risk, therefore, of deepening socio-spatial inequalities which could endanger the achievement of environmental and employment goals if left unchecked. The growth of global 'back offices' clearly illustrates this danger. Henkel (1994), for example, cites the cases of Swissair, which has switched its ticketing arrangement for flights worldwide to Bombay and an American insurance company that employs nurses at home in Ireland to process medical claims 'overnight'. More generally, the new global centres for 'back office' work are Russia, India and China, which offer masses of technologically skilled labour at very low cost. This illustrates the need for global regulation of labour market conditions if employment creation in such places is to avoid the worst excesses of a neo-liberal approach to economic and environmental management.

In summary, the scale and scope of economic and social changes outlined above presuppose a major qualitative and quantitative expansion in the scope and form of the state regulatory activity as neo-liberalism gives way to eco-Keynesianism. This more sustainable new regime of accumulation would incorporate a highly competitive and efficient formal economy producing tradeable commodities. Problems of competitiveness and economic efficiency, unemployment and environmental sustainability would be simultaneously tackled through a revised system of taxes and benefits that penalizes value-diminishing activities and rewards value-adding activities. Governments would have a key – though not sole – role to play in defining these new regulatory and governance arrangements. They in turn would help guide and shape choices of technologies, patterns of resource use and patterns of employment, production and consumption.

14.4 What would be an appropriate territorial level of state involvement in an eco-Keynesian mode of regulation in Europe?

The argument in the previous section is that the transition to a new sustainable regime of accumulation will require the creation of an enabling regulatory framework by a strong and socially progressive state. There is also, however, a question of the appropriate territorial level at which state power should be deployed in constructing this new mode of regulation, especially in the light of claims as to the increasing importance of global and sub-national spatial scales in the operation of economic and political processes. There is undoubtedly a general recognition of clear limits to the regulatory capacities of all national states. Recognition of such limits was important in the demise of the Fordist mode of regulation in the advanced capitalist states. A variety of crisis tendencies (Habermas, 1976), in particular those associated with fiscal crisis (O'Connor, 1973) could no longer be contained. As a result, national states could no longer pursue Keynesian policies to maintain full employment. In their place came neo-liberal policies, led by Thatcherism and Reaganomics. While often represented as deregulatory as national governments both acknowledged and encouraged the play of global market forces, these were in practice strongly re-regulatory as the relations between economy, society and state were redefined. There is growing evidence that they are subject to the same structural and systemic forces that underlay the demise of the previous Keynesian–Fordist regimes. It is precisely recognition of this that – *inter alia* – is encouraging exploration of links between the global and the local and supra-national and sub-national approaches to regulation (although there is no guarantee that these will not be subject to the same limits).

There is certainly evidence of political authority being recast at the supra-national level within Europe in the evolution of organizations such as the EU, which is both widening its spatial reach and deepening the degree of integration among its member states, and also sub-nationally via regional decentralization (Dunford and Hudson, 1996). This points within Europe to a degree of erosion of the powers of national states as a result of processes of 'hollowing out' (Jessop, 1994). The relationships of the EU to both globalization and regionalization processes are, however, ambiguous and the processes of policy formation and implementation within the EU are complex and contradictory (Hudson, 1996a). From one point of view, it can be seen as facilitating globalization. From another point of view, however, it can be seen as a barrier, a site of resistance, to it. At the same time, there have been growing pressures from various sub-national regionalist movements within Europe, in part because they see a greater potential for political-economic viability within a 'Europe of the Regions'. There are undoubtedly some regional economic 'winners' within the EU and many other cities and regions seek to emulate this success (Dunford and Hudson, 1996). There is also a deep ambivalence within the EU towards these processes. Claims about a 'Europe of the Regions' are open to competing interpretations which, on the one hand, emphasize deepening inequalities and competition between regions and, on the other hand, greater social cohesion as a result of strengthened Structural and Cohesion Fund policies (European Commission, 1996).

There is therefore no doubt that a restructuring of the territoriality of state power is ongoing and that further changes will be needed as part of the process of facilitating the emergence of a more sustainable regime of accumulation.

The relationship of national states within Europe to the supra- and sub-nationalization of state power is also an ambivalent one, however. Some see the demise of the national state as state power moves both upwards to emergent supra-national states (such as the EU, though it is important to recall that this has evolved as a result of decisions *of* national states) and downwards to sub-national units, representing, respectively, larger or smaller versions of current national state forms. Ruggie (1993) argues that in the EU '. . . the process of unbundling territoriality has gone further than anywhere else', but nonetheless state power remains strongly territorially, and nationally, based. Others caution against a too-ready acceptance of reports of 'the exaggerated death of the nation-state' (Anderson, 1995) and argue that what is emerging is a much more complex form of regulation involving supra-national, national and sub-national scales. Mann (1993) stresses that 'European nation states are neither dying nor retiring; they have merely shifted functions, and they may continue to do so in the future.' There are therefore strong grounds for believing that, for the foreseeable future, national states will continue to have a central role in processes of policy innovation, formation and implementation.

Globalization and regionalism are, therefore, both powerful, and to a degree related, tendencies but they are by no means uncontested ones. Thus while there continues to be a significant degree of unbundling of the relations between national territory and regulation, of 'hollowing out' of national states, they are still and will remain key actors in this emergent restructured regulatory system. Consequently, there are good grounds for believing that national states do and will retain important regulatory powers, both directly and indirectly, in so far as they play a key role in the construction of both supra-national and sub-national regulatory mechanisms. There are, however, considerable differences in the ways in which national states in Europe are seeking to construct regulatory frameworks to promote greater environmental sustainability (European Environmental Agency, 1996). While it therefore also remains an open question as to whether national states will choose to deploy their powers in pursuit of employment creation allied to environmental sustainability, regulatory systems that insist upon high environmental standards may well be compatible with economic success. For example, at regional level the economy of the Ruhr is being successfully restructured around clusters of environmental protection industries (Refeld, 1995), while at national level the German economy has prospered because of, rather than despite, stringent environmental standards. Compliance with these standards within the national territory has often given German companies a competitive edge by pioneering new environmentally friendly products and processes.

More generally, there is a growing recognition of the importance of environmental sustainability at the local and regional levels (Roberts, 1995). The transition to a more sustainable regime of accumulation will also necessitate regulatory changes at these territorial levels and a greater emphasis on socio-spatial justice and inclusion. For example, there could be policy changes to promote environmental restoration and improvement, including ways of using the environment and local heritage as a basis for improved recreational and leisure opportunities and, as a result, of generating income and jobs locally. There could also be policies to increase investment in 'soft' and ICT infrastructures that support individual and local economic initiatives for greater economic self-reliance and 'closure' of local and regional economies (while recognizing the difficulties of so doing: see Hudson and Plum, 1986). Without a doubt, however, the transition to a more sustainable regime of accumulation will need to create and guarantee 'room for manoeuvre' at local and regional levels, to allow local knowledge and creativity to be expressed and help shape local policy agendas, and to seek to combine employment creation with environmental valorization in ways that are sensitive to local and regional contexts (Hudson, 1996a).

At supra-national level, there are certainly themes in recent EU environmental policy documents which resonate with arguments as to the potential compatibility of environmental valorization and employment creation as policy goals. Any supra-national European regulatory framework would need to incorporate the sort of incentives and disincentives outlined there (see, for example, European Commission, 1993). For example, the need to switch taxation onto the use of environmental services (such as materials and land used, or wastes emitted) is recognized as part of an 'incentive-based' approach to the integration of environmental and economic policies (European Commission, 1994).

At the same time, differences in international and national environmental standards and regulations have allowed companies to escape strict regulation within their 'home' territories by seeking 'spatial fixes' by 'pollution dumping', particularly to Third World countries in which jobs take precedence over environmental protection (Leonard, 1988). There is no doubt that the emergence of a global regulatory authority (or authorities) would greatly ease the widespread diffusion and adoption of more stringent environmental regulation. There is some evidence of such international organizations and movements emerging (Yearley, 1995). It would be naive, however, to believe that considerable variations will not remain at the national, or perhaps macro-regional (for example, the EU or NAFTA), level in environmental standards, permitted pollution levels and regulatory regimes, not least as a result of the past impress of uneven development. This raises broader questions regarding the transition to 'sustainability' at global scale. Nevertheless, within the First World the transition to more sustainable production and employment creation 'at home' through restructuring which focuses on more technologically sophisticated and environmentally friendly products and processes can be compatible with both corporate and territorial competitive advantage. That said, the imperatives of capitalist production will continue to impose limits both on company strategies and on national states' 'room for manoeuvre' in policy formation – with longer-term implications for 'sustainability', a point developed in the final section.

14.5 Conclusions and implications

Establishing a new, more sustainable regime of accumulation will require radical structural economic change. Many of the elements that might form part of a policy package within an eco-Keynesian mode of regulation are already evident: the need for a facilitating meta-policy framework at international, national, regional and local levels; the identification and removal of barriers to change; and the need to steer and harness technological and cultural developments in ways that support the attainment of economic, social and environmental goals. The development of a new system of transfer payments and public expenditures which, unlike earlier redistribution systems, contributes to rather than detracts from the achievement of economic, social and environmental 'goods' such as efficient production, job creation and the performance of welfare-enhancing work, is central to this approach. The lynchpin of the system is its guaranteeing a citizen's income while providing incentives and flexibilities necessary to ensure that welfare-enhancing work is accomplished and a balance struck in personal and societal time allocations between work, education and leisure.

Although some elements of past economic practice may remain in a new, environmentally sustainable economy, a return to the certainties of Keynesian 'full employment' is not a feasible option. 'Full employment' in its Keynesian partial sense of formal-sector employment for men will no longer occupy a pivotal role in the equating of production and consumption as the route to increased economic growth rates. Indeed, increasing growth rates *per se* will no longer be the sole – or even main – goal of public policies. Even if a switch to environmentally more sustainable production increased employment, it is unlikely that overall unemployment levels would be significantly reduced. Nevertheless, there are compelling reasons why a switch to a more equitable distribution of work, of the costs and benefits of production, and to a more environmentally sustainable economy is necessary.

Unemployment and environmental stress can both be related to the current neo-liberal development paradigm. Radical revision of the contemporary policy framework is needed to tackle these problems. Nonetheless, a revision intended to internalize environmental costs will not *necessarily* and *automatically* also resolve the unemployment problem (and vice versa), although such a revision would undoubtedly have employment implications (both positive and negative). Such a review should seek to *create* major synergies between the achievement of greater environmental protection, the creation of more equitable and satisfying employment opportunities and the delivery of enhanced individual and collective welfare. Changes should be steered in ways that reduce the materials-intensiveness of production and economic growth. Complementary measures are needed to enhance social inclusion, guarantee minimum living standards, and reduce the materials-intensity of delivering these. Policy changes should facilitate the welfare-enhancing use of human resources, including 'work' that contributes to welfare by enhancing environmental quality or reducing the stresses that comfortable living place on the environment.

In summary, there is a need for a switch to a more sustainable regime of accumulation shaped and sustained by an eco-Keynesian mode of regulation. The distribution of, and entitlements to, society's output need to be shifted from their current close relationship to inherited wealth and position in the waged labour market to one which relates more closely to the socially useful work – waged and unwaged – that people carry out and to their rights as citizens. The required changes should be designed to facilitate greater environmental protection, improved economic performance (through productivity improvements of the currently unemployed) and greater social equity and cohesion. There are opportunities for win–win outcomes *if* the challenges faced are seen as systemically linked and a coherent strategy for economic restructuring is formulated and implemented accordingly. Conversely, a failure to do so will lead to continuing environmental destruction and permanent unemployment for many.

In the final analysis, however, there are unavoidable questions which must be faced as to the possibilities for and limits to state involvement in steering the trajectory of economic and social development within capitalism. The win–win scenarios referred to above assume that state regulation and involvement can successfully contain, if not abolish, crisis tendencies that are inherent to a capitalist economy. But there are powerful reasons to believe that an eco-Keynesian approach would be no less susceptible to the forces that drive a capitalist economy than any other mode of regulation and regime of accumulation. The maturing contradictions of state involvement were decisive in bringing about the end of the Fordist regime of accumulation and mode of regulation within the advanced capitalist countries. The emergent crises of one mode of state regulation spelled the end for Keynesianism and ushered in its neo-liberal successors, but these proved to be no more immune to crises. Similar longer-term dangers may await their eco-Keynesian successor within the parameters of a capitalist economy, with the resultant threat of a future of further environmental destruction and permanent unemployment for many. The best that may be expected – or hoped for – is to contain the problem for a while and buy some time in which more radical solutions might, perhaps, be explored.

References

Aglietta, M., 1979, *A Theory of Capitalist Regulation: the US Experience*, New Left Books, London

Aglietta, M., 1982, 'World capitalism in the Eighties', *New Left Review*, **136**, 5–42

Albert, M., 1993, *Capitalism against Capitalism*, Whurr, London

Allaert, G., 1994, 'Towards a sustainable Scheldt region', in Voogd, H. (ed.), *Issues in Environmental Planning*, Pion, London, 131–144

Allen, J., 1997, 'Economies of power and space', in Lee, R. and Wills, J. (eds), *Geographies of Economies*, Arnold, London, 59–70

Allen, J., Massey, D. and Cochrane, A., 1998, *Re-thinking the Region*, Routledge, London

Amin, A., 1989, 'Flexible specialisation and small firms in Italy; myths and realities', *Antipode*, **21**(1), 13–34

Amin, A., 1998, 'An institutionalist perspective on regional economic development', paper presented to the RGS Economic Geography Research Group Seminar, Institutions and Governance, 3 July, UCL, London

Amin, A. and Cohendet, P., 1997, 'Learning and adaptation in decentralised business networks', paper presented to the Final ESF EMOT Conference, Stresa, Italy, 11–13 September

Amin, A. and Hausner, J. (eds), 1997, *Beyond Market and Hierarchy: Interactive Governance and Social Complexity*, Edward Elgar, Aldershot

Amin, A. and Sadler, D., 1992, ' "Europeanisation" in the automotive components sector and its implications for state and locality', paper presented to the ESF RURE meeting, Copenhagen, 3–6 September

Amin, A. and Smith, I., 1990, 'Decline and restructuring in the UK motor vehicle components industry', *Scottish Journal of Political Economy*, **37**(3), 109–140

Amin, A. and Smith, I., 1991, 'Vertical integration or disintegration? The case of the UK car parts industry', in Law, C.M. (ed.), 1991, *Restructuring the Global Automobile Industry: National and Regional Impacts*, Routledge, London, 169–199

Amin, A. and Thomas, A., 1996, 'The negotiated economy: state and civic institutions in Denmark', *Economy and Society*, **25**, 255–281

Amin, A. and Thrift, N., 1994, *Globalization, Institutions and Regional Development in Europe*, Oxford University Press, Oxford

Amin, A., Cameron, A. and Hudson, R., 1998, 'Welfare as Work? The Potential of the UK Social Economy', *mimeo*, University of Durham, Department of Geography, 33

Anderson, J., 1995, 'The exaggerated death of the nation state', in Anderson, J., Brook, C. and Cochrane, A., *A Global World?*, Oxford University Press, 65–112

Anderson, J., Duncan, S. and Hudson, R., 1983, 'Uneven development, redundant spaces', in Anderson, J., Duncan, S. and Hudson, R. (eds), *Redundant Spaces in Cities and Regions?*, Academic Press, London, 1–15

Ardagh, J., 1982, *France in the 1980s*, Penguin, Harmondsworth

Arrow, K., 1962, 'The economic implications of learning by doing', *Review of Economic Studies*, **29**, 155–173

Asheim, B., 1996, ' "Learning regions" in a globalised world economy: towards a new competitive advantage of industrial districts?', paper presented to the First European Urban and Regional Studies Conference, 11–14 April, University of Exeter

Ashton, D.N. and Maguire, M.J., 1987, 'Young adults in the labour market', *Research Paper 55, Department of Employment*, HMSO, London

Athreye, S., 1998, 'On markets in knowledge', *ESRC Centre for Business Research Working Paper 83*, University of Cambridge

Atkinson, J.H., 1986, 'Opencast coal working', paper presented at a conference convened by CPRE, Stafford, October 1985

Austrin, T. and Beynon, H., 1979, 'Global Outpost: The Working Class Experience of Big Business in North East England, 1964–79', *mimeo*, University of Durham, Department of Sociology

Ayres, R.U., 1989, 'Industrial metabolism and global change', *International Social Science Journal*, **121**, 363–373

Ayres, R.U., 1991, *Computer Integrated Manufacturing: Revolution in Progress*, Chapman and Hall, London

Ayres, R.U., Norberg-Bohm, V., Prince, J., Stigliani, W. and Yanowitz, J., 1988, *Industrial Metabolism, the Environment and Applications of Materials-Balance Principles for Selected Chemicals*, RR-89-11, IIASA, Laxenberg, Austria

Ayres, R.U., Frankl, P. and Lee, H., 1994, *Life Cycle Analysis of Semiconductor*, CMER Working Paper, INSEAD, Fontainebleau, France

Bagnasco, A., 1977, *Tre Italie: La Problematica Territoriale dello Sviluppo Italiano*, Il Mulino, Bologna

Balassa, B., 1981, *The Newly Industrialising Countries in the World Economy*, Pergamon Press, Oxford

Baran, P. and Sweezy, P., 1968, *Monopoly Capitalism*, Penguin, Harmondsworth

Barnes, T., 1996, *Logics of Dislocation*, Guilford, New York

Beattie, C., Fothergill, S., Gore, T. and Herrington, A., 1997, *The Real Level of Unemployment*, Centre for Regional Economic and Social Research, Sheffield Hallam University

Beck, U., 1992, *Risk Society: Towards a New Modernity*, Sage, London

Bell, F., 1985, *At the Works*, Virago, London (1st edition 1907)

Bellet, M., Colletis, G. and Lung, Y., 1993, 'Economie de proximités', *Revue d'Economie Régionale et Urbaine*, **3**, 357–361

Benwell Community Development Project, 1978a, *Permanent Unemployment*, BCDP, 85–87 Adelaide Terrace, Newcastle upon Tyne NE4 8BB

Benwell Community Development Project, 1978b, *Slums on the Drawing Board*, BCDP, 85–87 Adelaide Terrace, Newcastle upon Tyne NE4 8BB

Beynon, H., 1984, *Working for Ford*, Penguin, Harmondsworth

Beynon, H. (ed.), 1985, *Digging Deeper*, Verso, London

Beynon, H., 1995, 'The changing experience of work: Britain in the 1990s', paper presented to the Conference on Education and Training for the Future Labour Markets of Europe, 21–24 September 1995, University of Durham

Beynon, H. and Austrin, T., 1994, *Masters and Servants*, Rivers Oram, London

Beynon, H. and Hudson, R., 1993, 'Place and space in contemporary Europe: some lessons and reflections', *Antipode*, **25**(3), 177–190

Beynon, H., Hudson, R., Lewis, J., Sadler, D. and Townsend, A., 1985, 'Middlesbrough: contradictions in economic and cultural change', *Middlesbrough Locality Study Working Paper 1*, Department of Geography and Sociology, University of Durham, Durham

Beynon, H., Hudson, R. and Sadler, D., 1986a, 'The growth and internationalization of Teesside's chemicals industry', *Middlesbrough Locality Study, Working Paper 3*, Departments of Geography and Sociology, University of Durham, Durham

Beynon, H., Hudson, R. and Sadler, D., 1986b, 'Nationalised industries and the destruction of communities: some evidence from north east England', *Capital and Class*, **29**, 29–58

Beynon, H., Cox, A. and Hudson, R., 1989, 'Opencast coal mining and the politics of coal production', *Capital and Class*, **40**, 89–114

Beynon, H., Hudson, R. and Sadler, D., 1991, *A Tale of Two Industries: The Contraction of Coal and Steel in North East England*, Open University Press, Milton Keynes

Beynon, H., Hudson, R. and Sadler, D., 1994, *A Place Called Teesside: a Locality in a Global Economy*, Edinburgh University Press, Edinburgh

Beynon, H., Cox, A. and Hudson, R., 1999, *Digging up Trouble: the Environment, Protest and Opencast Coal Mining*, Rivers Oram, London (forthcoming)

Blackaby, R. (ed.), 1979, *De-Industrialisation*, Heinemann, London

Bleitrach, D. and Chenu, A., 1982, 'Regional planning – regulation or deepening of social contradictions? The example of Fos-sur-Mer and the Marseilles Metropolitan Area', in Hudson, R. and Lewis, J.R. (eds), *London Papers in Regional Science 11: Regional Planning in Europe*, Pion, London, 148–178

Blyton, P. and Turnbull, P. (eds), 1992, *Reassessing Human Resource Management*, Sage, London

Board of Trade, 1963, *The North East: a Programme for Regional Development and Growth*, Cmnd 2206, HMSO, London

Borzaga, C. and Goglio, S., 1979, 'Economic development and regional imbalances: the case of Italy, 1945–76', in Hudson, R. (ed.), *New Regional Problems in Britain and Italy, Volume 2*, Final Report 415-78-12-UK to the Directorate of Research, Science and Education of the Commission of the European Communities, Brussels, 37–78

Boston Consulting Group and PRS Consulting International, 1991, *The Competitive Challenge Facing the European Automotive Components Industry*, Devonshire House, London W1X 5FH

Boulding, K. 1985, *The World as a Total System*, Sage, London

Boulding, P., 1988, 'Reindustrialization Strategies in Steel Closure Areas: A Comparison of Corby, Ebbw Vale and Hartlepool', unpublished PhD thesis, Department of Geography, University of Durham, Durham

Boulton, L., 1996, 'Energy efficiency', *Financial Times Survey*, 11 November

Bowden P.J., 1965, 'Regional problems and policies in the North-East of England', in Wilson, T. (ed.), *Papers on Regional Development*, Blackwell, Oxford, 20–39

Boyer, R. and Drache, D. (eds), 1995, *States against Markets*, Routledge, London

Braczyk, H.-J., Cooke, P. and Heidenreich, M. (eds), 1998, *Regional Innovations Systems*, UCL Press, London

Braverman, H., 1974, *Labor and Monopoly Capital: The Degradation of Work in the Twentieth Century*, Monthly Review Press, New York

Breathnach, P., 1982, 'The demise of growth-centre policy: the case of the Republic of Ireland', in Hudson, R. and Lewis, J.R. (eds), *London Papers in Regional Science 11: Regional Planning in Europe*, Pion, London, 35–56

Brenner, R. and Glick, M., 1991, 'The Regulation approach: theory and history', *New Left Review*, **188**, 45–120

Bridge, J., 1993, 'The regional impacts of inward investment in Europe: a study of the North of England', *Regional Prospects in Europe*, **3**, 53–57

British Coal, 1988, *Annual Report and Accounts* (1987/88), London

British Steel Corporation, 1973, *Ten Year Development Strategy*, Cmmd. 5226, HMSO, London

British Steel Corporation, 1974–1982, *Annual Report and Accounts*, British Steel Corporation, Sheffield

Britton, F.E.K., 1994, *Rethinking Work: New Concepts of Work in a Knowledge Society*, Report of the RACE Program, EcoPlan International, Paris

BSC, 1980, 'The case for Consett closure', *mimeo*, British Steel Corporation, Sheffield

Building Services Research and Information Association, 1994, *Environmental Code of Practice for Buildings and Their Services*, London

Bulmer, M. (ed.), 1978, *Mining and Social Change*, Croom Helm, Beckenham

Burawoy, M., 1985, *The Politics of Production*, Verso, London

Burns, W., 1967, *Newcastle: A Study in Replanning at Newcastle Upon Tyne*, Leonard Hill, Newcastle upon Tyne

Business Council for Sustainable Development, 1994, *Internalising Environmental Costs to Promote Eco-Efficiency*, BCSD, Geneva

Bussell, L., 1990, 'Inward investment: impact on the Northern Region', *Business Review North*, **2**, 24–32

Byrne, D., 1990, *Beyond the Inner City*, Open University Press, Milton Keynes

Byrne, D. and Parson, D., 1983, 'The state and the reserve army: the management of class relations in space', in Anderson, J., Duncan, S. and Hudson, R. (eds), *Redundant Spaces in Cities and Regions?*, Academic Press, London, 127–154

Camagni, R., 1991, 'Local "milieu", uncertainty and innovation networks: towards a new dynamic theory of economic space', in Camagni, R. (ed.), *Innovation Networks: Spatial Perspectives*, Belhaven, London, 121–142

Cameron, G.C., 1979, 'The national industrial strategy and regional policy', in Maclennan, D. and Parr, J.B. (eds), *Regional Policy: Past Experience and New Directions*, Martin Robertson, Oxford, 297–322

Carney, J., 1980, 'Regions in crisis: accumulation, regional problems and crisis formation', in Carney, J., Hudson, R. and Lewis, J. (eds), *Regions in Crisis: New Perspectives in European Regional Theory*, Croom Helm, London, 28–59

Carney, J., 1988, *The Derwentside Industrial Development Programme: Background, Progress and Future Objectives*, DIDA, Consett

Carney, J. and Hudson, R., 1974, 'Ideology, public policy and underdevelopment in the North East', *North East Area Study Working Paper 6*, University of Durham, Durham

Carney, J. and Hudson, R., 1978, 'Capital, politics and ideology: the North East of England, 1870–1946', *Antipode*, **10**(2), 64–78

Carney, J., Hudson, R., Ive, G. and Lewis, J., 1976, 'Regional underdevelopment in late capitalism: a study of North East England', in Masser, I. (ed.), *London Papers in Regional Science 6: Theory and Practice in Regional Science*, Pion, London, 11–29

Carney, J., Hudson, R. and Lewis, J., 1977, 'Coal combines and inter-regional uneven development in the UK', in Massey, D.B. and Batey, P.J. (eds), *London Papers in Regional Science 7: Alternative Frameworks for Analysis*, Pion, London, 52–67

Carney, J., Hudson, R. and Lewis, J. (eds), 1980, *Regions in Crisis: New Perspectives in European Regional Theory*, Croom Helm, London

Cassell, M., 1995, 'Inward looking perspective', *Financial Times*, 5 December

Castells, M. and Godard, F., 1974, *Monopoville: l'Entreprise, l'Etat, l'Urbaine*, Mouton, Paris

Caulkins, D., 1992, 'The unexpected entrepreneurs: small high technology firms and regional development in Wales and North East England', in Rothstein, F.A. and Blim, M. (eds), *Anthropology and the Global Factory*, Bergin and Garvey, New York, 119–135

Central Statistical Office, 1975–81, *Regional Trends*, Nos 11–17, HMSO, London

Chisholm, M., 1962, *Rural Settlement and Land Use: an Essay in Location*, Hutchinson, London

Clark, G., 1987, 'Enhancing competitiveness in the US steel industry – the steel-workers and National Steel's Cooperative partnership', unpublished manuscript, Center for Labor Studies, School of Urban and Public Affairs, Carnegie Mellon University, Pittsburgh, PA15213-3890, USA

Clark, G.L. and Dear, M., 1984, *State Apparatus: Structures and Language of Legitimacy*, Allen and Unwin, London

Clark, G.L. and Wrigley, N., 1995, 'Such costs: a framework for economic geography', *Transactions, Institute of British Geographers*, N.S., **20**(2), 204–223

Cleveland County Council, 1975, *The Economic Impact of North Sea Oil in Cleveland*, Middlesbrough

Cleveland County Council, 1979, *County Structure Plan Re-assessment Programme, Discussion Paper 1: Employment*, Middlesbrough

Cleveland County Council, 1985, *Cleveland Structure Plan: Issues Report*, Middlesbrough

Cleveland County Council, 1987, *Unemployment Strategy*, Middlesbrough

Cockerill, A. and Silbertson, A., 1974, *The Steel Industry: International Comparisons of Industrial Structure and Performance*, Cambridge University Press, Cambridge

Commission of the European Communities, 1977, *Bulletin of the EC*, 11.1.3.1–1.3.5 and 12.1.1.1–1.1.4., Luxembourg

Commission of the European Communities, 1980, *Basic Statistics of the Community*, Luxembourg

Commission of the European Communities, 1992, *Towards Sustainability: a European Community Programme of Policy and Action in Relation to the Environment and Sustainable Development*, COM 92(23), Brussels

Commission of the European Communities, 1996, *Employment in Europe*, Luxembourg

Conti, S. and Enrietti, A., 1995, 'The Italian automobile industry and the case of Fiat: one country, one company, one market', in Hudson, R. and Schamp, E.W. (eds), *Towards a New Map of Automobile Manufacturing in Europe? New Production Concepts and Spatial Restructuring*, Springer, Berlin, 117–146

Cooke, P., 1990, *Back to the Future: Modernity, Postmodernity and Locality*, Unwin Hyman, London

Coriat, B., 1991, 'Technical flexibility and mass production: flexible specialisation and dynamic flexibility', in Benko, G. and Dunford, M. (eds), *Industrial Change and Regional Development: the Transformation of New Industrial Spaces*, Belhaven, London, 134–157

Cousins, J. and Brown, R., 1970, 'Shipbuilding in the North-East', in Dewdney, J.C. (ed.), *Durham County and City with Teesside*, British Association for the Advancement of Science, Durham, 313–329

Cox, K.R., 1997, 'Globalisation and geographies of workers' struggles in the late twentieth century', in Lee, R. and Wills, J. (eds), *Geographies of Economies*, Arnold, London, 177–185

Crang, M., 1998, *Cultural Geography*, Routledge, London

Crang, P., 1997, 'Introduction: cultural turns and the (re)constitution of economic geography', in Lee, R. and Wills, J. (eds), *Geographies of Economies*, Arnold, London, 3–15

Crewe, L. and Davenport, E., 1992, 'The puppet show: changing buyer–supplier relations within clothing retailing', *Transactions of the Institute of British Geographers*, N.S., **17**, 183–197

Crossman, R., 1976, *The Diaries of the Cabinet Minister*, Vol. 2, Hamish Hamilton, London

Curry, J., 1993, 'The flexibility fetish: a review essay on flexible specialisation', *Capital and Class*, **50**, 99–126

Cusumano, M.A., 1985, *The Japanese Automobile Industry*, Harvard University Press, Cambridge, Massachusetts

Damette, E., 1980, 'The regional framework of monopoly exploitation: new problems and trends', in Carney, J., Hudson, R. and Lewis, J. (eds), *Regions in Crisis: New Perspectives in European Regional Theory*, Croom Helm, London, 76–92

Danset, D., Ugolini, E., Corbellotti, R., Sacher, F. and Straniero, R., 1979, 'A Usinor-Longwy, le sens d'une lutte', *Economie et Politique*, September, 13–15

Daysh, G.H.J. and Symonds, J.S., 1953, *West Durham: A Problem Area in North-Eastern England*, Blackwell, Oxford

de Vroey, M., 1984, 'A regulation interpretation of the contemporary crisis', *Capital and Class*, **23**, 45–66

Dennis, N., Henriques, F. and Slaughter, C., 1956, *Coal is Our Life*, Eyre and Spottiswoode, London

Department of Economic Affairs, 1965, *The National Plan*, HMSO, London

Department of Employment, 1978, *Department of Employment Gazette*, HMSO, London

Department of the Environment, 1964–1973, *Local Housing Statistics*, HMSO, London (prior to 1971, Ministry of Housing and Local Government)

Derwentside Industrial Development Agency, n.d., *Derwentside Fact File*, Consett, County Durham

Dicken, P., 1990, 'Seducing foreign investors – the competitive bidding strategies of local and regional agencies in the United Kingdom', in Hebbert, M. and Hansen, J.C. (eds), *Unfamiliar Territory*, Avebury, Aldershot, 162–186

Dicken, P., 1992a, *Global Shift*, Paul Chapman, London

Dicken, P., 1992b, 'Europe 1992 and strategic change in the international automobile industry', *Environment and Planning A*, **24**, 11–31

Dicken, P. and Oberg, S., 1996, 'The global context: Europe in a world of dynamic economic and population change', *European Urban and Regional Studies*, **3**(2), 101–120

Dicken, P. and Thrift, N., 1992, 'The organization of production and the production of organization: why business enterprises matter in the study of geographical organization', *Transactions, Institute of British Geographers*, N.S., **17**(3), 279–291

Dobson, R., 1996, 'Revealed: scandal of children's illegal jobs', *Independent on Sunday*, 12 May

Dore, R.P., 1983, 'Goodwill and the spirit of market capitalism', *British Journal of Sociology*, **34**, 3

Douglass, D. and Krieger, J., 1983, *A Miner's Life*, Routledge and Kegan Paul, Andover

Dunford, M., 1990, 'Theories of regulation', *Society and Space*, **8**, 297–321

Dunford, M. and Hudson, R., 1996, *Successful European Regions: Northern Ireland Learning from Others?*, Northern Ireland Economic Council, Belfast

Dunford, M.F. and Perrons, D., 1986, 'The restructuring of the post-war British space economy', in Rowthorne, B. and Martin, R. (eds), *The Geography of Deindustrialisation*, Macmillan, London, 53–105

Durand, C., 1981, *Chômage et Violence: Longwy en Lutte*, Galilée, Paris

Durham County Council, 1951, *County Development Plan: Draft written analysis*, Durham County Council

Durham County Council Community Education Department, 1991, *Making Music Work – Project Report, 1991–2*, County Hall, Durham

Durham University Business School, 1995, *Profitability Trends in Northern Industries*, Lloyds Bowmaker

Economist Intelligence Unit, 1992, *The European Automotive Components Industry – 1992 Edition*, Special Report No. 2158, 40 Duke Street, London W1A 1DW

Ellerbeck West Public Inquiry, 1989, *Inspector's Report (Mr J. Dunlop), Ellerbeck West Opencast Public Inquiry (February 1989)*, Ref. APP/C2300/A/87/67269

European Commission, 1993, *Growth, Competitiveness, Employment: The Challenges and Ways Forward into the 21st Century*, Brussels

European Commission, 1994, *The Potential Benefits of Integration of Environmental and Economic Policies: An Incentive Based Approach*, Brussels

European Commission, 1996, *First Cohesion Report*, Brussels

European Environmental Agency, 1996, *Environmental Taxes, Implementation and Environmental Effectiveness*, Copenhagen

Ferrao, J. and Vale, M., 1993, 'People carriers: a new opportunity for the periphery? Lessons from the Ford/VW project in Portugal', paper presented to the ESF RURE meeting, Stockholm, 15–18 April

Ferrao, J. and Vale, M., 1995, 'Multi-purpose vehicles, a new opportunity for the periphery?', in Hudson, R. and Schamp, E.W. (eds), *Towards a New Map of Automobile Manufacturing in Europe? New Production Concepts and Spatial Restructuring*, Springer, Berlin, 195–217

Financial Times, 1978, 'French steel rescue: control without nationalisation', 22 September, 18

Financial Times, 1979, 'Steel closure leaves Consett without hope', 13 December, 5

Financial Times, 1980a, 'Goodwill is not enough for Consett', 19 September, 6

Financial Times, 1980b, 'How the French Government bought change in the steel industry', 13 March, 3

Financial Times, 1980c, 'Northern group in talks on Consett', 2 September, 6

Financial Times, 1981, 'French steel groups see deficits', 28 October, 24

Financial Times, 1982, 'French steel workers burn down company HQ in jobs protest', 16 July, 2

Fincher, R., 1983, 'The inconsistency of eclecticism', *Environment and Planning A*, **15**, 607–622

Firn, J.R., 1975, 'External control and regional development: the case of Scotland', *Environment and Planning A*, 7, 393–414

Florida, R., 1995, 'The industrial transformation of the Great Lakes Region', in Cooke, P. (ed.), *The Rise of the Rustbelt*, University of London Press, London, 162–176

Fontela, E., 1994, 'The long-term outlook for growth and employment', in *OECD Societies in Transition: The Future of Work and Leisure*, OECD, Paris, 25–40

Foray, D., 1993, 'Feasibility of a single regime of intellectual property rights', in Humbert, M. (ed.), *The Impact of Globalisation on Europe's Firms and Regions*, Pinner, London, 85–95

Forbes, R. (ed.), 1987, *Polisses and Candymen*, The Tommy Armstrong Trust, Old Miners' Hall, Delves Lane, Consett, County Durham

Foss, N., 1966, 'Introduction: the emerging competence perspective', in Foss, N. and Knudsen, J. (eds), *Towards a Competence Theory of the Firm*, Routledge, London, 1–12

Fothergill, S. and Gudgin, G., 1978, *Regional Employment Statistics on a Comparable Basis 1952–75*, Centre for Environmental Studies, Occasional Paper No. 5, London

FPCFMM, 1978, 'Acier: Bruxelles programme des liquidations', Fédération du Parti Communist Français du Meurthe et Moselle, *Economie et Politique*, September, 27–35

Friedman, A., 1977, *Industry and Labour*, Macmillan, London

Friedman, M. and Friedman, R., 1980, *Free to Choose: A Personal Statement*, Harcourt Brace, New York

Friis, P., 1980, 'Regional problems in Denmark: myth or reality?', *Dunelm Translations* 4, Department of Geography, University of Durham

Fröbel, F., Heinrichs, J. and Kreye, O., 1980, *The New International Division of Labour*, Cambridge University Press, Cambridge

Garofoli, G. (ed.), 1992, *Endogenous Development and Southern Europe*, Avebury, Basingstoke

Garrahan, P. and Stewart, P., 1992, *The Nissan Enigma: Flexibility at Work in a Local Economy*, Mansell, London

Gauche-Cazalis, C., 1979, 'Sidérurgie étatiser pour mieux casser', *Economie et Politique*, January, 28–33

Geddes, M., 1978, 'Regional policy and crisis and the cuts', paper presented to a meeting of the Conference of Socialist Economists Regionalism Group, London

George, S., 1992, *The Debt Boomerang: How Third World Debt Harms Us All*, Pluto Press, London

Gertler, M., 1997, 'The invention of regional culture', in Lee, R. and Wills, J. (eds), *Geographies of Economies*, Arnold, London, 67–58

Gibbs, D., 1996, 'Integrating sustainable development and economic restructuring: a role for regulation theory?', *Geoforum*, **27**(1), 1–10

Gibson-Graham, J.K., 1996, *The End of Capitalism (As We Know It)*, Blackwell, Oxford

Gibson-Graham, J.K., 1997, 'Re-placing class in economic geographies: possibilities for a new class politics', in Lee, R. and Wills, J. (eds), *Geographies of Economies*, Arnold, London, 87–97

Giddens, A., 1973, *The Class Structure of Advanced Societies*, Hutchinson, London

Giddens, A., 1979, *Central Problems in Sociological Theory*, Macmillan, London

Giddens, A., 1981, *A Contemporary Critique of Historical Materialism*, Macmillan, London

Giddens, A., 1984, *The Constitution of Society*, Polity, Cambridge

Giddens, A., 1990, *The Consequences of Modernity*, Polity, Cambridge

Gladstone, F., 1976, *The Politics of Planning*, Temple Smith, London

Goodin, R., 1992, *Green Political Theory*, Polity, Cambridge

Goodwin, M. and Painter, J., 1996, 'Local governance, the crisis of Fordism and the changing geographies of regulation', *Transactions, Institute of British Geographers*, N.S., **21**, 635–648

Gordon, D., 1988, 'The global economy: new edifice or crumbling foundations?', *New Left Review*, **168**, 124–165

Grabher, G., 1990, 'On the weakness of strong ties – The ambivalent role of inter-firm relations in the decline and reorganisation of the Ruhr', *WZB Discussion Paper FS I 90-4*, Berlin

Grabher, G., 1993, 'The weakness of strong ties: the lock-in of regional development in the Ruhr area', in Grabher, G. (ed.), *The Embedded Firm: on the Socio-economics of Industrial Networks*, Routledge, London, 255–277

Granovetter, M., 1985, 'Economic action and social structure: the problem of embeddedness', *American Journal of Sociology*, **91**(3), 481–510

Gregory, D., 1975, 'The Cow Green Reservoir', in Smith, P.J. (ed.), *The Politics of Physical Resources*, Penguin, Harmondsworth, 144–201

Gregory, D., 1994, 'Social theory and human geography', in Gregory, D., Martin, R. and Smith, G. (eds), *Human Geography: Society, Space and Social Science*, Macmillan, Basingstoke, 78–112

Gregory, D., Martin, R. and Smith, G., 1994, 'Introduction: human geography, social change and social science', in Gregory, D., Martin, R. and Smith, G. (eds), *Human Geography: Society, Space and Social Science*, Macmillan, Basingstoke, 1–18

Gregson, N., 1986, 'On duality and dualism: the case of time-geography and structuralism', *Progress in Human Geography*, **10**, 184–205

Habermas, J., 1976, *Legitimation Crisis*, Heinemann, London

Habermas, J., 1979, 'Conservatism and capitalist crisis', *New Left Review*, **115**, 73–84

Hadjimichalis, C. and Sadler, D. (eds), 1995, *Europe on the Margins: New Mosaics of Inequality*, Wiley, London

Haggett, P., 1965, *Locational Analysis in Human Geography*, Arnold, London

Halford, S. and Savage, M., 1997, 'Rethinking restructuring: embodiment, agency and identity in organisational change', in Lee, R. and Wills, J. (eds), *Geographies of Economies*, Arnold, London, 108–117

Harris, D., 1988, *The Arts in Derwentside – Towards Development*, Consultant's Report commissioned by Derwentside District Council, Civic Centre, Consett, County Durham

Harvey, D., 1973, *Social Justice and the City*, Arnold, London

Harvey, D., 1982, *The Limits to Capital*, Blackwell, Oxford

Harvey, D., 1989, *The Condition of Postmodernity*, Blackwell, Oxford

Harvey, D., 1993, 'The nature of the environment: the dialectics of social and environmental change', *Socialist Register*, 1–51

Harvey, D., 1996, *Justice, Nature and the Geography of Difference*, Blackwell, Oxford

Harvey, D. and Scott, A.J., 1987, 'The nature of human geography: theory and empirical specificity in the transition from Fordism to flexible accumulation', paper presented to the Quantitative Methods Study Group of the Institute of British Geographers, 10 April

Harvey, G., 1917, *Capitalism in the Northern Coalfield*, City Library, Newcastle upon Tyne

Hassink, R., 1992a, *Regional Innovation Policy*, Doctoral thesis, University of Utrecht

Hassink, R., 1992b, *Regional Innovation Policy: Case Studies from the Ruhr Area, Baden-Württemberg and the North East of England*, The Netherlands Geographical Studies, No. 145, Den Haag

Henkel, H.O., 1994, 'Risks and opportunities of telework for the individual, the environment and society at large', *Proceedings of the European Assembly on Teleworking and New Ways of Working*, 3–4 November, Berlin, 11–17

Herod, A., 1997, 'From a geography of labor to a Labor geography: Labor's spatial fix and the geography of capitalism', *Antipode*, **29**(1), 1–31

Hewison, R., 1987, *The Heritage Industry*, Methuen, London

Hines, C. and Searle, G., 1979, *Automatic Unemployment: Discussion of the Impact of Microelectronic Technology on UK Unemployment and the Responses this Demands*, Earth Resources Research, London

Hobsbawm, E.J., 1977, 'Some reflections on the break-up of Britain', *New Left Review*, **105**, 3–24

Hodgson, G., 1988, *Economics and Institutions: A Manifesto for Modern Institutional Economics*, Polity, London

Hodgson, G.M., 1993, *Economics and Evolution*, Polity, Cambridge

Hole, V.W., Adderson, I.M. and Pountney, M.T., 1979, *Washington New Town: The Early Years*, Department of the Environment, HMSO, London

Holloway, J. and Piciotto, S. (eds), 1978, *State and Capital: a Marxist Debate*, Arnold, London

House, J. and Fullerton, B., 1960, *Tees-side at Mid-Century: an Industrial and Economic Survey*, Macmillan, London

Hudson, R., 1974, 'Images of the retailing environment: an example of the use of the repertory grid methodology', *Environment and Behaviour*, **6**, 470–494

Hudson, R., 1976a, *New Towns in North East England*, Report HR 1734, Final Report to the Social Science Research Council, London, Volumes 1 and 2

Hudson, R., 1976b, 'Preliminary Notes on Restructuring Production in the Northern Region', *mimeo*, Department of Geography, University of Durham

Hudson, R., 1979a, 'New Towns and spatial policy: the case of Washington New Town', in *The Production of the Built Environment: Proceedings of the Bartlett Summer School*, Bartlett School of Architecture and Planning, University College London

Hudson, R., 1979b, 'Space, place and placelessness: some questions concerning methodology', *Progress in Human Geography*, **3**, 169–174

Hudson, R., 1980a, 'Regional development policies and female employment', *Area*, **12**(3), 229–234

Hudson, R., 1980b, 'Women and work: a study of Washington New Town', *Occasional Publications (New Series) 16*, Department of Geography, University of Durham, Durham

Hudson, R., 1981a, 'Capital accumulation and regional problems: a study of North East England, 1945–80', in Plum, V. (ed.), *North East England*, Publikationer frå Institut för Geografi, Samfundsanalyse og Datalogi, Kompendium 25, Roskilde Universitetscenter, Roskilde, Denmark, 1–54

Hudson, R., 1981b, 'The development of the chemicals industry on Teesside', in Plum, V. (ed.), *North East England*, Publikationer frå Institut för Geografi, Samfundsanalyse og Datalogi, Kompendium 25, Roskilde Universitetscenter, Roskilde, Denmark, 54–94

Hudson, R., 1983a, 'Capital accumulation and chemicals production in Western Europe in the postwar period', *Environment and Planning A*, **15**, 105–122

Hudson, R., 1983b, 'Capital accumulation and regional problems: a study of North East England 1945–80', in Hamilton, F.E.I. and Linge, G. (eds), *Regional Industrial Systems*, John Wiley, Chichester, 75–102

Hudson, R., 1985, 'The paradoxes of state intervention: the impact of nationalised industry policies and regional policy on employment in the Northern Region in the post-war period', in Chapman, R. (ed.), *Public Policy Studies: North East England*, Edinburgh University Press, Edinburgh, 57–79

Hudson, R., 1986a, 'Producing an industrial wasteland: capital, labour, and the State in North East England', in Martin, R.L. and Rowthorne, B. (eds), *Deindustrialization and the British Space Economy*, Macmillan, London, 169–213

Hudson, R., 1986b, 'Nationalised industry policies and regional policies; the role of the state in the deindustrialisation and reindustrialisation of regions', *Society and Space*, 4(1), 7–28

Hudson, R., 1987, 'Changing spatial divisions of labour in manufacturing and their impacts on localities', paper presented to the 8th Conference of Nordic Radical Geographers, Elmsta, Sweden, 24–27 September

Hudson, R., 1988a, 'Producing a divided society: state policies and the management of change in housing and labour markets in a peripheral region', in Allen, J. and Hamnett, C. (eds), *Studies in Housing and Labour Market Change*, Hutchinson, London, 214–236

Hudson, R., 1988b, 'Uneven development in capitalist societies: changing spatial divisions of labour, forms of spatial organisation of production and service provision, and their impact on localities', *Transactions, Institute of British Geographers*, N.S., **13**, 484–496

Hudson, R., 1989a, *Wrecking a Region: State Policies, Party Politics and Regional Change*, Pion, London

Hudson, R., 1989b, 'Labour market changes and new forms of work in old industrial regions: maybe flexibility for some but not flexible accumulation', *Society and Space*, 7, 5–30

Hudson, R., 1990, 'Re-thinking regions: some preliminary considerations on regions and social change', in Johnston, R.J., Hoekveld, G. and Hauer, J. (eds), *Regional Geography: Current Developments and Future Prospects*, Routledge, London, 67–84

Hudson, R., 1991a, *Making Music Work – Final Project Report, volumes 1 and 2*, University of Durham

Hudson, R., 1991b, 'The North in the 1980s: new times in the "Great North" or just more of the same?', *Area*, **23**(1), 47–56

Hudson, R., 1992a, 'The Japanese, the United Kingdom automobile industry and the automobile industry in the United Kingdom', *Change in the Automobile Industry: An International Comparison*, Discussion Paper No. 9, Department of Geography, University of Durham

Hudson, R., 1992b, 'Institutional change, cultural transformation and economic regeneration; myths and realities from Europe's old industrial areas', *Occasional Publication No. 26*, Department of Geography, University of Durham

Hudson, R., 1992c, 'Industrial restructuring and spatial change: myths and realities in the changing geography of production in the 1980s', *Scottish Geographical Magazine*, **2**, 74–81

Hudson, R., 1992d, *British Energy Policy and the Market for Coal*, Memorandum submitted to the House of Commons Select Committee on Trade and Industry on behalf of Easington District Council

Hudson, R., 1993, 'Spatially uneven development, and the production of spaces and places: some preliminary considerations, and a case study of Consett', in Hauer, J. and Hoekveld, G.J. (eds), *Moving Regions*, Netherlands Geographical Studies, 43–68

Hudson, R., 1994a, 'New production concepts, new production geographies? Reflections on changes in the automobile industry', *Transactions, Institute of British Geographers*, N.S., **19**, 331–345

Hudson, R., 1994b, 'Institutional change, cultural transformation and economic regeneration: myths and realities from Europe's old industrial regions', in Amin, A. and Thrift, N.J. (eds), *Globalization, Institutions and Regional Development in Europe*, Oxford University Press, Oxford, 331–345

Hudson, R., 1995a, 'Making music work? Alternative regeneration strategies in a deindustrialised locality: the case of Derwentside', *Transactions, Institute of British Geographers*, N.S., **20**(4), 460–473

Hudson, R., 1995b, 'The Japanese, the European market and the automobile industry in the United Kingdom', in Hudson, R. and Schamp, E.W. (eds), *Towards a New Map of Automobile Manufacturing in Europe? New Production Concepts and Spatial Restructuring*, Springer, Berlin, 63–92

Hudson, R., 1995c, 'The end of mass production, the end of the mass collective worker and their respective geographies? Or more old wine in new bottles?', paper presented to the Annual Conference of the Institute of British Geographers, Newcastle upon Tyne, 3–6 January

Hudson R., 1995d, 'Towards sustainable industrial production: but in what sense sustainable?', in Taylor, M. (ed.), *Environmental Change: Industry, Power and Place*, Avebury, Winchester, 37–56

Hudson, R., 1995e, 'The role of foreign investment', in Evans, L., Johnson, P. and Thomas, B. (eds), *Northern Region Economy: Progress and Prospects*, Cassell, London, 79–95

Hudson, R., 1996a, 'Putting policy into practice: policy implementation problems, with special reference to the European Mediterranean', Keynote Address to the International Conference on Desertification in the Mediterranean, Sissi, Crete, 28–30 October

Hudson, R., 1996b, 'The learning economy, the learning firm and the region: a sympathetic critique of the limits to learning', Paper prepared for an International Seminar on Learning and Territoriality, Université Montesquieu-Bordeaux XIV, 28–30 November

Hudson, R., 1997a, 'Regional futures: industrial restructuring, new production concepts and spatial development strategies in Europe', *Regional Studies*, **31**(5), 467–478

Hudson, R., 1997b, 'The end of mass production and of the mass collective worker? Experimenting with production, employment and their geographies', in Lee, R. and Wills, J. (eds), *Geographies of Economies*, Arnold, London, 302–310

Hudson, R., 1999a, '"The Learning Economy, the Learning Firm and the Learning Region": a sympathetic critique of the limits to learning', *European Urban and Regional Studies*, **6**(1), 59–72

Hudson, R., 1999b, 'Globalisation and the restructuring of the UK space-economy: national state policies and regional differentiation', in Vellinga, M. (ed.), *Globalisation, National States and Regional Responses* (forthcoming)

Hudson, R. and Plum, V., 1986, 'Deconcentration or decentralisation? Local government and the possibilities for local control of local economies', in Goldsmith, M. and Villadsen, S. (eds), *Urban Political Theory and the Management of Fiscal Stress*, Gower, Aldershot

Hudson, R. and Sadler, D., 1983a, 'Region, class and the politics of steel closures in the European Community', *Society and Space*, **1**, 405–428

Hudson, R. and Sadler, D., 1983b, 'Anatomy of disaster: closure of Consett steel-works', *Northern Economic Review*, **6**, 2–17

Hudson, R. and Sadler, D., 1985, 'The development of Middlesbrough's iron and steel industry, 1841–1985', *Middlesbrough Locality Study Working Paper 2*, Departments of Geography and Sociology, University of Durham

Hudson, R. and Sadler, D., 1986a, 'Contesting works closures in Western Europe's old industrial regions: defending place or betraying class?', in Scott, A.J. and Storper, M. (eds), *Production, Territory, Work*, Allen and Unwin, London, 172–193

Hudson, R. and Sadler, D., 1986b, *Forging a United Engineering and Steels Industry?*, Sheffield City Council, Sheffield

Hudson, R. and Sadler, D., 1986c, 'Communities in crisis: the social and political effects of steel closures in France, West Germany, and the United Kingdom', *Urban Affairs Quarterly*, **21**, 171–186

Hudson, R. and Sadler, D., 1989, *The International Steel Industry: Restructuring, State Policies and Localities*, Routledge, London

Hudson, R. and Sadler, D., 1990, 'State policies and the changing geography of the coal industry in the UK in the 1980s and 1990s', *Transactions, Institute of British Geographers*, N.S., **15**(4), 435–454

Hudson, R. and Sadler, D., 1991, 'Manufacturing success? Reindustrialisation policies in Derwentside in the 1980s', *Occasional Publication No. 25*, Department of Geography, University of Durham

Hudson, R. and Schamp, E.W. (eds), 1995, *Towards a New Map of Automobile Manufacturing in Europe? New Production Concepts and Spatial Restructuring*, Springer, Berlin

Hudson, R. and Townsend, A., 1992, 'Trends in tourism employment and resulting policy choices for local government', in Johnson, P. and Thomas, B. (eds), *Perspectives on Tourism Policy*, Mansell, London, 49–68

Hudson, R. and Weaver, P., 1997, 'In search of employment creation via environmental valorisation: exploring a possible eco-Keynesian future for Europe', *Environment and Planning A*, **29**, 1647–1661

Hudson, R. and Williams, A., 1995, *Divided Britain*, Wiley, Chichester (2nd edition; 1st edition, 1989, Belhaven, London)

Hudson, R., Krogsgaard Hansen, L. and Schech, S., 1992, 'Jobs for the girls? The new service sector economy of Derwentside in the 1980s', *Occasional Publication No. 26*, Department of Geography, University of Durham

Humbert, M. (ed.), 1993, *The Impact of Globalisation on Europe's Firms and Industries*, Pinter, London

Ingham, G., 1982, 'Divisions within the dominant class and "British exceptionalism"', in Giddens, A. and Mackenzie, G. (eds), *Social Class and the Division of Labour*, Cambridge University Press, Cambridge, 209–227

Inward, 1987, *Labour Performance and Productivity in North West England*, Duxbury Park, Chorley, Lancashire PR7 4AT

Isard, W., 1956, *Location and the Space-economy*, Wiley, New York

Jackson, T., 1996, *Material Concerns: Pollution, Profit and Quality of Life*, Routledge, London

Jackson, T. and Marks, N., 1994, *Measuring Sustainable Economic Welfare – A Pilot Index: 1950–1990*, Stockholm Environment Institute

Jacobs, J., 1961, *The Death and Life of the Great American Cities*, Penguin, Harmondsworth

Jacobs, M., 1991, *The Green Economy*, Pluto, London

Jenkins, T. and McLaren, D., 1994, *Working Future? Jobs and the Environment*, Friends of the Earth, London

Jensen-Butler, C., 1981, 'A critique of behavioural geography: an epistemological analysis of cognitive mapping and of Hagerstrans's time-space model', *Working Paper 12*, Geographical Institute, University of Aarhus, 8000 Aarhus C, Denmark

Jessop, B., 1982, *The Capitalist State*, Martin Robertson, Oxford

Jessop, B., 1990, *State theory: Putting Capitalist States in their Place*, Cambridge University Press, Cambridge

Jessop, B., 1993, 'Towards a Schumpeterian workfare state? Preliminary remarks on a post-Fordist political economy', *Studies in Political Economy*, **40**, 7–39

Jessop, B., 1994, 'Post-Fordism and the State', in Amin, A. (ed.), *Post-Fordism: A Reader*, 57–84, Blackwell, Oxford

Jessop, B., 1997, 'Capitalism and its future: Remarks on regulation, government and governance', *Review of International Political Economy*, **4**, 561–581

Johnson, B., 1992, 'Institutional learning', in Lundvall, B.-A. (ed.), 1992, *National Systems of Innovation: Towards a Theory of Innovation and Interactive Learning*, Pinter, London, 23–44

Johnson, P. and Conway, C., 1995, 'Entrepreneurship and new firm formation', in Evans, L., Johnson, P. and Thomas, B. (eds), *Northern Region Economy: Progress and Prospects*, Cassell, 97–114

Johnston, R.J., 1997, *Geography and Geographies: Anglo-American Human Geography Since 1945*, Arnold, London (5th edition)

Joint Trades Unions for Consett Steelworkers (JTUCS), 1980, 'No case for closure', *mimeo*, Consett Branch Library, Victoria Road, Consett, County Durham

Jones, B., 1988, 'Doctors join Greens against opencast mines', *The New Statesman*, 3 June

Judd, D. and Parkinson, M. (eds), 1990, *Leadership and Urban Regeneration*, Sage, London

Kearney, J., 1990, *A Report on the Decade Waltz: the Consett Music Festival*, Making Music Work, Old Miners' Hall, Delves Lane, Consett, County Durham

Kearney, T., 1990, *Painted Red: A Social History of Consett, 1840–1990*, DCA, Consett, County Durham

Keeble, D., 1971, 'Employment mobility in Britain', in Chisholm, M. and Manners, G. (eds), *Spatial Policy Problems of the British Economy*, Cambridge University Press, Cambridge

Keeble, D. and Walker, S., 1994, 'New firms, small firms and dead firms: spatial patterns and determinants in the United Kingdom', *Regional Studies*, **28**(4), 411–428

Kenney, M. and Florida, R., 1988, 'Beyond mass production: production and the labour process in Japan', *Politics and Society*, **16**(1), 121–158

Kirby, D., 1995, 'The development of the services sector', in Evans, L., Johnson, P. and Thomas, B. (eds), *Northern Region Economy: Progress and Prospects*, Cassell, London, 43–58

Kistler, E. and Strech, K.-D., 1992, 'Die Sonne der Arbeit: Arbeitseinstellung als Forschungsgegenstand im Transformationsprozess', in Jaufmann, D., Kistler, E., Meier, K. and Strech, K.-D. (eds), *Empirische Sozialforschung im vereinten Deutschland*, Frankfurt

Kitson, M. and Michie, J., 1998, 'Markets, competition and innovation', *ESRC Centre for Business Research Working Paper 86*, University of Cambridge

Krieger, J., 1979, 'British colliery closure programmes in the North East: from para-dox to contradiction', in Cullen, J.G. (ed.), *Analysis and Decision in Regional Policy.* Papers in Regional Science 9, Pion, London, 219–232

Läpple, D. and van Hoogstraten, P., 1980, 'Remarks on the spatial structure of capitalist development: the case of the Netherlands', in Carney, J., Hudson, R. and Lewis, J.R. (eds), *Regions in Crisis*, Croom Helm, Beckenham, 117–166

Lash, D. and Urry, J., 1987, *The End of Organised Capitalism*, Polity, Cambridge

Lash, S. and Urry, J., 1994, *Economies of Signs and Space*, Sage, London

Le Figaro, 1979, 6 February, 9

Leborgne, D. and Lipietz, A., 1988, 'New technologies, new modes of regulation: some spatial implications', *Society and Space*, **6**(3), 263–280

Lee, R., 1996, 'Moral money? LETS and the social construction of economic geographies in south east England', *Environment and Planning A*, **28**, 1377–1394

Lee, R. and Wills, J. (eds), 1997, *Geographies of Economies*, Arnold, London

Lenk, H., 1994, 'Value changes and the achieving society: a social–philosophical perspective', in *OECD Societies in Transition: The Future of Work and Leisure*, OECD, Paris, 81–94

Leonard, H.J., 1988, *Pollution and the Struggle for the World Product: Multinational Corporations, Environment and International Competitive Advantage*, Cambridge University Press, Cambridge

Linge, G.J.R. and Hamilton, F.E.I., 1981, 'International industrial systems', in Ham-ilton, F.E.I. and Linge, G.J.R. (eds), *Spatial Analysis, Industry and the Industrial Environment Progress in Research and Applications*, Vol. 2, Wiley, Chichester, 1–117

Lipietz, A., 1979, *Crise et Inflation: pourquoi?*, Maspero, Paris

Lipietz, A., 1980a, 'Inter-regional polarisation and the tertiarisation of society', *Papers and Proceedings of the Regional Science Association*, **44**, 3–17

Lipietz, A., 1980b, 'The structuration of space, the problem of land, and spatial policy', in Carney, J., Hudson, R. and Lewis, J. (eds), *Regions in Crisis: New Perspectives in European Regional Theory*, Croom Helm, London, 60–75

Lipietz, A., 1984, 'Imperialism or the beast of the apocalypse', *Capital and Class*, **22**, 81–110

Lipietz, A., 1986, 'New tendencies in the international division of labour: regimes of accumulation and modes of regulation', in Scott, A.J. and Storper, M. (eds), *Production, Territory, Work*, Unwin Hyman, London, 16–40

Lipietz, A., 1987, *Mirages and Miracles: the Crises of Global Fordism*, Verso, London

Lipietz, A., 1992, *Towards a New Economic Order: Postfordism, Ecology and Democ-racy*, Polity, Cambridge

Llewelyn-Davies, R., Weeks and Partners, 1966, *Washington New Town Master Plan and Report*, Washington Development Corporation, Washington New Town, Tyne-Wear, England

Löfstedt, R.E., 1995, *Times Higher Educational Supplement*, 31 March, 18

Lundvall, B.-A. (ed.), 1992, *National Systems of Innovation: Towards a Theory of Innovation and Interactive Learning*, Pinter, London

Lundvall, B.-A., 1995, 'The learning economy – challenges to economic theory and policy', revised version of a paper presented to the EAEPE Conference, Copenha-gen, 27–29 October 1994

Lundvall, B.-A. and Johnson, B., 1994, 'The learning economy', *Journal of Industry Studies*, **1**(2), 23–42

Lutz, C., 1994, 'Prospects of social cohesion in OECD countries', in *OECD Societies in Transition: The Future of Work and Leisure*, OECD, Paris, 95–119

MacDonald, R., 1991, 'Runners, fallers and plodders: youth and the enterprise culture', Paper presented to the 14th National Small Firms and Policy Conference, Blackpool

MacGregor, I., 1987, *The Enemies Within: the Story of the Miners' Strike, 1984–5*, Fontana, London

Mair, A., 1991, 'Just-in-time manufacturing and the spatial structure of the automobile industry', *Change in the Automobile Industry: An International Comparison*, Discussion Paper No. 4, Department of Geography, University of Durham

Malmberg, A., 1991, 'Restructuring the Swedish manufacturing industry – the case of the motor vehicles industry', in Law, C.M. (ed.), *Restructuring the Global Automobile Industry: National and Regional Impacts*, Routledge, London, 200–214

Malmberg, A., 1992, 'The restructuring of the Swedish car production system', paper presented to the ESF RURE meeting, Copenhagen, 3–6 September

Mandel, E., 1963, 'The dialectic of class and region in Belgium', *New Left Review*, **20**, 5–31

Mandel, E., 1975, *Late Capitalism*, New Left Books, London

Mandel, E., 1978, *The Second Slump*, New Left Books, London

Mann, M., 1993, 'Nation-states in Europe and other continents: diversifying, developing, not dying', *Proceedings of the American Academy of Arts and Sciences*, **122**(3), 115–140

Manpower Services Commission, 1979, *Annual Report 1978–9*, MSC, London

Manpower Services Commission, 1981, *Regional Employment Market Intelligence Trends*, No. 10, MSC, Newcastle Upon Tyne

Manwairing, T., 1981, 'Labour productivity and the crisis at BSC', *Capital and Class*, **14**, 61–97

Marsh, F., 1983, *Japanese Overseas Investment – the New Challenge*, SR 142, Economist Intelligence Unit, London

Martin, R., 1988, 'Industrial capitalism in transition: the contemporary reorganisation of the British space-economy', in Massey, D. and Allen, J., *Uneven Redevelopment: Cities and Regions in Transition*, Hodder and Stoughton, London, 202–231

Martin, R., Sunley, P. and Wills, J., 1994, 'Unions and the politics of deindustrialisation: comments on how Geography complicates class analysis', *Antipode*, **26**(1), 59–76

Maskell, P., 1998, 'Low-tech competitive advantage and the role of proximity', *European Urban and Regional Studies*, **5**(2), 99–118

Maskell, P. and Malmberg, A., 1995, 'Localised learning and industrial competitiveness', *BRIE Working Paper Number 80*, Berkeley Roundtable on the International Economy, University of California, Berkeley

Maskell, P., Eskelinen, H., Hannibalsson, I., Malmberg, A. and Vatne, E., 1998, *Comprehensive, Localised Learning and Regional Development*, Routledge, London

Massey, D., 1978, 'Survey regionalism: some current issues', *Capital and Class*, **6**, 106–125

Massey, D., 1979, 'In what sense a regional problem?', *Regional Studies*, **13**, 233–243

Massey, D., 1995, *Spatial Divisions of Labour: Social Structures and the Geography of Production*, Macmillan, London (2nd edition, 1st edition, 1984)

Mayer, M., 1992, 'The shifting local political system in European cities', in Dunford, M. and Kafkalas, G. (eds), *Cities and Regions in the New Europe*, Belhaven, London, 255–276

McDowell, L., 1997, 'A tale of two cities? Embedded organization and embodied workers in the City of London', in Lee, R. and Wills, J. (eds), *Geographies of Economies*, Arnold, London, 18–29

McDowell, L. and Massey, D., 1984, 'A woman's place?', in Massey, D. and Allen, J. (eds), *Geography Matters*, Cambridge University Press, Cambridge, 128–147

Metcalfe, J.S., 1996, 'Technology strategy in an evolutionary world', The Honeywell/Sweatt Lecture, Centre for Development of Technological Leadership, University of Minnesota, Minnesota

Metcalfe, J.S., 1997, *Evolutionary Economics and Creative Destruction*, The Graz Schumpeter Lectures, University of Graz, Austria

Metcalfe, J.S., 1998, 'Evolutionary concepts in relation to evolutionary economics', *CRIC Working Paper 4*, University of Manchester, 39

Miller, P., Pons, J.N. and Naude, P., 1996, 'Global teams', *Financial Times*, 14 June, 10

Mingione, E., 1985, 'Social reproduction and the labour force: the case of Southern Italy', in Redclift, N. and Mingione, E. (eds), *Beyond Employment: Household, Gender and Subsistence*, Blackwell, Oxford, 14–55

Moore, B., Rhodes, J. and Tyler, P., 1977, 'The impact of regional policy in the 1970s', *Centre for Environmental Studies Review*, 1, 67–77

Morgan, K., 1981, 'Regional crises, the European steel industry and decline in Britain', research paper, *mimeo*, Department of Urban and Regional Studies, University of Sussex

Morgan, K., 1983, 'Restructuring steel: the crises of labour and locality in Britain', *International Journal of Urban and Regional Research*, 7, 175–200

Morgan, K., 1995, 'The Learning Region: institutions, innovation and regional renewal', *Papers in Planning Research Number 157*, Department of City and Regional Planning, University of Wales, Cardiff

Morris, L., 1986, 'Redundant populations: deindustrialization in a northeast English town', in Lee, R.M. (ed.), *Redundancies, Layoffs and Plant Closures*, Croom Helm, Beckenham, 40–61

Mulberg, J., 1995, *Social Limits to Economic Theory*, Routledge, London

Murdoch, J., 1997, 'Towards a geography of heterogeneous associations', *Progress in Human Geography*, 21(3), 321–337

Murdoch, J., Banks, J. and Marsden, T., 1998, 'An economy of conventions? Some thoughts on conventions theory and its application to the agro-food sector', paper presented to the RGS EGRG Conference 'Geographies of Commodities', University of Manchester, 1–2 September

Murray, F., 1983, 'The decentralization of production – the decline of the mass-collective worker', *Capital and Class*, 19, 74–99

Nairn, T., 1977, *The Break-up of Britain: Crisis and Neo-nationalism*, New Left Books, London

National Economic Development Office, 1991, *The Experience of Nissan Suppliers: Lessons for the UK Engineering Industry*, NEDO, Millbank Tower, London

Nelson, R.R. and Winter, S.G., 1982, *An Evolutionary Theory of Economic Change*, Harvard University Press, Cambridge, MA

Newby, H., 1980, *Green and Pleasant Land? Social Change in Rural England*, Penguin, Harmondsworth

Nicholson, M., 1970, *The Environmental Revolution*, Pelican, Hardmondsworth

Noiriel, E., 1980, *Vivre et Lutter à Longwy*, Maspéro, Paris

Nordhaus, W.D. and Tobin, J., 1972, 'Is growth obsolete?', *National Bureau of Economic Research, General Series 9*, Columbia University Press, New York

North East Area Study, 1975, *Social Consequences and Implications of the Teesside Structure Plan*, University of Durham

North East Trades Union Studies Information Unit, 1976, *The Crisis Facing the UK Power Plant Manufacturing Industry*, NETUSIU, Newcastle upon Tyne

North East Trades Union Studies Information Unit, 1977, *Multinationals in Tyne and Wear*, NETUSIU, Newcastle upon Tyne

Northern Economic Planning Council, 1966, *Challenge of the Changing North*, Newcastle upon Tyne

Northern Region Joint Monitoring Team, 1980, *Second 'State of the Region' Report*, Department of the Environment, Northern Region Office, Newcastle upon Tyne

Northern Region Strategy Team, 1975, 'Change and efficiency in the Northern Region, 1948–73', *NRST Technical Report No. 3*, Newcastle upon Tyne

Northern Region Strategy Team, 1976a, 'Linkages in the Northern Region', *Working Paper 6*, published for North East County Councils Association by Cleveland County Council, Middlesbrough

Northern Region Strategy Team, 1976b, 'Public expenditure in the Northern Region', *NRST Technical Report No. 12*, Newcastle upon Tyne

Northern Region Strategy Team, 1977, *Strategic Plan for the Northern Region*, Volumes 1–5, Newcastle upon Tyne

North of England Assembly, 1995, *Profile 1995*, Newcastle upon Tyne

North Tyneside Community Development Project, 1978a, *North Shields, Living with Industrial Change*, Home Office, London

North Tyneside Community Development Project, 1978b, *North Shields: Women's Work*, Home Office, London

North Tyneside Trades Council, 1979, *Shipbuilding – the Cost of Redundancy*, Newcastle upon Tyne

O'Connor, J., 1973, *The Fiscal Crisis of the State*, St Martin's Press, New York

Odgaard, M. and Hudson, R., 1998, 'The misplacement of learning in economic geography', Universities of Durham and Roskilde, *mimeo*, 20

Offe, C., 1975a, 'The theory of the capitalist state and the problem of policy formation', in Lindberg, L.N., Alford, R., Crouch, C. and Offe, C. (eds), *Stress and Contradiction in Modern Capitalism*, DC Heath, Lexington, MA, 125–144

Offe, C., 1975b, 'Introduction to Part III', in Lindberg, L.N., Alford, R., Crouch, C. and Offe, C. (eds) *Stress and Contradiction in Modern Capitalism*, DC Heath, Lexington, MA, 245–259

Ohmae, K., 1995, *The End of the Nation State: The Role of Regional Economies*, The Free Press, New York

Okamura, C. and Kawahito, H., 1990, *Karoshi*, Mado-Sha, Tokyo

Olin Wright, 1978, *Class, Crisis and the State*, New Left Review, London

Organization for Economic Cooperation and Development, 1976, *The 1974–5 Recession and the Employment of Women*, OECD, Paris

Organization for Economic Cooperation and Development, 1980, *Steel in the 1980s*, OECD, Paris

Osterland, M., 1986, 'Deregulation and erosion of the standard employment pattern: the spread of grey-zone work in the Federal Republic of Germany', unpublished manuscript available from Zentrale Wissenschaftliche Einrichtung Arbeit und Betrieb, Universität Bremen, Postfach 330440, 2800 Bremen 33, Germany

Osterland, M., 1987, 'Declining industries plant closing and local markets: a case study of the city of Bremen', unpublished manuscript available from Zentrale Wissenschaftliche Einrichtung Arbeit und Betrieb, Universität Bremen, Postfach 330440, 2800 Bremen 33, Germany

Painter, J. and Goodwin, M., 1995, 'Local governance and concrete research: investigating the uneven development of regulation', *Economy and Society*, **24**, 334–356

Parkes, C., 1991, 'Dying heart still beats', *Financial Times*, 6 December

Pearce, D., Markandya, A. and Barbier, E., 1989, *Blueprint for a Green Economy*, Earthscan, London

Peck, F. and Stone, T., 1993, 'Japanese inward investment in the north east of England; reassessing "Japanization"', *Environment and Planning C*, **11**, 56–67

Peck, J., 1994, 'Regulating labour: the social regulation and reproduction of local labour markets', in Amin, A. and Thrift, N. (eds), *Globalization, Institutions and Regional Development in Europe*, Oxford University Press, Oxford, 147–176

Peck, J., 1996, *Workplace: the Social Regulation of Labour Markets*, Guilford, New York

Peet, R., 1997a, *Modern Geographical Thought*, Blackwell, Oxford

Peet, R., 1997b, 'The cultural production of economic forms', in Lee, R. and Wills, J. (eds), *Geographies of Economies*, Arnold, London, 37–46

Pepler, G. and MacFarlane, P.W., 1949, *The North East Area Development Plan*, unpublished Interim Report presented to the Minister of Town and Country Planning

Perrons, D., 1979, 'The role of Ireland in the new international division of labour: a proposed framework for regional analysis', *Urban and Regional Studies Working Papers 15*, School of Urban and Regional Studies, University of Sussex

Pickvance, C.G., 1976, 'On the study of urban social movements', in Pickvance, C.G. (ed.), *Urban Sociology: Critical Essays*, Methuen, London, 198–218

Pine, B.J., 1993, *Mass Customization: the New Frontier in Business Competition*, Harvard University Press, Cambridge, MA

Piore, M. and Sabel, C., 1984, *The Second Industrial Divide*, Basic Books, New York

Pocock, D. (ed.), 1981, *Humanistic Geography and Literature*, Croom Helm, Beckenham

Pocock, D. and Hudson, R., 1978, *Images of the Urban Environment*, Macmillan, London

Pollard, J., 1995, 'The contradictions of flexibility: labour control and resistance in the Los Angeles industry', *Geoforum*, **26**(2), 121–123

Pollert, A., 1988, 'Dismantling flexibility', *Capital and Class*, **34**, 42–75

Polyani, K., 1957, *The Great Transformation: the Political and Economic Origins of Our Time*, Beacon Press, Boston, MA (2nd edition; 1st edition, 1944)

Pred, A., 1967, *Behaviour and Location*, University of Lund, Department of Human Geography, Lund, Sweden

Priestley, J.B., 1994, *English Journey*, Mandarin, London (1st edition, 1934)

Pugliese, E., 1985, 'Farm workers in Italy: agricultural working class, landless peasants or clients of the welfare state?', in Hudson, R. and Lewis, J. (eds), *Uneven Development in Southern Europe*, Methuen, Andover, 123–139

Pugliese, E., 1991, 'Restructuring of the labour market and the role of Third World migrations in Europe', paper presented to the Conference on Undefended Cities and Regions Facing the New European Order, Lemnos

Purdy, D., 1994, 'Citizenship, basic income and the state', *New Left Review*, **208**, 30–48

Putnam, R., 1993, *Making Democracy Work*, Princeton University Press, Princeton, NJ

Rainnie, A., 1993, 'The reorganization of large firm subcontracting', *Capital and Class*, **49**, 53–76

Rankin, S., 1987, 'Exploitation and the labour theory of value: a neo-Marxian reply', *Capital and Class*, **32**, 104–116

Redclift, N. and Mingione, E. (eds), 1985, *Beyond Employment: Household, Gender and Subsistence*, Blackwell, Oxford

Rees, G. and Thomas, M., 1989, 'From coalminers to entrepreneurs? A case study in the sociology of reindustrialisation', paper presented to the British Sociological Association Annual Conference, Plymouth

Refeld, D., 1995, 'Disintegration and reintegration of production clusters in the Ruhr area', in Cooke, P. (ed.), *The Rise of the Rustbelt*, UCL Press, London, 85–102

Regional Policy Research Unit, 1979, 'State regional policies and uneven development: the case of North East England', *Final Report RP270*, Centre for Environmental Studies, London, Parts 1–9

Relph, E., 1976, *Place and Placelessness*, Pion, London

Repetto, R., 1992, *Green Fees: How a Tax Shift Can Work for the Environment and the Economy*, World Resources Institute, Washington DC

Rich, M., 1996, 'Britain's textile manufacturers cotton on to cheap labour', *Financial Times*, 16 April, 23

Roberts, B., 1987, 'Marx after Steebman: separating Marxism from "surplus theory"', *Capital and Class*, **32**, 84–103

Roberts, P., 1995, *Environmentally Sustainable Business: A Local and Regional Perspective*, Paul Chapman, London

Robertson, J., 1995, *Electronics, Environment and Employment: Harnessing Private Gain to the Common Good*, Oxford Green College Centre for Environmental Policy and Understanding

Robinson, F., Shaw, K. and Lawrence, M., 1993, 'It takes two to Quango: UDCs and their local communities in the north east', *Northern Economic Review*, **21**, 47–60

Robinson, J.F.F., 1990, *The Great North?*, Centre for Urban and Regional Development Studies, University of Newcastle upon Tyne and BBC TV North East

Robinson, J.F.F. and Sadler, D., 1984, 'Consett after the closure', *Occasional Publication New Series Number 19*, Department of Geography, University of Durham

Robinson, J.F.F. and Storey, D., 1979, 'Employment change in manufacturing industry in Cleveland, 1965–76', paper presented to a meeting of the Conference of Socialist Economists Regionalism Group, Durham

Rodwin, L. and Sazanami, H. (eds), 1989, *Deindustrialization and Regional Economic Transformation: the Experience of the United States*, Unwin Hyman, London

Rosenberg, N., 1982, *Inside the Black Box: Technology and Economics*, Cambridge University Press, Cambridge

Rowthorne, B., 1983, 'The past strikes back', in Hall, S. and Jacques, M. (eds), *The Politics of Thatcherism*, Lawrence and Wishart, London, 63–78

Rubinstein, J., 1992, *The Changing US Auto Industry*, Routledge, London

Ruggie, J.G., 1993, 'Territoriality and beyond: problematizing modernity in international relations', *International Organization*, **47**(1), 139–174

Sadler, D., 1982a, 'Capital accumulation, state intervention and regional response: the re-structuring of the EEC steel industry', *Graduate Discussion Paper No. 1*, Department of Geography, University of Durham

Sadler, D., 1982b, 'The closure of Consett steel works: isolation and division within region and class', *mimeo*, Department of Geography, University of Durham

Sadler, D., 1991a, 'Beyond 1992: the evolution of European Community policies towards the automobile industry', *Change in the Automobile Industry: An International Comparison*, Discussion Paper No. 7, Department of Geography, University of Durham

Sadler, D., 1991b, 'Strategic change in the west European automotive components industry', *Change in the Automobile Industry: An International Comparison*, Discussion Paper No. 8, Department of Geography, University of Durham

Sadler, D., 1992, *Global Region*, Pergamon, Oxford

Sadler, D., 1997, 'The role of supply chain management in the "Europeanisation" of the automobile production system', in Lee, R. and Wills, J. (eds), *Geographies of Economies*, Arnold, London, 311–320

Sadler, D. and Amin, A., 1994, ' "Europeanisation" in the automotive components sector and its implications for state and locality', in Hudson, R. and Schamp, E.W. (eds), *Towards a New Map of Automobile Manufacturing in Europe? New Production Concepts and Spatial Restructuring*, Springer, Berlin, 39–62

Sadler, D., Swain, A. and Hudson, R., 1993, 'The automobile industry and eastern Europe', *Area*, **25**(4), 339–349

Salais, M. and Storper, M., 1992, 'The four worlds of contemporary industry', *Cambridge Journal of Economics*, **16**, 169–193

Sassen, S., 1991, *The Global City*, Princeton University Press, Princeton, NJ

Savary, J., 1993, 'Strategies of European automobile manufacturers, with special reference to Renault', paper presented to the ESF RURE meeting, Stockholm, 15–18 April

Save Scotswood Campaign Committee, 1979, *Economic Audit on Vickers Scotswood*, SSCC, Newcastle upon Tyne

Sayer, A., 1984, *Method in Social Science*, Hutchinson, London

Sayer, A., 1986, 'New developments in manufacturing: the just-in-time system', *Capital and Class*, **30**, 43–72

Sayer, A., 1989, 'Postfordism in question', *International Journal of Urban and Regional Research*, **13**, 666–695

Sayer, A. and Walker, R., 1992, *The New Social Economy: Reworking the Division of Labour*, Blackwell, Oxford

Schamp, E., 1991, 'Towards a spatial reorganisation of the German car industry? The implications of new production concepts', in Benko, G. and Dunford, M. (eds), *Industrial Change and Regional Development: the Transformation of New Industrial Spaces*, Belhaven, London, 159–171

Schamp, E., 1992, 'The German car production system going European', paper presented to the ESF RURE meeting, Copenhagen, 3–6 September

Schamp, E.W., 1995, 'The German automobile industry going European', in Hudson, R. and Schamp, E.W. (eds), *Towards a New Map of Automobile Manufacturing in Europe? New Production Concepts and Spatial Restructuring*, Springer, Berlin, 93–116

Schmidt-Bleek, F., 1994, 'Work in a sustainable economy', in *Proceedings of the European Assembly on Teleworking and New Ways of Working*, 3–4 November, Berlin, 19–34

Schoenberger, E., 1994, *The Cultural Crisis of the Firm*, Blackwell, Oxford

Scott, A.J., 1988a, *New Industrial Spaces*, Pion, London

Scott, A.J., 1988b, *Metropolis: From the Division of Labour to Urban Form*, University of California Press, Berkeley and Los Angeles

Seers, D., Schaffer, B. and Kiljunen, M.L. (eds), 1979, *Underdeveloped Europe: Studies in Core–Periphery Relations*, Harvester, Brighton

Sengenberger, W. and Loveman, G., 1987, *Smaller Units of Employment New Industrial Organization Programme, DP3/1987*, International Institute for Labour Studies, International Labour Organization, Geneva

Semmens, P., 1970, 'The chemical industry (of Teesside and South Durham)', in Dewdney, J.C. (ed.), *Durham County and City with Teesside*, British Association for the Advancement of Science, Durham, 330–340

Shanks, M., 1977, *Planning and Politics: the British Experience, 1960–76*, Allen and Unwin, London

Sheppard, E. and Barnes, T., 1990, *The Capitalist Space Economy: Geographical Analysis after Ricardo, Marx and Sraffa*, Unwin Hyman, London

Simmie, J. (ed.), 1997, *Innovation, Networks and Learning Regions*, Jessica Kingsley, London

Simon, H.A., 1959, 'Theories of decision making in economics and behavioural science', *American Economic Review*, **XLIX**(3), 253–283

Simonis, U.E., 1994, 'Industrial restructuring in industrial countries', in Ayres, R.U. and Simonis, U.E. (eds), *Industrial Metabolism: Restructuring for Sustainable Development*, United Nations University Press, Tokyo, 31–54

Smith, I. and Stone, I., 1989, 'Foreign investment in the North: distinguishing fact from hype', *Northern Economic Review*, **18**, 50–61

Smith, N., 1984, *Uneven Development: Nature, Capital and the Production of Space*, Blackwell, Oxford

Soja, E., 1980, 'The socio-spatial dialectic', *Annals of the Association of American Geographers*, **70**, 207–225

Spellman, J., 1991, *Attracting Inward Investment: The Strategies of a Regional Development Organisation*, Unpublished MA thesis, University of Manchester

Sraffa, P., 1960, *The Production of Commodities by Means of Commodities*, Cambridge University Press, Cambridge

Steedman, I., 1977, *Marx after Sraffa*, New Left Books, London

Stohr, W. (ed.), 1990, *Global Challenge and Local Response*, Mansell, London

Stohr, W. and Fraser Taylor, D.R. (eds), 1981, *Development from Above or Below?*, Wiley, London

Stone, I. and Peck, F., 1996, 'The foreign-owned manufacturing sector in UK peripheral regions, 1978–93: restructuring and comparative performance', *Regional Studies*, **30**(1), 55–68

Storey, D., 1990, 'Evaluation of policies and measures to create local employment', *Urban Studies*, **27**, 669–684

Storper, M., 1995, 'The resurgence of regional economies, ten years after: the region as a nexus of untraded dependencies', *European Urban and Regional Studies*, **2**(3), 191–223

Storper, M., 1997, *The Regional World*, Guilford, New York

Storper, M. and Scott, A.J., 1989, 'The geographical foundations and social regulation of flexible production systems', in Dear, M. and Wolch, J. (eds), *The Power of Geography*, Unwin Hyman, London, 21–40

Storper, M. and Walker, R., 1989, *The Capitalist Imperative: Territory, Technology and Industrial Growth*, Blackwell, Oxford

Strange, S., 1988, *States and Markets*, Pinter, London

Swain, A., 1992, 'Eastern Europe and the global strategies of automobile producers', *Change in the Automobile Industry: An International Comparison*, Discussion Paper No. 11, Department of Geography, University of Durham

Taylor, G., 1979, 'The restructuring of capital in the Teesside chemical and steel industries', paper presented to a meeting of the Conference of Socialist Economists Regionalism Group, Durham

Taylor, M.J., 1994, 'Industrialisation, enterprise power and environmental change: an exploration of concepts', paper presented to the IGU Commission on the Organisation of Industrial Space, Budapest, 16–20 August

Taylor, M.J., 1995, 'Linking economy, environment and policy', in Taylor, M.J. (ed.), *Environmental Change: Industry, Power and Policy*, Avebury, Aldershot, 1–12

Taylor, P.J., 1979, '"Difficult to let", "Difficult to live in", and sometimes "Difficult to get out of": an essay on the provision of council housing, with special reference to Killingworth', *Environment and Planning A*, **11**, 1305–1320

Taylor, R., 1997, 'Call centres expected to employ 1m by 2000', *Financial Times*, 24 June

Teesside County Borough Council, 1972, *Teesside Structure Plan: Report of Survey*, Middlesbrough

The Ecologist, 1972, *A Blueprint for Survival*, Penguin, Harmondsworth

The Sunday Times, 1982, 'The spectre behind Longwy's SOS sign', 30 October, 12

The Times, 1981, 'Long, slow haul back for Consett', 2 November, 13

Théret, B., 1994, 'To have or to be: on the problem of the interaction between state and economy in its "solidaristic" mode of regulation', *Economy and Society*, **23**, 1–46

Thomas, B., 1995, 'Tourism: is it underdeveloped?', in Evans, L., Johnson, P. and Thomas, B. (eds), *Northern Region Economy: Progress and Prospects*, Cassell, London, 59–78

Thompson, E.P., 1968, *The Making of the English Working Class*, Penguin, Harmondsworth

Thrift, N. and Olds, K., 1996, 'Refiguring the economic in economic geography', *Progress in Human Geography*, **20**(3), 311–337

Thrift, N., 1994, 'On the social and cultural determinants of international financial centres: the case of the City of London', in Cambridge, S., Martin, B. and Thrift, N. (eds), *Money, Space and Power*, Blackwell, Oxford, 327–355

Tickell, A. and Peck, J., 1992, 'Accumulation, regulation and the geographies of post-Fordism: missing links in regulationist research', *Progress in Human Geography*, **16**(2), 190–218

Toft Jensen, H., 1982, 'The role of the state in regional development, planning and managment', in Hudson, R. and Lewis, J. (eds), *London Papers in Regional Science 11: Regional Planning in Europe*, Pion, London

Townsend, A.R., 1986, 'Spatial aspects of the growth of part-time employment in Britain', *Regional Studies*, **20**, 313–330

Townsend, P., Phillimore, P. and Beattie, A., 1988, *Health and Deprivation: Inequality and the North*, Croom Helm, Beckenham

Tuan, Y.F., 1977, *Space and Place: The Perspective of Experience*, Edward Arnold, London

Turok, I., 1993, 'Inward investment and local linkages: how deeply embedded is Silicon Glen?', *Regional Studies*, **27**(5), 401–418

United Nations Development Program, 1993, *Human Development Report 1993*, Oxford University Press, New York

United Nations Economic Commission for Europe and International Robotics Federation, 1994, *World Industrial Robots 1994: Statistics 1983–93 and Forecasts to 1997*, United Nations, Geneva

United Nations World Commission on Environment and Development, 1987, *Our Common Future*, Oxford University Press

Upham, M., 1980, 'The BSC: retrospect and prospect', *Industrial Relations Journal*, **11**, 5–21

Urry, J., 1981, 'Localities, regions and social class', *International Journal of Urban and Regional Research*, **5**, 455–473

Urry, J., 1985, 'Social relations, space and time', in Gregory, D. and Urry, J. (eds), *Social Relations and Spatial Structures*, Macmillan, London, 20–48

van Tulder, R. and Ruigrok, W., 1993, 'Regionalisation, globalisation or glocalisation: the case of the world car industry', in Humbert, M. (ed.), *The Impact of Globalisation on Europe's Firms and Industries*, Pinter, London, 22–33

Vaughan, R., 1976, *Post-war Integration in Europe*, Edward Arnold, London

Veltz, P., 1991, 'New models of production organisation and trends in spatial development', in Benko, G. and Dunford, M. (eds), *Industrial Change and Regional Development: the Transformation of New Industrial Spaces*, Belhaven, London, 193–204

von Weizsäcker, E.U., 1994, *Earth Politics*, Zed Books, London

Wagstyl, S., 1997, 'Re-engineering a grand old name', *Financial Times*, 10 April

Walker, R., 1985, 'Is there a service economy? The changing capitalist division of labour', *Science and Society*, **49**, 42–83

Ward, N. and Almäs, R., 1997, 'Explaining change in the international agro-food system', *Review of International Political Economy*, **4**(4), 611–629

Warren, K., 1990, *Consett Iron, 1840 to 1980*, Clarendon, Oxford

Watson, M.W., Thomas, J., Temple, J.M.F. and Rees, H.G., 1986, 'Opencast Mining and Health – the Effects of Opencast Mining on Health in the Upper Neath Valley', *mimeo*

Washington Development Corporation, 1964–1979, 'Annual reports', in *Reports of the Development Corporations*, HMSO, London

Washington Development Corporation, 1974, 'Report on migration, January–December 1973', unpublished document, Washington Development Corporation, Washington New Town, Tyne-Wear, England

Weaver, C., 1982, 'The limits to economism: towards a political approach to regional development and planning', in Hudson, R. and Lewis, J.R. (eds), *London Papers in Regional Science 11: Regional Planning in Europe*, Pion, London, 184–202

Weaver, P.M., 1993, 'Synergies of association: Ecorestructuring, scale, and the industrial landscape', paper presented to the Symposium on Ecorestructuring, United Nations University, Tokyo, 5–7 July

Weaver, P.M., 1994, 'How life-cycle analysis and operational research methods could help clarify environmental policy: the case of fibre recycling in the pulp/paper sector', paper presented to the IGU Commission on the Organisation of Industrial Space, Budapest, 16–20 August

Weaver, P.M., 1995a, 'Steering the Eco-Transition: A Materials Accounts Based Approach', *mimeo*, Department of Geography, University of Durham

Weaver, P.M., 1995b, *Implementing Change: A Resource Paper for the Factor-10 Club Meeting*, Carnoules, September

Weaver, P.M. and Hudson, R., 1995, 'Economic restructuring and public expenditure for sustainable development: an eco-Keynesian model', in McClaren, D. (ed.), *Working Futures*, Friends of the Earth, London, 60–74

Weiss, L., 1997, 'Globalization and the myth of the powerless state', *New Left Review*, **225**, 3–27

Wickens, P., 1986, *The Road to Nissan*, Macmillan, London

Wilkenson, D., 1939, *The Town that was Murdered*, Left Book Club, London

Williams, K., Hasam, C., Williams, J., Cutler, T., Adcroft, A. and Johal, S., 1992, 'Against Lean Production', *mimeo*, Business Policy Unit, East London Business School, University of East London, 48

Williams, R., 1960, 'Advertising: the magic system', *New Left Review*, **4**

Williams, R., 1983, *Keywords: a Vocabulary of Culture and Society*, Fontana, London

Williams, R., 1989, *The Politics of Modernism*, Verso, London

Williamson, B., 1982, *Class, Culture and Community*, Routledge and Kegan Paul, Henley-on-Thames

Wills, J., 1998a, 'Taking on the Cosmo Corp? Experiments in trans-national labour organisation', *Economic Geography*, **74**, 111–130

Wills, J., 1998b, 'Building labour institutions to stage the world order? International trade unionism and European Works Councils', paper presented to the RGS EGRG Seminar on Institutions and Governance, 3 July, UCL, London

Wilson, A., 1970, *Entropy in Urban and Regional Modelling*, Pion, London

Winterton, J., 1985, 'Computerized coal: new technology in the mines', in Beynon, H. (ed.), *Digging Deeper*, Verso, London, 231–244

Womack, J.P., Jones, D.T. and Roos, D., 1990, *The Machine that Changed the World*, Macmillan, New York

World Bank, 1994, *Annual Report 1994*, Washington DC

Wrigley, N. and Lowe, M. (eds), 1995, *Retailing, Consumption and Capital: Towards the New Retail Geography*, Longman, London

Wuppertal Institute for Climate, Environment and Energy: Factor 10 Club, 1994, *Carnoules Declaration*, Wuppertal

Yates, C., 1998, 'Defining the fault lines: new divisions in the working class', *Capital and Class*, **66**, 119–147

Yearley, S., 1995, 'The transnational politics of the environment', in Anderson, J., Brook, C. and Cochrane, A. (eds), *A Global World?*, Oxford University Press, Oxford, 209–248

Zukin, S. and di Maggio, P., 1990, 'Introduction', in Zukin, S. and di Maggio, P. (eds), *Structures of Capital: the Social Organisation of the Economy*, Cambridge University Press, Cambridge, 1–36

Index

Printed and bound by CPI Group (UK) Ltd, Croydon, CR0 4YY

01/11/2024

01782615-0015